Antisense Therapeutics

METHODS IN MOLECULAR MEDICINE™

John M. Walker, SERIES EDITOR

Antisense Therapeutics

Second Edition

Edited by

M. Ian Phillips, PhD, DSc

Vice President for Research
University of South Florida, Tampa, FL

Foreword by

Stanley T. Crooke, MD, PhD

Isis Pharmaceuticals Inc., Carlsbad, CA

Humana Press ☀ Totowa, New Jersey

Cover illustration: "The principle of antisense inhibition," Figure 1 from chapter 1, Antisense Therapeutics: *A Promise Waiting to be Fulfilled,* by M. Ian Philips

Cover design by Patricia F. Cleary.

Printed in the United States of America. 10 9 8 7 6 5 4 3 2 1

Library of Congress Cataloging in Publication Data
Antisense therapeutics / edited by M. Ian Phillips.— 2nd ed.
        p. ; cm. — (Methods in molecular medicine ; 106)
    Includes bibliographical references and index.
    ISBN 1-58829-205-3 (alk. paper); eISBN 1-59259-854-4
    1. Antisense nucleic acids—Therapeutic use. [DNLM: 1. Oligonucleotides, Antisense—therapeutic use. 2. Oligonucleotides, Antisense—pharmacology. QU 57 A6332 2005] I. Phillips, M. Ian. II. Series.
    RM666.A564A585 2005
      615'.31—dc22   2004006680

Foreword

We are now more than 15 years into a large-scale experiment to determine the viability of antisense technology. The challenges of creating a new pharmacological drug discovery platform are prodigious, requiring sizeable investments, long-term commitment, insight, and perseverance. For antisense technology to progress, advances in understanding the behavior of the receptor, RNA, and the behavior of the drugs, oligonucleotide analogs, were necessary. A new medicinal industry, the medicinal industry of oligonucleotides, had to be invented, and numerous drug development challenges—such as creating efficient manufacturing and analytical processes and formulations—had to be overcome. All of those advances then needed to be focused in drug candidates designed to interact with specific targets and to be effective in patients with specific diseases. This has taken time and a good bit of money and although the progress in the technology has been gratifying, there have, of course, been failures of individual clinical trials and individual drugs along the way.

What have we learned? Antisense technology works. Oligonucleotide analogs with a reasonable drug-dependent property can be synthesized and used to inhibit gene function through a variety of antisense mechanisms. Antisense drugs distribute to a wide range of tissues and reduce the expression of targets in a dose fashion consistent with the pharmaceutics of the drugs. First-generation antisense drugs are sufficient for relatively severe indications and second-generation drugs are performing significantly better. Moreover, these drugs are effective by a wide variety of routes including intravenous, subcutaneous, intradermal, rectal, and aerosol, and progress in oral delivery has been reported. Today numerous clinical trials in a wide range of diseases using a variety of oligonucleotide chemistries and antisense mechanisms are in progress.

In this year alone, positive clinical data in rheumatoid arthritis, diabetes, hyperlipidemia, cancer, and other diseases have been reported.

In this edition of *Antisense Therapeutics*, a number of approaches to antisense and therapeutic areas are discussed, as well as specific diagnostic opportunities. That the breadth of activities presented in this volume is as impressive as it is and yet does not begin to cover all of the work in progress, underscores the range of utility and potential value of antisense technology.

Nevertheless, despite antisense being an accepted tool that has facilitated better understanding of biological systems, much remains to be done before the true potential of the technology for therapeutic purposes can be defined. What this volume emphasizes, however, is that exponential progress in defining the long-term roles and value of antisense-based therapeutics is being made.

We look forward to the continued evolution of the technology.

Stanley T. Crooke, MD, PhD

Preface

This is the second edition of *Antisense Therapeutics*. The first edition was edited by Sudhir Agrawal and published in 1996. At that time there was no therapy based on antisense, but plenty of promise for the highly specific targeting of genes that cause disease. Antisense oligonucleotides were first reported as viral replication inhibitors by Paul Zamecnik and Mary Stephenson in 1978. Although this was excellent work, nothing much happened until new procedures for synthesizing DNA sequences were developed. Once oligonucleotides were easy to make, more and more studies were published in the 1980s, most of which were directed to cells in culture. In the early 1990s antisense oligonucleotides were increasingly tested in vivo. There were many controversies and a great deal of concern about backbone modification of the phosphodiester bridges that link the DNA bases. To protect against breakdown by nucleases in cells or blood, phosphorothioate oligonucleotides were adopted. In 1998 a phosphorothioated antisense agent was the first FDA-approved antisense therapy. Vitravene™, developed by Isis Pharmaceuticals, made antisense therapeutics a reality.

Since then, the complete sequencing of the human genome in April, 2003 has demonstrated the presence of a vast number of targets for antisense oligonucleotides. So we now have thousands of targets, hundreds of preclinical animal studies, and some 20 clinical trials ongoing. Any successful trial with an antisense compound will open a floodgate of new therapies for a panoply of diseases.

This second edition of *Antisense Therapeutics* deals less with the basic science of antisense and more with the actual therapeutic applications. For that reason it is organized into disease states.

I thank the authors for their patience and their strong contributions. Since this book was being edited at a time when I moved from the University of Florida to the University of South Florida, I ended up with two secretaries. I would like to thank Ms. Gayle Butters at the University of Florida and Mr. Eric J. Wheeler at the University of South Florida for their essential help. I am also grateful to Craig Adams at Humana Press for his patience.

M. Ian Phillips, PhD, DSc

Contents

Contributors

NARIMAN V. AMIRKHANOV • *Departments of Biochemistry and Molecular Pharmacology, Kimmel Cancer Center, Thomas Jefferson University, Philadelphia, PA*

KAZUNORI AOKI • *Section for Studies on Host-Immune Response, National Cancer Center Research Institute, Tokyo, Japan*

VIKRAM ARORA • *Research and Development, AVI BioPharma, Corvallis, OR*

MOHAN R. ARUVA • *Department of Radiology, Kimmel Cancer Center, Thomas Jefferson University, Philadelphia, PA*

WILLIAM A. BANKS • *GRECC, VA Medical Center St. Louis, Department of Internal Medicine, St. Louis University, St. Louis, MO*

RHONDA M. BRAND • *Division of Emergency Medicine, Evanston Northwestern Healthcare, and Department of Medicine, Feinberg School of Medicine, Northwestern University, Evanston, IL*

ATIS CHAKRABARTI • *Departments of Biochemistry and Molecular Pharmacology, Kimmel Cancer Center, Thomas Jefferson University, Philadelphia, PA*

STANLEY T. CROOKE • *Chairman and CEO, ISIS Pharmaceuticals Inc., Carlsbad, CA*

FEDERICA DEL MONTE • *Cardiovascular Research Center, Massachusetts General Hospital and Harvard Medical School, Boston, MA*

F. ANDREW DORR • *Salmedix Inc., San Diego, CA*

TADAO FUNATO • *Division of Molecular Diagnostics, Tohoku University School of Medicine, Sendai, Japan*

ROGER J. HAJJAR • *Cardiovascular Research Center, Massachusetts General Hospital and Harvard Medical School, Boston, MA*

SIAN E. HARDING • *National Heart and Lung Institute, Imperial College, London, UK*

PATRICK L. IVERSEN • *AVI BioPharma, Corvallis, OR*

LAURA B. JAEGER • *Department of Pharmacological and Physiological Science, St. Louis University, St. Louis, MO*

RONALD JUBIN • *Department of Antiviral Therapy, Schering Plough Research Institute, Kenilworth, NJ*

BIRGITTA KIMURA • *Department of Anthropology, University of Florida, Gainesville, FL*

NICHOLAS KIPSHIDZE • *Lenox Hill Heart and Vascular Institute, Cardiovascular Research Foundation, Lenox Hill Hospital, New York, NY*

MARTIN B. LEON • *Lenox Hill Heart and Vascular Institute, Cardiovascular Research Foundation, Lenox Hill Hospital, New York, NY*

JEFFREY W. MOSES • *Lenox Hill Heart and Vascular Institute, Cardiovascular Research Foundation, Lenox Hill Hospital, New York, NY*

SHUMPEI OHNAMI • *Central RI Laboratory, National Cancer Center Research Institute, Tokyo, Japan*

ROSANNE M. ORR • *Cancer Research UK Centre for Cancer Therapeutic, The Institute of Cancer Research, Sutton, Surrey, UK*

M. IAN PHILLIPS • *Vice President for Research, Office of Research, University of South Florida, Tampa, FL*

WENYI QIN • *Department of Surgery, University of Missouri, Columbia, MO*

PONUGOTI S. RAO • *Department of Radiology, Kimmel Cancer Center, Thomas Jefferson University, Philadelphia, PA*

MANUEL RIEBER • *Tumor Cell Biology Laboratory, Center of Microbiology and Cell Biology, IVIC, Caracas, Venezuela*

WILLIS K. SAMSON • *Department of Pharmacological and Physiological Science, St. Louis University, St. Louis, MO*

EDWARD R. SAUTER • *Department of Surgery, University of Missouri, Columbia, MO*

MARY STRASBERG-RIEBER • *Tumor Cell Biology Laboratory, Center of Microbiology and Cell Biology, IVIC, Caracas, Venezuela*

INGO TAMM • *Department of Hematology and Oncology, Charite, Campus Virchow, Humboldt University of Berlin, Berlin, Germany*

MEGHAN M. TAYLOR • *Department of Pharmacological and Physiological Science, St. Louis University, St. Louis, MO*

MATHEW L. THAKUR • *Department of Radiology, Kimmel Cancer Center, Thomas Jefferson University, Philadelphia, PA*

XIAOBING TIAN • *Departments of Biochemistry and Molecular Pharmacology, Kimmel Cancer Center, Thomas Jefferson University, Philadelphia, PA*

ERIC WICKSTROM • *Departments of Biochemistry and Molecular Pharmacology, Kimmel Cancer Center, Thomas Jefferson University, Philadelphia, PA*

BRUCE R. YACYSHYN • *Louis Stokes VA Hospital and Case Western Reserve University, Cleveland, OH*

TERUHIKO YOSHIDA • *Genetics Division, National Cancer Center Research Institute, Tokyo, Japan*

Y. CLARE ZHANG • *Department of Pediatrics, University of South Florida, St. Petersburg, FL*

WEIZHU ZHU • *Department of Surgery, University of Missouri, Columbia, MO*

I

INTRODUCTION

1

Antisense Therapeutics

A Promise Waiting to Be Fulfilled

M. Ian Phillips

1. Introduction

During the past decade, only one antisense-based therapy has received full Food and Drug Administration (FDA) approval. Vitravene™, developed by Isis Pharmaceuticals, was the first drug based on antisense technology to be successfully commercialized and used in treatment *(1)*. The therapeutic area it is used in is a small niche related to the treatment of preventing blindness in acquired immunodeficiency syndrome (AIDS) patients by inhibiting cytomegalovirus-induced retinitis. The success of Vitravene, however, showed that antisense could be taken all the way through the FDA approval process and provide those patients taking it with a vitally important effect. With Vitravene we saw the first breakthrough in antisense therapy, and, yet, euphoria has turned to disappointment without a second breakthrough. Subsequent trials of Affinitak (Isis), an antisense inhibitor of protein kinase C α, failed to show statistically significant benefits as an antisense therapy for the treatment of non–small cell carcinoma of the lung better than the median survival with control treatments. The results nevertheless proved that antisense was well tolerated and tended toward greater benefit to the survival of patients ($p < 0.054$). The promise of antisense therapy is so attractive that some 20 trials continue.

The appeal of antisense is that it potentially provides highly specific, nontoxic effects for safe and effective therapeutics of an enormous number of diseases including AIDS, Crohn's disease, pouchitis, psoriasis, cancers, diabetes, mulitiple sclerosis, muscular dystrophy, restenosis, asthma, rheumatoid arthritis, hepatitis, skin diseases, polycystic kidney disease, and chronic cardiovas-

From: *Methods in Molecular Medicine, Vol. 106: Antisense Therapeutics, Second Edition*
Edited by: I. Phillips © Humana Press Inc., Totowa, NJ

cular disease, such as hypertension, restenosis, and heart failure. Successes in phase I have shown that antisense therapy consistently has excellent safety results. With each trial we learn more, and this makes each new antisense drug candidate more easy to test. We are hampered by a lack of understanding of the theoretical considerations for optimal antisense inhibition. Failures in the past have been the result of incorrect design and use of unmodified backbones causing instability, overly long oligonucleotides leading to unpredictable targeting, and aptermeric or nonantisense effects. However, with each experiment we learned more. For example, high doses of antisense in monkeys triggered cardiovascular collapse *(2)*. This result was a setback until it was found that the reaction could be accounted for by the extremely high doses and a sensitivity to complement activation unique to nonhuman primates *(3)*. Human trials, by contrast, have shown how well antisense is tolerated and how few side effects are encountered. The number of trials is increasing, and more than 2000 patients have received antisense. Isis is the leader with 11 phase I, 7 phase II, and 3 phase III trials. Genta is active with Genasense, and antisense to Bcl 2 for antitumor cell treatment is in phase III. AVI Biopharm has a third generation antisense platform, and around this it is testing four phase I, five phase II, and two phase III trials. Hybridon has conducted two phase I and has two phase II trials planned.

2. Mechanism of Antisense Inhibition

Antisense oligonucleotides (AS-ODNs) are designed to bind and inactivate specific mRNA sequences inside cells. The potential uses for AS-ODNs is vast because RNA is so ubiquitous and abundant. With the publication of the human genome sequence, we now have such a wide open access to the sequences of genes that antisense can in theory be applied to almost every known gene to inhibit its mRNA. Inhibiting mRNA prevents specific proteins from being produced. Although routine human therapy may have been difficult to achieve, at a scientific level, antisense gene knockdown has become one of the fastest ways to study new therapeutic targets.

AS-ODNs are synthetically made, single-stranded short sequences of DNA bases designed to hybridize to specific sequences of mRNA forming a duplex. This DNA-RNA coupling attracts an endogenous nuclease, RNase H, that destroys the bound RNA and frees the DNA antisense to rehybridize with another copy of mRNA *(2)*. In this way, the effect is not only highly specific but prolonged because of the recycling of the antisense DNA sequence. The reduction in mRNA reduces the total amount of protein specified by mRNA. It is also theorized that hybridization sterically prevents ribosomes from translating the message of the mRNA into protein. Therefore, there are at least two

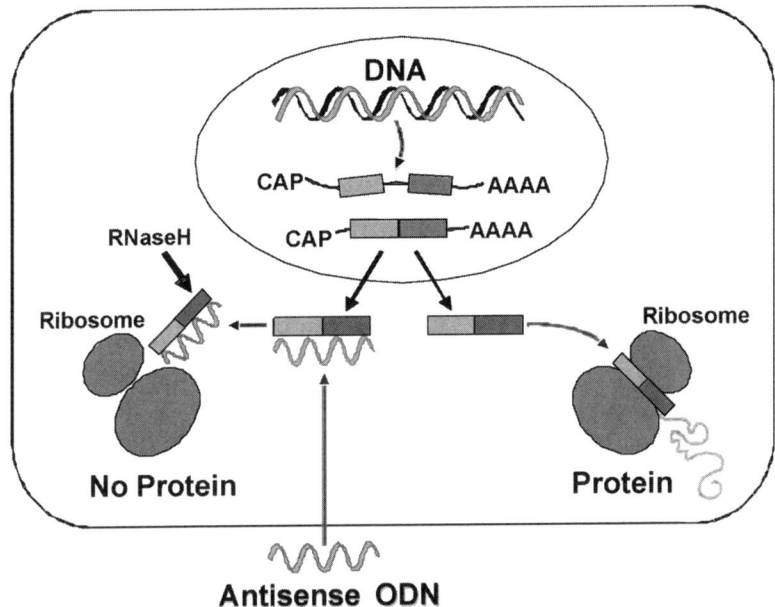

Fig. 1. Mechanism of AS-ODN posttranscriptional inhibition. AS-ODN enters the cell by an unknown uptake mechanism and hybridizes with a copy of a specific mRNA. The ODN-RNA duplex then prevents protein translation by (1) attracting RNase H to degrade the RNA and (2) steric hindrance of the ribosomal access and/or assembly. Note that the extent of inhibition depends on the AS-ODN competing with endogenous copies of RNA.

ways in which antisenses can work to effectively reduce the amount of protein being elaborated: RNase H degradation of RNA and hindering of ribosomal assembly and translation (**Fig. 1**). However, unless the antisense is designed to inhibit transcription, antisense would rarely be 100% inhibitory because the antisense inhibition of RNA does not shut down the transcription of endogenous copies of mRNA. It competes with the RNA being produced by the cell, and the effect is a gene knockdown rather than knockout. This has the advantage of being more physiological as a therapeutic agent, since antisense does not cause a mutation and does not prevent a protein that is involved in normal physiology from assuming its role. What antisense therapy does very effectively is reduce overexpression of proteins, and it is the overexpression of proteins that can cause disease states.

3. Stability

One of the problems that dogged early attempts to achieve a therapy with antisense was the question of stability. This is largely being answered by numerous ways to modify backbones of the DNA sequence in an AS-ODN. Native DNA has a phosphodiester bridge between each successive base of the DNA sequence. It was quickly learned that unmodified AS-ODNs were very short lasting, because they were unprotected from breakdown by nucleases, which break apart the nuclear acids. A very successful modification was phosphorothioate in which a sulfur atom replaces one oxygen atom in the phosphorate group of the phosphodiester bond. Phosphorothioate oligonucleotides are resistant to nucleases and are stable. This extends the life of the AS-ODN to several days instead of a few hours. Many variations on this theme have been tested and patented so that there is now a range of second- and even third-generation backbone modifications available *(2–4)*. Each company appears to favor its own particular modification. Isis uses phosphorothivates with 2'-*O*-methyl modification. Hybridon favors its IMO™ backbone modification, which can increase or decrease immunomodulation. AVI Biopharm has used NeuGene® as a platform of third-generation antisense for its nine clinical trials. A factor in developing backbone modifications such as these and others, including peptide nucleic acid, is the cost.

4. Cellular Uptake

Another area that has required time (and money) to investigate is the optimal conditions for uptake and distribution. This is particularly important when it comes to systemic injection as opposed to the early experiments in which antisenses were simply applied to cells in culture. There is both uptake and efflux of intact AS-ODNs in cells *(5)*.The backbone modifications become extremely important when systemic injections are used because of nucleases and the binding of oligonucleotides to proteins. The backbone modification can alter cell uptake, distribution, metabolism, and excretion. Nonantisense effects are a concern because they may alter the interpretation of whether the antisense effect is truly through an antisense mechanism or not. Mechanisms for the uptake of oligonucleotides into cells are still not clearly understood. The lack of a theory of the uptake and kinetic effects on oligonucleotides has required a lot of trial-and-error studies. This affects how to determine the optimal length of the oligonucleotide, the optimal concentration for effective treatment, and the frequency of treatments to maintain constant therapy. Despite these complications and holes in the study of antisense, phosphorothioated oligonucleotides are surprisingly easy to work with. In our own studies, which were in vivo applications of AS-ODN, we aimed injections into the brain and into the blood at receptor targets involved in cardiovascular disease. We found highly significant effects using AS-ODNs of 15–18 bases in length delivered in the brain without any vehicle *(6)* and in the blood delivered with liposomes *(7)*. Call it science or dumb luck, we nevertheless were able to show significant physiological effects of antisense delivery in models of hypertension. Because hypertension is a chronic disease, the findings were remarkable because of the long-lasting efficacy of a single antisense treatment. Reductions in blood pressure lasted weeks with a single systemic injection of antisense targeting β-1 receptors *(8)*.

The distribution of AS-ODNs injected systemically is to all parts of the body except the brain. The lipophobicity and/or negative change appear to prevent AS-ODNs from crossing the blood-brain barrier. However, the oligonucleotides accumulate in liver, kidney, and spleen. The lack of entry into the brain probably translates into few side effects. With the antisense to β-1 receptors, this could be a definite advantage *(8)*. For treating liver or kidney disease, however, AS-ODNs might have a built-in advantage in terms of delivery.

5. The Target

Clearly, the target protein for antisense inhibition is crucially important for a therapeutic effect. To reach the target, the antisense therapy must enter the cell through an uptake mechanism and escape from endosomes and lysosomes within the cell in sufficient amounts to avoid intracellular degradation. If the target mRNA is shielded or coiled, it may be difficult for AS-ODNs to hybridize. DNA and RNA are folded and studded with regulated proteins. Predicting how RNA folds and its secondary structures in a living cell is still very difficult. Once again, trial and error must be used. The stability of the oligos also depends on the interactions of the G-C proportions because of the three hydrogen bonds instead of the two hydrogen bonds that are in the A-T interaction. Having sufficient length of bases is necessary to make a specific match, but having too long a sequence can overlap the coding regions and inhibit more than single-target RNA.

Even when everything is successful and there is good uptake—good inhibition of the target—it does not necessarily lead to a therapeutic effect, because the target may not be the only player in the disease. If knocking down one gene leads to an increase in a compensatory gene, there may be little or no effect. Alternatively, a target gene may have been involved in starting the disease, but once the disease is present that target is no longer necessary, and, therefore, inhibiting it does not alter the disease state. Targeting transcription factors or signaling pathway proteins important in regulating cells may not be specific enough. If the target protein is overexpressed only in the disease state, then antisense should be efficacious, but if the target is similarly expressed in both normal and malignant cells, antisense treatment may cause both types of cells to undergo apoptosis. Then the therapy becomes a question of benefit vs risk. Because of the competition for RNA inhibition with antisense vs endogenous production of copies of mRNA in a cell, antisense for cancer is not a cell killer and, therefore, will not destroy all cancerous cells. However, it can be used with other treatments for cancer, and that is the protocol proposed for Affinitak and for Genasense.

6. Alternative to Oligonucleotides

In recent years, there has been a tremendous increase in interest in morpholinos *(9)*, small inhibitory RNA (siRNA) *(10)*, as well as ribozymes *(11)*. Morpho-linos are assembled from four different morpholino subunits each of which contains one of the four genetic bases linked to a six-sided morpholine ring. Morpholinos are supposed to have complete resistance to nucleases, high sequence specificity, and predictable targeting because they invade the RNA secondary structure and are fast and easy to deliver to the nucleus without liposome delivery systems. siRNAs are double-stranded RNA (dsRNA) molecules of 21–25 bp in length. They mediate RNA interference, an antiviral response initially identified in *Caenorhabditis elegans* and subsequently found active in specific gene silencing in many other organisms including mammalian cells. The sense and antisense strands of an siRNA first unwind, and the antisense strand binds to the target mRNA and recruits RNA-induced silencing complex (RISC) (**Fig. 2**). The sense strand is released from RISC, and RISC catalyzes the mRNA cleavage. The gene silencing efficiency of siRNA has reportedly been greater than antisense in general, typically reaching 80–90%. However, the maximal effects of optimal AS-ODNs and siRNAs targeting the same mRNA sequence are comparable. siRNAs are being used because of their stability and specificity, but it is not clear how effective they will be in systemic injections or oral delivery. Vickers et al. *(12)* conducted a comparative study of single-stranded AS-ODNs vs siRNA. Examination of 80 siRNA oligonucleotide duplexes designed to bind human RNA showed that both strategies are valid in terms of potency, maximal effects, specificity, and duration of action, at least in vitro.

The design of AS-ODNs and siRNAs follows different rules. Unlike AS-ODNs, the selection of an effective siRNA does not depend on the secondary mRNA structure or sequence accessibility. Instead, nucleotide composition and the release rate of the sense strand from RISC seem to play major roles. Several siRNA molecules targeting the same mRNA can be used in combination to achieve greater effects and to avoid cellular resistance to siRNA. An independent combinatorial effect of AS-ODNs and siRNAs has also been observed

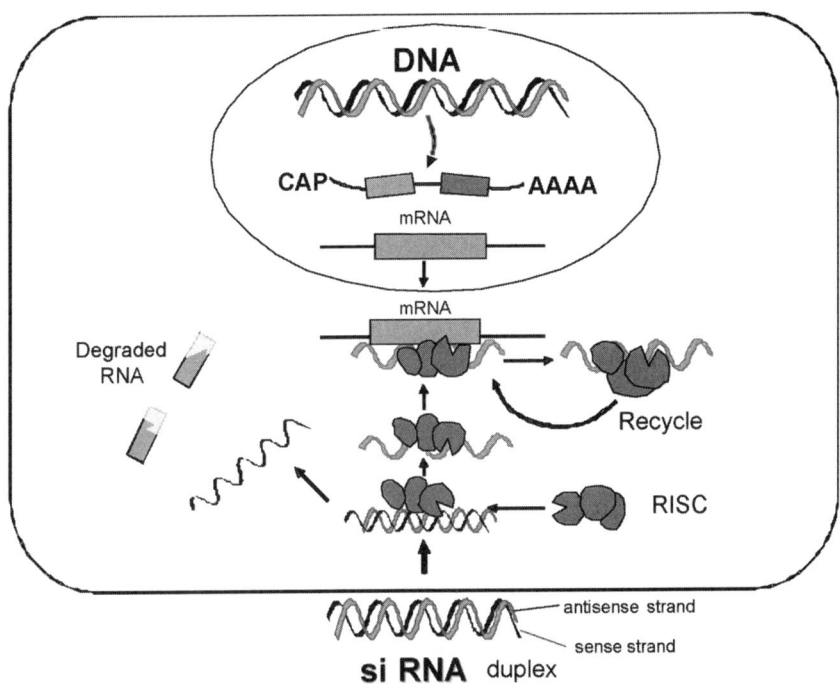

Fig. 2. Mechanism of siRNA. Synthetic siRNA enters the cell as a dsRNA with sense and antisense strands. RISC multiprotein made up of helicase, RNase III, and an activating protein unwinds the two strands of RNA and uses the antisense to recognize the chosen sequence of RNA. The RNase cleaves the sequence of mRNA, which is degraded by cellular nucleases. The RISC-antisense complex can then recycle and silence more copies of mRNA.

when siRNA was coadministered with nonhomologous AS-ODNs, targeting distant regions of the same mRNA. As alternative therapeutics, development of siRNA has covered a wide variety of disease models in a short time. The most studied fields of siRNA application are cancer and infectious diseases. siRNA has been administered in vivo in unmodified states. Following iv injection into mice, the highest inhibition of target mRNA was found in liver, kidney, spleen, lung, and pancreas. If both strategies are equally effective, then the deciding factor in choosing one over the other would depend on the price of production. In addition, experience with AS-ODNs will count for some time against the newness of siRNA molecules. However, a lot will depend on whether there are side effects that are not due to the antisense mechanism, or if one approach is associated with more side effects than the other.

7. Conclusion

The brief history of antisense therapeutics has been characterized by cycles of success and disappointment. However, through it all, the promise of antisense therapy has been so appealing that hope remains for that block-buster breakthrough that will open the doors for so many potential treatments. There are now thousands of targets available with known genomic sequences. There are hundreds of preclinical studies pointing to new treatments with antisense. And there are a score of human trials that are paving the way. Once one major treatment is accepted, each new antisense therapy will be more easily and quickly brought to those who suffer from diseases that are not yet satisfactorily treated with drugs.

References

1. Crooke, S. T. (2004) Progress in antisense technology. *Annu. Rev. Med.* **55,** 61–95.
2. Crooke, S. T. (1998) Molecular mechanisms of antisense drugs: RNase H. *Antisense Nucleic Acid Drug Dev.* **8(2),** 133–134.
3. Wickstrom, E. and Smith, J. B. (1998) DNA combination therapy to stop tumor growth. *Cancer J. Sci. Am.* **4(Suppl. 1),** S43–S47.
4. Agrawal, S., Kandimalla, E. R., Yu, D., et al. (2002) GEM 231, a second-generation antisense agent complementary to protein kinase A alpha subunit, potentiate antitumor activity of irinotecan in human colon, pancreas, prostrate and lung cancer xenografts. *Int. J. Oncol.* **21(1),** 65–72.
5. Li, B., Hughes, J. A., and Phillips, M. I. (1997) Uptake and efflux of intact antisense phosphorothioate deoxyoligonucleotide directed against angiotensin receptors in bovine adrenal cells. *Neurochem. Int.* **31(3),** 393–403.
6. Gyurko, R., Wielbo, D., and Phillips, M. I. (1993) Antisense inhibition of AT1 receptor mRNA and angiotensinogen mRNA in the brain of spontaneously hypertensive rats reduces hypertension of neurogenic origin. *Regul. Pept.* **49(2),** 167–174.
7. Phillips, M. I. (2001) Gene therapy for hypertension: sense and antisense strategies. *Expert Opin. Biol. Ther.* **(4),** 655–662.
8. Zhang, Y. C., Bui, J. D., Shen, L., and Phillips, M. I. (2000) Antisense inhibition of beta(1)-adrenergic receptor mRNA in a single dose produces a profound and prolonged reduction in high blood pressure in spontaneously hypertensive rats. *Circulation* **101(6),** 682–688.
9. Summerton, J. (1999) Morpholino antisense oligomers: the case for an RNase H-independent structural type. *Biochim. Biophys. Acta.* **1489(1),** 141–158.
10. Zamore, P. D. and Aronin, N. (2003) siRNAs knocks down hepatitis. *Nat. Med.* **9(3),** 266–267.
11. Fedor, M. J. and Westhof, E. (2002) Ribozymes: the first 20 years. *Mol. Cell.* **10(4),** 703–704.
12. Vickers, T. A., Koo, S., Bennett, C. F., Crooke, S. T., Dean, N. M., and Baker, B. F. (2003) Efficient reduction of target RNAs by small interfering RNA and RNase H–dependent antisense agents: a comparative analysis. *J. Biol. Chem.* **278(9),** 7108–7118.

2

Antisense Inhibition

Oligonucleotides, Ribozymes, and siRNAs

Y. Clare Zhang, Meghan M. Taylor, Willis K. Samson, and M. Ian Phillips

1. Introduction

Over a span of more than two decades, antisense strategies for gene therapy have expanded from antisense oligonucleotides (AS-ODNs) solely, to the addition of ribozymes and, more recently, to the inclusion of small interfering RNAs (siRNAs). Antisense therapeutics has also experienced its phases of high expectation, sudden disappointment, and meticulous rediscovery, while maintaining its status as a viable and effective gene therapy approach. With the discovery of RNA interference (RNAi) and development in delivery of these gene drugs, more preclinical and clinical investigations are anticipated to take place in the near future to finally fulfill the promise of antisense therapeutics in humans.

2. Antisense Oligonucleotides

AS-ODNs are typically 18–25 bases in length, consisting of sequences that are complementary to the target RNA. They can be injected directly into tissues or delivered systemically. Once delivered into cells, oligonucleotide binds to its RNA counterpart and suppresses expression of the proteins encoded by target RNA. The specificity of this approach is based on the probability that any sequence longer than a minimal number of nucleotides (nt)—13 for RNA and 17 for DNA—occurs only once within the human genome The idea of antisense therapy for inhibiting disease-associated proteins has become par-

From: *Methods in Molecular Medicine, Vol. 106: Antisense Therapeutics, Second Edition*
Edited by: I. Phillips © Humana Press Inc., Totowa, NJ

ticularly appealing since Zamecnik and Stephenson *(1)* first demonstrated in 1978 the reduction of Rous sarcoma viral RNA translation by a specific oligo-nucleotide.

2.1. Mechanisms of Antisense Inhibition

Gene expression can be altered by oligonucleotides by means of either posttranscriptional inhibition or splicing shift. Posttranscriptional inhibition is accomplished by several mechanisms including sterical blockade of ribosomal access to the target mRNA, induction of RNase H cleavage of mRNA, and inhibition of ribosomal assembly. The net outcome of this process is the diminished translation of target proteins. Oligonucleotides chemically modified by phosphorothioation are especially effective in activating RNase H, resulting in sequence-specific digestion of the target mRNA molecules. This destruction of RNA while leaving the DNA oligonucleotide intact allows the oligonucleotide to be recycled, which makes AS-ODNs long lasting. A majority of antisense studies so far, including most clinical trials, are aimed at reducing undesired disease-associated proteins by virtue of translational inhibition. Alternatively, oligonucleotides that are RNase H inactive and designed toward a certain exon–intron junction can prevent the pre-mRNA splicing at the targeted site and redirect the splicing to a more favored site. The therapeutic potential of this approach has been exemplified in the correction of the expression of β-globin and the breast cancer gene *BCL-X* in related diseases. Certain forms of β-thalassemia are caused by aberrant splicing of β-globin pre-mRNA that leads to abrogation of the protein production *(2)*. AS-ODNs designed to the untoward splice site have been proven effective at inhibiting aberrant splicing and at restoring β-globin expression in thalassemic patients *(3)*. Likewise, alternative splicing of *BCL-X* pre-mRNA gives rise to two isoforms, *BCL-XL* and *BCL-XS*, with opposing antiapoptotic and proapoptotic activities. Targeting the *BCL-XL* splice site with oligonucleotides favored production of the proapoptotic *BCL-XS* protein that enhances cell death in prostate and breast tumor cells *(4)*.

2.2. Targeting Antisense

Although antisense can be designed against any region of the target RNA in theory, different sequences vary markedly in efficiency of gene inhibition. The accessibility of oligonucleotides to RNA is considered the most important factor in choosing the optimal antisense sequences. Computational analysis of the secondary structure of RNA by programs such as mfold or RNAstructure has been used to facilitate selection of target sites for antisense action *(5)*; however, it does not take into account the three-dimensional structures as well as the instant interaction of RNA molecules with other factors. More commonly taken routes involve evaluation of accessible sites by use of RNase H mapping *(6)* or scanning oligonucleotide arrays for the best hybridization signals *(7)*. Nevertheless, in general, targeting the start codon AUG, where mRNA is supposedly open for ribosomal entry, has been a successful strategy, although in many cases other sequences turned out to be more effective. Despite these predictive approaches, the selection of optimal antisense sequences still requires trial-and-error testing initially and, in the end, needs to be confirmed in vivo.

2.3. Chemical Modifications

Stability and efficient delivery, prerequisites for oligonucleotides to achieve observable therapeutic effects, have been obstacles due to their macromolecular nature. Numerous chemical modifications and delivery approaches have been developed to overcome this problem (**Fig. 1**). The first generation of antisense agents contains backbone modifications such as replacement of oxygen atom of the phosphate linkage by sulfur (phosphorothioates), methyl group (methylphosphonates), or amines (phosphoramidates). Of these, the phosphorothioates have been the most successful and used for gene silencing because of their sufficient resistance to nucleases and ability to induce RNase H functions. However, their profiles of binding affinity to the target sequences, specificity, and cellular uptake are less satisfactory. The second generation of antisense modifications was aimed at improving these properties, among which substitutions of position 2' of ribose with an alkoxyl group (e.g., methyl or methoxyethyl groups) were most successful. 2'-*O*-methyl and 2'-*O*-methoxyethyl derivatives can be further combined with phosphorothioate linkage *(8)*. The third generation contains structural elements, such as zwitterionic oligonucleotides (possessing both positive and negative charges in the molecule); locked nucleic acids (LNAs)/bridged nucleic acids (BNAs) *(9)*; morpholino *(10)*; peptide nucleic acids (PNAs) (with a pseudopeptide backbone) *(11)*; and, more recently, hexitol nucleic acids (HNA) *(12)*. All of the modifications enhanced AA-ODNs in terms of nuclease resistance; specific binding; and with agents such as PNA and morpholino, cellular uptake. However, the ability of oligonucleotides to induce RNase H cleavage was abolished by these alterations. Therefore, chimeric oligonucleotides with an unmodified RNase H–susceptible core flanked by modified nuclease-resistant nucleotides have recently been proposed to address this issue and applied in a number of investigations *(13)*, including clinical trials.

2.4. Delivery of Antisense

Oligonucleotides are primarily taken up by cells via endocytosis. Only a portion of oligonucleotides are able to escape endosome/lysome, enter the nucleus, and bind to its RNA complement. Because of the hydrophilic and

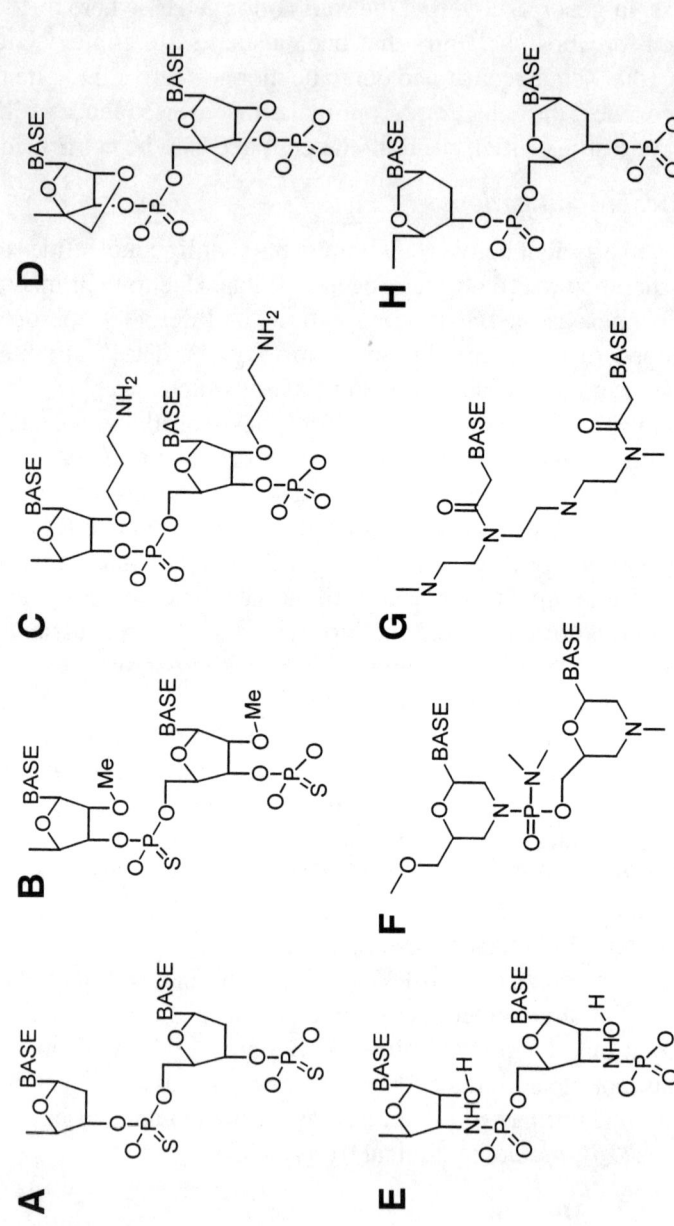

Fig. 1. Structures of synthetic oligonucleotides: (**A**) phosphodiester; (**B**) 2'-*O*-methyl phosphorothioate; (**C**) 2'-*O*-aminopropyl phosphorothioate. (**B**) 2'-*O*-methyl phosphorothioate; (**C**) 2'-*O*-aminopropyl phosphodiester; (**D**) locked/bridged nucleic acids (LNA/BNA); (**E**) phosphoramidate; (**F**) morpholino; (**G**) peptide nucleic acid (PNA); (**H**) hexitol nucleic acid (HNA).

macromolecular nature, permeation of oligonucleotides across cell membrane is relatively difficult. Even after two decades of research, safe and efficient delivery of oligonucleotides in vivo still remains a major barrier to the clinical success of antisense therapies. Cationic liposomes and electroporation are commonly used carriers. A large variety of liposomal formulas have been developed to facilitate antisense delivery, some of which have entered clinical trials *(14)*. More recently, nanoparticles and oligonucleotide conjugates have shown improved cellular uptake, biodistribution, and targeted delivery, especially in cancer treatment *(15,16)*. A hydrodynamic tail vein injection has proven very effective in delivering oligonucleotides into liver of rodents *(17)*. Inhalable and topical applications of oligonucleotides in patients have shown satisfactory profiles of uptake and distribution *(18,19)*. However, interestingly, most AS-ODNs that are therapeutically valuable in animal models and in patients have been administered in the form of naked compounds, despite the progress in antisense delivery.

2.5. Antisense in Therapies

Antisense therapeutics has seen its ups and downs since the first antisense trial was planned in leukemia in 1992 *(20)*, followed by the excitement over the FDA approval of the first antisense drug, Fomivirsen, for the treatment of cytomeglovirus (CMV) retinitis in 1998 *(21)*. In addition, more recently, a phase III trial reported disappointing results for Affinitak (an antisense inhibitor of protein kinase C-α [PKC-α]) for the treatment of non–small cell lung cancer (NSCLC). Cancer is the major target of ongoing clinical trials using antisense therapies, followed by human immunodeficiency virus (HIV) and other immune-related diseases (**Table 1**). The targets of antisense for cancer treatment include genes involved in cell growth, apoptosis, angiogenesis, and metastasis. A limitation for antisense as a therapy for cancer may be the single-target approach. Even if the target is successfully inhibited by antisense, other targets may be activated and compensate for the antisense inhibition. Another potential problem is that for successful suppression of cancer growth, the inhibition should be 100%. However, the mechanism of antisense inhibition is always in competition with constitutive copies of mRNA, making a 100% knockdown difficult to achieve. It is noteworthy that after extensive efforts at endogenous expression of antisense RNA by plasmids and viral vectors in a variety of disease models, viral delivery of antisense has recently advanced to human patients; VRX 496 (a lentivirus vector encoding antisense to HIV-1 env protein) started its phase I trial in 2003. Cancer vaccine, a cell therapy using NSCLC cell lines genetically engineered to express transforming growth factor-β (TGF-β) antisense, has also been tested in patients with lung cancer. With the emergence of new generations of modified oligonucleotides and delivery

Table 1
Ongoing Clinical Trials for Antisense Therapy

mRNA target	Drug	Company	Diseases[a]	Phase	Notes[b]
BCL-2	G3139 (Genasense)	Genta/Aventis	Melanoma, MM, CLL, NSCLC	III	18mer/PS
Ha-Ras	ISIS 2503	ISIS	Solid tumors	II	20mer/PS
PKC-α	ISIS 3521 (Affinitak)	ISIS/Eli Lilly	NSCLC, solid tumors	III	20mer/PS
c-RAF	ISIS 5132	ISIS	Solid tumors	Discontinued	20mer/PS
PKC	PKC412	Novartis	Solid tumors, eye infection	II	
PKA-R1α	GEM 231	Hybridon	Solid tumors	I–II	18mer/AC
Ribonucleotide reductase	GTI 2040, GTI 2501	Lorus	Solid tumors	I–II	21mer/PS
c-RAF	LErafAON	Neopharm	Solid tumors	I–II	
c-MYC	Oncomyc-NG	AVI BioPharma	Cancer	I–II	
c-MYC	AVI 4126	AVI BioPharma	Cancer, kidney disease	I–II	
Clusterin	OGX-011	OncoGenex	Prostate cancer	I	21mer/AC
TGF-β2	Cancer vaccine	NovaRx	NSCLC	II	NSCLC cells engineered to express TGF-β2 antisene
Cytochrome P450	AVI 4557 (Neugene)	AVI BioPharma	Drug adverse effects	I–II	
c-MYB	LR/INX-3001	Gewirtz et al.	CML	I–II	24mer/PS
DNA methyltransferase	MG-98	MethylGene	Solid tumors	I–II	20mer/AC
HIV-1	HGTV-43	Enzo Biochem	HIV	I	
HIV gag	GEM-92	Hybridon	HIV	I	
HIV env	VRX496	VIRxSYS	HIV	I	Antisense in lentivirus

c-MYC	Resten-NG	AVI BioPharm	CAD, kidney disease, cancer	I–II
HCV	ISIS 14803	ISIS	HCV	II
ICAM-1	ISIS 2302 (alicaforsen)	ISIS	Crohn disease, psoriasis	II–III
Adenosine A1R	EPI-2010	EpiGenesis	Asthma	II
TNF-α	ISIS 104838	ISIS	Arthritis, Crohn disease, psoriasis	II

[a]MM, multiple myeloma; CLL, chronic lymphatic leukemia; CML, chronic myelogenous leukemia; CAD, coronary artery disease. [b]PS, phosphorothioates; AC, advanced chemistry oligonucleotides.

technologies, antisense therapeutics is closer to fulfilling its promise in the clinic for diseases other than cancer, such as cardiovascular disease, psoriasis, and Crohn's disease.

3. Ribozymes
3.1. What Are Ribozymes?

It was discovered in the early 1980s that some naturally occurring RNA molecules have enzymatic activity *(22,23)*. These enzymatic RNA molecules were termed *ribozymes*. Ribozymes recognize specific RNA sequences and then catalyze a site-specific phosphodiester bond cleavage within the target molecule. Following cleavage, the ribozyme releases itself and binds to another target molecule, repeating the process. The cellular consequence varies depending on the setting. There are many naturally occurring ribozymes, including in plant viroids, ribosomes, self-splicing introns, and the RNA portion of RNase P. In plant and animal cells, as well as in viruses, ribozymes are necessary for some normal cellular processes such as transcription. The goal of most synthetic ribozyme usage, however, is reduction in targeted RNA and, thus, lower levels of the protein encoded by the target RNA.

Ribozyme substrate recognition occurs in the same manner as antisense pairing, through strand complementarity. Therefore, any decrease in target protein following ribozyme treatment could in part be due to antisense inhibition of translation or the recruitment of cellular enzymes to the double-stranded RNA (dsRNA) molecules. However, the ability of each ribozyme molecule to rapidly cleave multiple target molecules gives this technology an advantage over classic antisense that can act only on a single RNA molecule. In fact, the rate constants of ribozyme cleavage reactions can approach and exceed those of protein enzymes, including enzymes with similar functions such as RNase A *(24,25)*.

There are multiple types of ribozymes; the two most commonly used for research and therapeutic purposes are the hammerhead ribozyme and the hairpin ribozyme (**Figs. 2** and **3**). One of the smallest and most well-understood ribozymes, the hammerhead ribozyme, is composed of 30–40 nt and was originally discovered as a common sequence found in plant viroids that undergo site-specific, self-catalyzed cleavage as part of their replication process *(26)*. All hammerhead ribozymes have a common structure consisting of three base-paired helices connected by two invariant single-stranded regions forming the catalytic core. Helices 1 and 3 contain the antisense arms of the ribozyme. Helix 3 also contains the cleavage triplet, the site that is cut by the catalytic core. The triplet most commonly found in naturally occurring hammerhead ribozymes is GUC; however, mutagenesis studies have shown that any cleav-

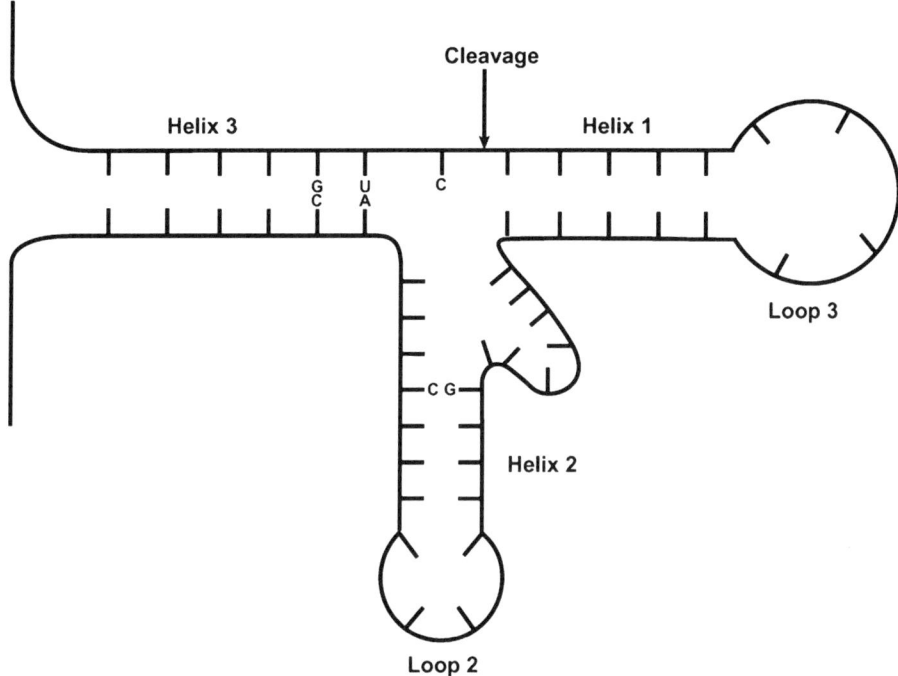

Fig. 2. Schematic of a natural hammerhead ribozyme. Hammerhead ribozymes consist of three helices, formed by complementary base pairing, which are connected by single-stranded regions. Loop 3 is removed to generate a *trans*-cleaving ribozyme; helices 1 and 3 then form the antisense arms. The most commonly found cleavage triplet, GUC, is indicated, as is the cleavage site. The single-stranded domain at the top of helix 2 is the catalytic core. Highly conserved GC residues in helix 2 are necessary for catalytic activity.

age triplet with the sequence NUH is tolerated, in which N is any nucleotide and H is A, U, or C *(27)*. Hammerhead ribozymes catalyze the hydrolysis of the phosphodiester bond at the 3' end of the cleavage triplet. The mechanism requires a divalent metal ion, usually Mg^{2+}, which plays two crucial roles in ribozyme function: it promotes proper folding of the catalytic core and also is a catalytic cofactor *(28)*.

Native hammerhead ribozymes are *cis*-cleaving enzymes, meaning that their targets lie within the same RNA molecule. The ribozyme structure can be engineered to create an intermolecular cleaving ribozyme consisting of two single-stranded antisense arms surrounding the catalytic core and helix 2 so that it will cleave within a different RNA molecule. Because RNA often folds into

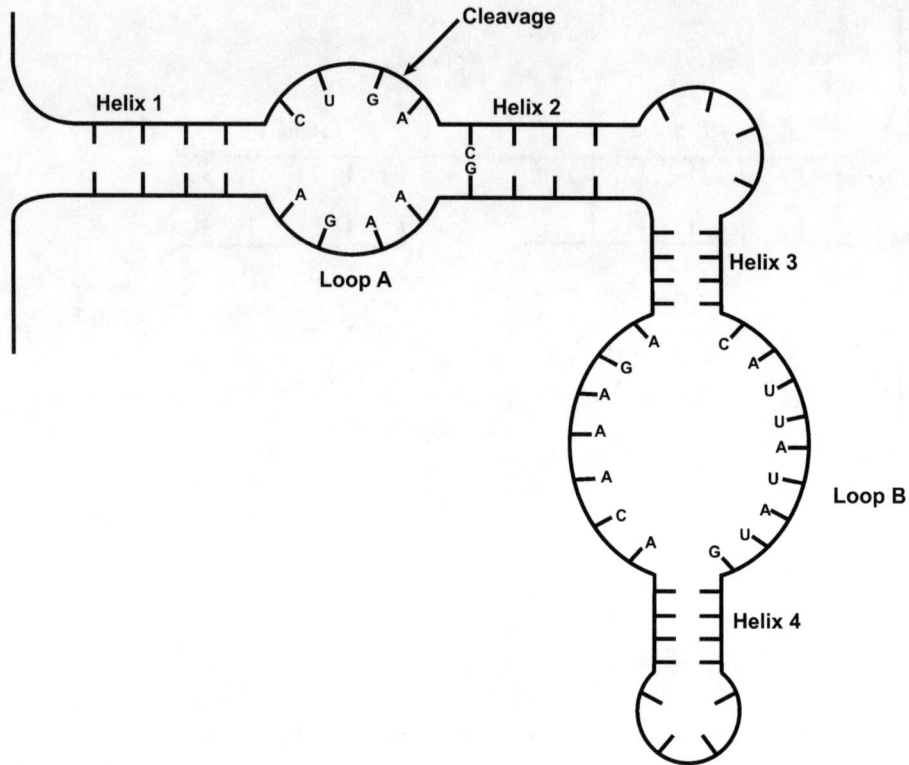

Fig. 3. Schematic of a natural hairpin ribozyme. Hairpin ribozymes consist of 4 helices, formed by complementary base pairing, which are connected by single stranded regions. The small loop at the base of the ribozyme is removed to generate a *trans*-cleaving hairpin ribozyme; helices 1–4 then form portions of the antisense arms. The cleavage site is indicated. Loops A and B comprise the catalytic core domains. The sequences of both loops are highly conserved, as are the GC residues in helix 2.

complex secondary structures, the accessibility of the target site to the annealing arms of the ribozyme must be considered when designing a ribozyme. Arm lengths of 7 to 8 nt are optimal to convey both specificity and access to most ribozymes *(29)*. These shorter annealing arms also aid in turnover of the ribozyme, enhancing the ability of each ribozyme molecule to cleave multiple target RNA molecules *(30)*.

Hairpin ribozymes, like hammerhead ribozymes, are found in some plant viroids that undergo self-catalyzed cleavage as part of their replication process. Hairpin ribozymes contain four base-paired helices and two unpaired loops. The ribozyme cleavage site resides within loop A. The helices can vary in length and will tolerate any sequence that maintains complementarity with the exception of a requirement for a guanine residue located at the beginning of helix 2, which is required for cleavage site recognition *(31)*. Nucleotides within the catalytic loop regions, however, must be highly conserved to ensure catalytic activity of the ribozyme *(31)*.

Hairpin ribozymes catalyze site-specific hydrolysis of the phosphodiester bond on the complementary strand of RNA that is one base upstream of the conserved guanine in helix 2. Hairpin ribozymes, like hammerhead ribozymes, require an Mg^{2+} ion to activate proper secondary structure. However, unlike the hammerhead ribozyme, Mg^{2+} does not play a direct role in the catalytic process *(32)*. The exact catalytic mechanism used by hairpin ribozymes is not yet fully understood. A greater understanding of how both hammerhead and hairpin ribozymes work and of methods to optimize their function will enhance their attractiveness as potential therapeutic agents.

3.2. Delivery of Ribozymes

Two major issues in the use of ribozymes for research and therapy are ensuring that the ribozyme is delivered to the target tissues and ensuring that the levels of ribozyme delivered are adequate to produce the desired effect. There are two methods for delivering the ribozyme to cells: exogenous delivery of a presynthesized ribozyme or endogenous expression of the ribozyme. Exogenous delivery is relatively easy and rapid; however, as with antisense, there are two main problems with this technique; cellular uptake of the ribozyme is often difficult to achieve, and once the ribozyme is taken up, it is quickly degraded. Cellular uptake of the ribozyme can be enhanced through the use of cationic liposomes. These cationic lipid micelles have the added benefit of protecting the ribozymes from RNase present in serum. To enhance further the lifespan of ribozymes they are frequently chemically modified. The addition of a 2'-*O*-methyl moiety on some or all of the bases is the most commonly used modification. Work is currently being done to engineer DNAzymes, which should be more stable than their ribozyme RNA counterparts *(33)*. One benefit of exogenous ribozyme delivery in vivo is that the immune system is fairly tolerant of foreign RNA molecules *(34)*.

The other method for delivering ribozymes, endogenous expression of the ribozyme, is most often accomplished using viral vectors; however, plasmid vectors may also be used. Both retroviral and DNA viral vectors have been used. Expression cassettes can be designed to carry cell type–specific or conditional transcription initiation sites, as well as to include reporter proteins. The big advantage of an endogenous ribozyme is that it can be continuously produced, allowing for the compromise of target protein production over a long period of time.

3.3. Research and Therapeutic Uses of Ribozymes

There are four main uses of ribozymes in the medical field: as a research tool, as a chemotherapeutic agent, as an antiviral agent, and as a method to overcome acquired dominant genetic diseases.

With the recent sequencing of the *Drosophila*, mouse, and human genomes, there was a surge of newly identified proteins whose role in the organism is currently unknown or not fully understood. The use of ribozymes to selectively target these new proteins offers an attractive method to rapidly screen their role in vivo. This method, along with other antisense techniques, offers several advantages over traditional methods of screening proteins. First, only a partial cDNA sequence is required to design a ribozyme. Second, ribozymes can be generated very rapidly, whereas both traditional and conditional knockout animals as well as transgenic overexpression animals require a significant amount of time to generate. Finally, ribozymes can lead to greater effects for longer periods of time when compared with antibody neutralization of the target protein.

In addition to rapid screening of new proteins, ribozyme technology can also be used to overcome problems with traditional protein function studies. For example, we use ribozymes to target a protein that when knocked out results in embryonic demise in mice and for which conditional knockouts have been unsuccessful *(35)*. Ribozymes can also be used to locally target a protein that is made in many tissues, such as to lower targeted protein levels in brain without altering protein expression in the periphery.

The specificity of ribozymes makes them very attractive as therapeutics in disease states in which a protein is overexpressed or is malfunctioning. Ribozymes have the capability to specifically recognize single nucleotide differences in their targets. This special feature has resulted in the development of ribozymes to target oncogenes that are frequently mutated in tumors. For instance, the oncogene H-ras is mutated at a high frequency in many cancers; therefore, a ribozyme that recognizes only the mutant H-ras transcript has the potential to be a very efficacious treatment. Several ribozymes have been developed that can discriminate between H-ras mutants and the normal H-ras transcript and initial studies have shown that stable expression of H-ras mutant ribozymes leads to reduced tumor formation in athymic mice *(36,37)*.

Alternative uses for ribozymes in cancer therapy are to block the elevation of normal gene products, such as c-fos, that occur in transformed cells or to block angiogenic pathways. One such antiangiogenic ribozyme is targeted to flt-1 mRNA, which encodes for the vascular endothelial growth factor receptor (VEGF-R). The ribozyme has been shown to be well tolerated when administered daily by sc injection, and this dosing schedule leads to prolonged elevation in the plasma levels of ribozyme *(38)*. This ribozyme is now in phase II clinical trials in which therapeutic efficacy in breast and colorectal cancers is being examined.

A different set of circumstances in which a ribozyme can offer great therapeutic potential is the treatment of acquired dominant genetic diseases. Retinitis pigmentosa is a genetic disease that causes carriers of the dominant P23H rhodopsin allele to slowly lose their vision. Hauswirth and Lewin have developed a ribozyme that recognizes only the dominant version of the gene transcript, which differs by two bases from the wild-type gene. Ribozyme treatment has resulted in a halt of disease progression in various species including rat, dog, and now monkey *(39,40)*. This treatment is currently being prepared to enter the first phase of clinical trials. Promising results from this study could open the door for the development of ribozymes to treat other dominant genetic disorders.

One final area where ribozyme therapy holds much promise is as antiviral agents, particularly in the treatment of retroviral infections. Many RNA viruses such as HIV have very high mutation rates throughout much of their genome that renders the mutated viruses resistant to current treatments. However, some sequences, including promoters and slicing signals, are highly conserved in HIV and among other RNA viruses. These regions provide excellent targets for ribozymes. In fact, some groups have designed ribozymes against conserved areas of HIV and have shown that ribozyme treatment can provide long-term HIV resistance and decrease HIV replication in infected cells *(41)*. Several companies now have ribozymes directed against HIV in various stages of clinical trials. Other viruses for which ribozyme treatments are also being designed include hepatitis B, hepatitis C, and the herpes viruses. **Table 2** summarizes ongoing clinical trials using ribozymes.

The use of ribozymes for the inhibition of gene expression holds great promise in both therapeutics and research; however, we have only begun to understand the potential of these molecules. Efforts to improve the stability and delivery of ribozymes will enhance their usefulness as therapeutic agents and lead to a greater recognition of the role of novel proteins in selected tissues and the body as a whole.

4. RNA Interference with siRNA

4.1. What Is siRNA?

RNAi is a form of antiviral immune response mounted by many higher eukaryotes—including plants, nematodes, and insects—on exposure to dsRNA. dsRNA molecules are key intermediates in the genomic replication of many viruses but are not normally found in eukaryotic cells. In contrast to the inter-

Table 2
Ongoing Clinical Trials for Ribozyme Therapy

mRNA target	Drug	Company	Diseases	Phase	Notes
VEGF-R	Angiozyme	Ribozyme	Solid tumors	II	
HCV genome	LY466700				
	Heptazyme	Ribozyme	Hepatitis C	II	
EGF-R type 2	Herzyme	Ribozyme	Breast, ovarian cancer	I	
Proliferating cell nuclear antigen	VitrenASE	Immusol	Retinal disorders	II	
HIV genome		Ribozyme	non-Hodgkin lymphoma in HIV patients	II	Stem cells transduced with retrovirus expressing anti-HIV ribozymes

feron responses of mammalian cells in the face of viral infection, RNAi is used by many other eukaryotes to defend against viruses through dsRNA-induced degradation of viral RNAs.

The first evidence that dsRNA could suppress gene functions came from the work in *Caenorhabditis elegans (42)*. In 1998, Fire et al. *(43)* found that sense RNA was as effective as antisense RNA for inhibiting genes. Subsequently, Zamore et al. *(44)* demonstrated that dsRNA was at least 10-fold more potent as a silencing trigger than was sense or antisense RNA alone. Since then, gene silencing by dsRNA has been termed RNAi, and its mechanisms have been elucidated vigorously. Our current mechanistic understanding of RNAi derives largely from work in the *Drosophila* system *(44,45)*. The first step of RNAi is to process longer dsRNA into 21- to 23-nt fragments that bear 3' overhangs by an RNase III–like enzyme called Dicer *(46)*. These approx 21 nt dsRNAs, which are termed as siRNA, are essential to form a large (approx 500-kDa) RNA-induced silencing complex (RISC) *(47)*. Through a yet-undefined mechanism, RISC cleaves the target mRNA that is complementary to the guide siRNA, whether the target RNA is a viral mRNA or a cognate gene.

4.2. Application in Mammalian Cells

The key characteristics of RNAi are its remarkable sequence specificity, and it can therefore be used to target gene expression. It was found in *Drosophila* that artificial siRNAs can be incorporated into RISC and induce degradation of target mRNA. However, previous efforts to induce RNAi in cultured mammalian cells had largely failed because long dsRNAs (>30 bp) could induce a potent, nonspecific interferon response and activation of the protein kinase PKR and 2',5'-oligoadenylate synthetase *(48,49)*. dsRNAs shorter than 30 bp do not activate the PKR and interferon pathways. In 2001, a pioneering work by Elbashir et al. *(50)* demonstrated that transfection of 21 nt synthetic siRNAs into cultured human cells can effectively inhibit gene expression in a sequence-specific manner with extremely high inhibitory efficiency. The advantages and great potential of RNAi technology, including high efficiency (typically >85%), have since raised tremendous interest in this field and generated rapidly emerging progress (reviewed in **refs. 51** and **52**). These advances have expanded RNAi technology from the use of synthetic siRNA for the endogenous production of small hairpin RNA (shRNA) by plasmid and viral vectors, and from transient inhibition in vitro to longer-lasting effects in vivo and in transgenic animals. This makes possible the utilization of RNAi in suppressing undesirable genes for human gene therapy.

siRNA-mediated gene silencing is sequence specific and dose dependent. As with antisense oligonucleotides and ribozymes, the efficient delivery of siRNA into cells of choice is currently the limiting factor to successful gene inhibition. The delivery of siRNA can be in the form of naked compounds *(53)*, or markedly improved by cationic liposomes *(54)* or electroporation *(55)*, depending on cell types. Structural variations and sequence mutations have been made to investigate the structural and sequence requirements for siRNA-induced gene silencing. It was found that the status of the 5' hydroxyl terminus of the antisense strand of an siRNA determines RNAi activity, while a 3' terminus block is tolerated *(56–58)*. Sequence mutations, on the other hand, are generally tolerated at the 5' end, but not the 3' end *(59)*. Chemical modifications such as phosphorothioation and 2'-*O*-methylation and 2'-*O*-allylation were developed to improve the nuclease resistance of synthetic siRNAs *(56,59,60)*. Certain chemical modifications at selected sites prolonged the siRNA activities, whereas others compromised the efficiency. Simultaneous knockdown of more than two genes is possible, as illustrated by the double suppression of the nuclear mitotic apparatus protein (NUMA1) and lamin in HeLa cells *(61)*. However, different siRNA species can possibly undergo reversible competition in a sequence-independent manner *(58,62)*, suggesting that the RNAi machinery might be titratable or limited in mammalian cells.

4.3. siRNA as an Antiviral Agent

The ability of RNAi to protect plants and insects from viral infection can be applied to mammals, although mammalian cells themselves do not possess inherent RNAi mechanisms. siRNAs designed against a variety of viruses, HIV *(63–68)*, hepatitis C virus (HCV) *(69–74)*, hepatitis B virus *(75–77)*, papillomavirus *(78)*, herpesvirus *(79)*, rotavirus *(80)*, and influenza virus *(81)*, have been tested in cell cultures and displayed high efficiency in inhibiting viral infection and replication. In experiments aimed at HIV suppression, siRNAs have been targeted to various regions of the HIV genome including the long-terminal repeats and all five encoded genes, and typically a 30- to 50-fold decrease was observed in the viral levels in cell lines as well as in primary T-lymphocytes *(63–68)*. Similar cases were reported in the siRNA-mediated HCV suppression *(69–74)*. Silencing of HCV RNAs was dose dependent and specific, resulting in a dramatic decrease in HCV RNAs and clearance of HCV in hepatoma cell lines bearing an HCV subgenome. The effects lasted for 3 to 4 d with synthetic siRNA and for more than 3 wk with expression vectors *(70,72)*. It is noteworthy that siRNA-resistant virus strains might already be present in the original viral population and thereby selectively survive the treatment with siRNA, as reported in an attempt to inhibit poliovirus with siRNAs *(82)*. These results suggest that a pool of different siRNA sequences should be used to avoid the selection pressure for favoring siRNA-immune virus.

4.4. Comparison of siRNA AS-ODNs

Direct comparisons of AS-ODNs and siRNAs to the same targets were made in cell culture and in vivo *(83–87)*. Dose-response experiments revealed that the IC_{50} value for the siRNA was about 100-fold lower than that of the AS-ODNs *(83)*. The effect of siRNA is also longer lasting than of AS-ODNs because siRNAs are more stable in fetal calf serum, human plasma, and cell cultures *(84)*. By contrast, Vickers et al. *(87)* reported that optimized RNase-H-dependent oligonucleotides and siRNAs are comparable in terms of potency, maximal effectiveness, sequence specificity, and duration of action. An independent combinatorial effect of AS-ODNs and siRNAs has been observed when siRNA was coadministered with nonhomologous antisense oligonucleotides, targeting distant regions of the same mRNA *(88)*.

4.5. Small Hairpin RNA

The silencing activity of synthetic siRNAs is transient, lasting 3–5 d in cell culture and 10 d in vivo *(53,70,84)*. This phenomenon can be attributed to siRNA degradation and dilution of siRNA concentrations over cell divisions. To achieve persistent inhibitory effects of RNAi, various plasmid- and virus-based vectors have been developed to express siRNA or shRNA endogenously. RNA polymerase III–dependent promoters such as U6, H1, and tRNA promoters are among the most commonly used, followed by RNA polymerase II–dependent promoters such as CMV promoter *(68,89–97)*. Pol III is ideal for transcribing small RNAs, and its transcripts are not modified posttranscriptionally. Vectors based on adenovirus, retrovirus, and lentivirus have been used to produce functional shRNA species, resulting in persistent and robust gene silencing in vitro and in vivo *(97–105)*. The structure, length, and composition of hairpins appeared to be crucial in determining siRNA activities; however, they varied greatly in these studies. So far it is unclear what makes the best shRNA. Furthermore, RNAi has provided a rapid and effective means to functionally silence genes in stem cells and transgenic animals *(99,100,106,107)*.

4.6. In Vivo Delivery

Efficient in vivo delivery of siRNAs has been reported in a number of mouse models *(53,54,74,97,108–110)*. After a rapid systemic injection into tail vein, the target gene expression was effectively inhibited in liver, kidney, spleen, lung, and pancreas *(111)*. Intravenous injection of siRNA targeting Fas specifically reduced Fas protein expression in mouse hepatocytes and protected mice from fulminant hepatitis induced by concanavalin A or an agonistic Fas antibody *(53)*. Moreover, siRNA is capable of gene silencing when administered into brain or retina *(108,110)*.

Although RNAi technology is still in the fledgling stage, exhaustive efforts in the past few years have advanced our knowledge of RNAi from an antiviral mechanism in higher eukaryotes, to a powerful tool in functional genetics in mammalian cells, and then to a promising therapeutic approach for gene therapy. While many mechanistic and functional questions await answers, the advantages of RNAi in terms of its extraordinary efficiency and specificity, coupled with extensive research for improving its stability, delivery, and duration of action, warrant further preclinical and clinical explorations in a wide variety of diseases.

References

1. Zamecnik, P. C. and Stephenson, M. L. (1978) Inhibition of Pous sarcoma virus replication and cell transformation by a specific oligodeoxynucleotide. *Proc. Natl. Acad. Sci. USA* **75**, 280–284.
2. Schwartz, E. and Benz, E. (1995) *Thalassemia Syndromes*. Churchill Livingston, New York.
3. Lacerra, G., Sierakowska, H., Carestia, C., et al. (2000) Restoration of hemoglobin A synthesis in erythroid cells from peripheral blood of thalassemic patients. *Proc. Natl. Acad. Sci. USA* **97**, 9591–9596.
4. Mercatante, D. R., Bortner, C. D., Cidlowski, J. A., and Kole, R. (2001) Modification of alternative splicing of Bcl-x pre-mRNA in prostate and breast cancer cells. Analysis of apoptosis and cell death. *J. Biol. Chem.* **276**, 16,411–16,417.
5. Mathews, D. H., Sabina, J., Zuker, M., and Turner, D. H. (1999) Expanded sequence dependence of thermodynamic parameters improves prediction of RNA secondary structure. *J. Mol. Biol.* **288**, 911–940.
6. Ho, S. P., Bao, Y., Lesher, T., et al. (1998) Mapping of RNA accessible sites for antisense experiments with oligonucleotide libraries. *Nat. Biotechnol.* **16**, 59–63.
7. Milner, N., Mir, K. U., and Southern, E. M. (1997) Selecting effective antisense reagents on combinatorial oligonucleotide arrays. *Nat. Biotechnol.* **15**, 537–541.
8. Manoharan, M. (1999) 2'-carbohydrate modifications in antisense oligonucleotide therapy: importance of conformation, configuration and conjugation. *Biochim. Biophys. Acta.* **1489**, 117–130.
9. Morita, K, Takagi M, Hasegawa C, et al. (2002) 2'-O,4'-C-ethylene-bridged nucleic acids (ENA): highly nuclease-resistant and thermodynamically stable oligonucleotides for antisense drug. *Bioorg. Med. Chem. Lett.* **12**, 73–76.
10. Summerton, J. (1999) Morpholino antisense oligomers: the case for an RNase—independent structure type. *Biochim. Biophys. Acta* **1489**, 141–158.
11. Elayadi, A. N. and Corey, D. R. (2001) Application of PNA and LNA oligomers to chemotherapy. *Curr. Opin. Investig. Drugs* **2**, 558–561.
12. Declercq, R., Van Aerschot, A., Read, R. J., Herdewijn, P., and Van Meervelt, L. (2002) Crystal structure of double helical hexitol nucleic acids. *J. Am. Chem. Soc.* **124**, 928–933.

13. Wang, H., Wang, S., Nan, L., Yu, D., Agrawal, S., and Zhang, R. (2002) Antisense anti-MDM2 mixed-backbone oligonucleotides enhance therapeutic efficacy of topoisomerase I inhibitor irinotecan in nude mice bearing human cancer xenografts: in vivo activity and mechanisms. *Int. J. Oncol.* **20,** 745–752.

14. Maurer, N., Fenske, D. B., and Cullis, P. R. (2001) Developments in liposomal drug delivery systems. *Expert Opin. Biol. Ther.* **1,** 923–947.

15. Brigger, I., Dubernet, C., and Couvreur, P. (2002) Nanoparticles in cancer therapy and diagnosis. *Adv. Drug Deliv. Rev.* **54,** 631–651.

16. Manoharan, M. (2002) Oligonucleotide conjugates as potential antisense drugs with improved uptake, biodistribution, targeted delivery, and mechanism of action. *Antisense Nucleic Acid Drug Dev.* **2,** 103–128.

17. Lecocq, M., Andrianaivo, F., Warnier, M. T., Wattiaux-De Coninck, S., Wattiaux, R., and Jadot, M. (2003) Uptake by mouse liver and intracellular fate of plasmid DNA after a rapid tail vein injection of a small or a large volume. *J. Gene Med.* **5,** 142–156.

18. Sandrasagra, A., Leonard, S. A., et al. (2002) Discovery and development of respirable antisense therapeutics for asthma. *Antisense Nucleic Acid Drug Dev.* **12,** 177–181.

19. Brand, R. M. (2001) Topical and transdermal delivery of antisense oligonucleotides. *Curr. Opin. Mol. Ther.* **3,** 244–248.

20. Reynolds, T. (1992) First antisense drug trials planned in leukemia. *J. Natl. Cancer Inst.* **84,** 288–290.

21. Roehr, B. (1998) Fomivirsen approved for CMV retinitis. *J. Int. Assoc. Physicians AIDS Care* **4,** 14–16.

22. Altman, S. (1990) Ribonuclease P. Postscript. *J. Biol. Chem.* **265,** 20,053–20,056.

23. Cech, T. R. (1987) The chemistry of self-splicing RNA and RNA enzymes. *Science* **236,** 1532–1539.

24. Esteban, J. A., Banerjee, A. R., and Burke, J. M. (1997) Kinetic mechanism of the hairpin ribozyme. *J. Biol. Chem.* **272,** 13,629–13,639.

25. Fedor, M. J. and Uhlenbeck, O. C. (1992) Kinetics of intermolecular cleavage by hammerhead ribozymes. *Biochemistry* **31,** 1242–1254.

26. Symons, R. H. (1992) Small catalytic RNAs. *Annu. Rev. Biochem.* **61,** 641–671.

27. Birikh, K. E., Heaton, P. A., and Eckstein, F. (1997) The structure, function and application of the hammerhead ribozyme. *Eur. J. Biochem.* **245,** 1–16.

28. Dahm, S. C. and Uhlenbeck, O. C. (1991) Role of divalent metal ions in the hammerhead RNA cleavage reaction. *Biochemistry* **30,** 9464–9469.

29. Lieber, A. and Strauss, M. (1995) Selection of efficient cleavage sites in target RNAs by using a ribozyme expression library. *Mol. Cell. Biol. 15,* 540–551.

30. Hendry, P. and McCall, M. (1996) Unexpected anisotropy in substrate cleavage rates by asymmetric hammerhead ribozymes. *Nucleic Acids Res.* **24,** 2679–2684.

31. Fedor, M. J. (2000) Structure and function of the hairpin ribozyme. *J. Mol. Biol.* **297,** 269–291.

32. Young, K. J., Gill, F., and Grasby, J. A. (1997) Metal ions play a passive role in the hairpin ribozyme catalyzed reaction. *Nucleic Acids Res.* **25,** 3760–3766.

33. Li, Y. and Breaker, R. R. (1999) Deoxyribozymes: new players in the ancient game of biocatalysis. *Curr. Opin. Struct. Biol.* **9,** 315–323.

34. Riddell, S. R., Elliott, M., Lewinsohn, D. A., et al. (1996) T-cell mediated rejection of gene-modified HIV-specific cytotoxic T lymphocytes in HIV infected patients. *Nat. Med.* **2,** 216–223.

35. Taylor, M. M. and Samson, W. K. (2002) Ribozyme compromise of adrenomedullin mRNA reveals a physiological role in the regulation of water intake. *Am. J. Physiol.* **282,** R1739–R1745.

36. Koizumi, M., Hayase, Y., Iwai, S., et al. (1989) Design of a RNA enzyme distinguishing a single base mutation in RNA. *Nucleic Acids Res.* **17,** 7059–7071.

37. Koizumi, M., Kmiya, H., and Ohtsuka, D. (1992) Ribozymes designed to inhibit transformation of NIH/3T3 cells by the activated c-Ha-ras gene. *Gene* **117,** 179–184.

38. Usman, N. and Blatt, L. M. (2000) Nuclease-resistant synthetic ribozymes: developing a new class of therapeutics. *J. Clin. Invest.* **106,** 1197–1202.

39. Lewin, A. S., Drenser, K. A., Hauswirth, W. W., et al. (1998) Ribozyme rescue of photoreceptor cells in a transgenic rat model of autosomal dominant retinitis pigmentosa. *Nat. Med.* **4,** 967–971.

40. Lavail, M. M., Yasumura, D., Matthes, M. T., et al. (2000) Ribozyme rescue of photoreceptor cells in P23H transgenic rats: long-term survival and late-stage therapy. *Proc. Nat. Acad. Sci. USA* **97,** 11,488–11,493.

41. Muotri, A. R., Pereira, L. V., Vasques, L. R., et al. (1999) Ribozymes and the antigene therapy: how a catalytic RNA can be used to inhibit gene function. *Gene* **237,** 303–310.

42. Guo, S. and Kemphues, K. J. (1995) Par-1, a gene required for establishing polarity in C. elegans embryos, encodes a putative Ser/Thr kinase that is asymmetrically distributed. *Cell* **81,** 611–620.

43. Fire, A., Xu, S., Montgomery, M. K., Kostas, S. A., Driver, S. E., and Mello, C. C. (1998) Potent and specific genetic interference by double-stranded RNA in *Caenorhabditis elegans. Nature* **391,** 806–811.

44. Zamore, P. D., Tuschl, T., Sharp, P. A., and Bartel, D. P. (2000) RNAi: double-stranded RNA directs the ATP-dependent cleavage of mRNA at 21 to 23 nucleotide intervals. *Cell* **101,** 25–33.

45. Elbashir, S. M., Harborth, J., Lendeckel, W., Yalcin, A., Weber, K., and Tuschl, T. (2001) RNA interference is mediated by 21- and 22-nucleotide RNAs. *Genes Dev.* **15,** 188–200.

46. Bernstein, D., Caudy, A. A., Hammond, S. M., and Hannon, G. J. (2001) Role for a bidentate ribonuclease in the initiation step of RNA interference. *Nature* **409,** 363–366.

47. Hammond, S. M., Bernstein, F., Beach, D., and Hannon, G. J. (2000) An RNA-directed nuclease mediates post-transcriptional gene silencing in Drosophila cells. *Nature* **404,** 293–296.

48. Stark, G. R., Kerr, I. M., Williams, B. R., et al. (1998) How cells respond to interferons. *Annu. Rev. Biochem.* **67,** 227–264.

49. Minks, M. A., West, D. K., Benvin, S., and Baglioni, C. (1979) Structural require-ments of double-stranded RNA for the activation of 2',5'-oligo(A) polymerase and protein kinase of interferon-treated Hela cells. *J. Biol. Chem.* **254,** 10,180–10,183.

50. Elbashir, S. M., Harborth, J., Lendeckel, W., et al. (2001) Duplexes of 21-nucle-otide RNAs mediate RNA interference in cultured mammalian cells. *Nature* **411,** 494–498.

51. McManus, M. T. and Sharp, P. A. (2002) Gene silencing in mammals by small interfering RNAs. *Nat. Rev.* **3,** 737–747.

52. Shi, Y. (2003) Mammalian RNAi for the masses. *Trends Genet.* **19,** 9–12.

53. Song, E., Lee, S. K., Wang, J., et al. (2003) RNA interference targeting Fas pro-tects mice from fulminant hepatitis. *Nat. Med.* **9,** 347–351.

54. Sorensen, D. R., Leirdal, M., and Sioud, M. (2003) Gene silencing by systemic delivery of synthetic siRNAs in adult mice. *J. Mol. Biol.* **327,** 761–766.

55. Herweijer, H. and Wolff, J. A. (2003) Progress and prospects: naked DNA gene transfer and therapy. *Gene Ther.* **10,** 453–458.

56. Czauderna, F., Fechtner, M., Dames, S., et al. (2003) Structural variations and stabilizing modifications of synthetic siRNAs in mammalian cells. *Nucleic Acid Res.* **31,** 2705–2716.

57. Chiu, Y. L. and Rana, T. M. (2002) RNAi in human cells: basic structural and functional features of small interfering RNA. *Mol. Cell* **10,** 549–561.

58. Holen, T., Amarzguioui, M., Wiiger, M. T., Babaie, E., and Prydz, H. (2002) Positional effects of short interfering RNAs targeting the human coagulation trig-ger tissue factor. *Nucleic Acids Res.* **30,** 1757–1766.

59. Amarzguioui, M., Holen, T., Babaie, E., and Prydz, H. (2003) Tolerance for muta-tions and chemical modifications in a siRNA. *Nucleic Acids Res.* **31,** 589–595.

60. Harborth, J., Elbashir, S. M., Vandenburgh, K., et al. (2003) Sequence, chemical, and structural variation of small interfering rnas and short hairpin RNAs and the effect on mammalian gene silencing. *Antisense Nucleic Acid Drug Dev.* **13,** 83–105.

61. Elbashir SM, Harborth J, Weber K, Tuschl T. Analysis of gene function in somatic mammalian cells using small interfering RNAs. Methods 2002; 26: 199–213.

62. MaManus, M. T., Haines, B. B., Chen, J,. and Sharp, P. A. (2002) SiRNA-medi-ated gene silencing in T-lymphocytes. *J. Immunol.* **169,** 5754–5760.

63. Jacque, J. M., Triques, K., and Stevenson, M. (2002) Modulation of HIV-1 repli-cation by RNA interference. *Nature* **418,** 435–438.

64. Novina, C. D., Murray, M. F., Dykxhoorn, D. M., et al. (2002) siRNA-directed inhibition of HIV-1 infection. *Nat. Med.* **8,** 681–686.

65. Surabhi, R. M. and Gaynor, R. B. (2002) RNA interference directed against viral and cellular targets inhibits human immunodeficiency virus type 1 replication. *J. Virol.* **76,** 12,963–12,973.

66. Capodici, J., Kariko, K., and Weissman, D. (2002) Inhibition of HIV-1 infection by small interfering RNA-mediated RNA interference. *J. Immunol.* **169,** 5196–5201.

67. Coburn, G. A. and Cullen, B. R. (2002) Potent and specific inhibition of human immunodeficiency virus type 1 replication by RNA interference. *J. Virol.* **76,** 9225–9231.

68. Lee, N. S., Dohjima, T., Bauer, G., Li, H., et al. (2002) Expression of small interfering RNAs targeted against HIV-1 rev transcripts in human cells. *Nat. Biotechnol.* **20,** *500–505.*

69. Yokota, T., Sakamoto, N., Enomoto, N., et al. (2003) Inhibition of intracellular hepatitis C virus replication by synthetic and vector-derived small interfering RNAs. *EMBO Rep.* **4,** 602–608.

70. Wilson, J. A., Jayasena, S., Khvorova, A., et al. (2003) RNA interference blocks gene expression and RNA synthesis from hepatitis C replicons propagated in human liver cells. *Proc. Natl. Acad. Sci. USA* **100,** 2783–2788.

71. Kapadia, S. B., Brideau-Andersen, A., and Chisari, F. V. (2003) Interference of hepatitis C virus RNA replication by short interfering RNAs. *Proc. Natl. Acad. Sci. USA* **100,** 2014–2018.

72. Randall, G., Grakoui, A., and Rice, C. M. (2003) Clearance of replicating hepatitis C virus replicon RNAs in cell culture by small interfering RNAs. *Proc. Natl. Acad. Sci. USA* **100,** 235–240.

73. Seo, M. Y., Abrignani, S., Houghton, M., and Han, J. H. (2003) Small interfering RNA-mediated inhibition of hepatitis C virus replication in the human hepatoma cell line Huh-7. *J. Virol.* **77,** 810–812.

74. McCaffrey, A. P., Meuse, L., Pham, T. T., Conklin, D. S., Hannon, G. J., and Kay, M. A. (2002) RNA interference in adult mice. *Nature* **418,** 38–39.

75. Hamasaki, K., Nakao, K., Matsumoto, K., Ichikawa, T., Ishikawa, H., and Eguchi, K. (2003) Short interfering RNA-directed inhibition of hepatitis B virus replication. *FEBS Lett.* **543,** 51–54.

76. McCaffery, A. P., Nakai, H., Pandey, K., et al. (2003) Inhibition of hepatitis B virus in mice by RNA interference. *Nat. Biotechnol.* **21,** 639–644.

77. Shlomai, A. and Shaul, Y. (2003) Inhibition of hepatitis B virus expression and replication by RNA interference. *Hepatology* **37,** 764–770.

78. Hall, A. H. and Alexander, K. A. (2003) RNA interference of human papillomavirus type 18 E6 and E7 induces senescence in Hela cells. *J. Virol.* **77,** 6066–6069.

79. Jia, Q. and Sun, R. (2003) Inhibition of gammaherpesvirus replication by RNA interference. *J. Virol.* **77,** 3301–3306.

80. Dector, M. A., Romero, P., Lopez, S., and Arias, F. (2002) Rotavirus gene silencing by small interfering RNAs. *EMBO Rep.* **3,** 1175–1180.

81. Ge, Q., McManus, M. T., Nguyen, T., et al. (2003) RNA interference of influenza virus production by directly targeting mRNA for degradation and indirectly inhibiting all viral RNA transcription. *Proc. Natl. Acad. Sci. USA* **100,** 2718–2723.

82. Gitlin, L., Karelsky, S., and Andino, R. (2002) Short interfering RNA confers intracellular antiviral immunity in human cells. *Nature* **418,** 430–434.

83. Miyagishi, M., Hayashi, T., and Taira, K. (2003) Comparison of the suppressive effects of antisense oligonucleotides and siRNAs directed against the same targets in mammalian cells. *Antisense Nucleic Acid Drug Dev.* **13,** 1–7.

84. Bertrand, J. R., Pottier, M., Vekris, A., et al. (2002) Comparison of antisense oligonucleotides and siRNAs in cell culture and in vivo. *Biochem. Biophys. Res. Commun.* **29,** 1000–1004.
85. Aoki, Y., Cioca, D. P., Oidaira, H., Kamiya, J., and Kiyosawa, K. (2003) RNA interference may be more potent than antisense RNA in human cancer cell lines. *Clin. Exp. Pharmacol. Physiol.* **30,** 96–102.
86. Garber, K. (2003) Better blocker: RNA interference dazzles research community. *J. Natl. Cancer Inst.* **95,** 500–502.
87. Vickers, T. A., Koo, S., Bennett, C. F., Crooke, S. T., Dean, N. M., and Baker, B. F. (2003) Efficient reduction of target RNAs by small interfering RNA and RNase H-dependent antisense agents. *J. Biol. Chem.* **278,** 7108–7118.
88. Hemmings-Mieszczak, M., Dorn, G., Natt, F. J., et al. (2003) Independent combinatorial effect of antisense oligonucleotides and RNAi-mediated specific inhibition of the recombinant rat P2X3 receptor. *Nucleic Acids Res.* **31,** 2117–2126.
89. Yu, J. Y., DeRuiter, S. L., and Turner, D. L. (2002) RNA interference by expression of short-interfering RNAs and hairpin RNAs in mammalian cells. *Proc. Natl. Acad. Sci. USA* **99,** 6047–6052.
90. Paddison, P. J., Caudy, A. A., Bernstein, E., et al. (2002) Short hairpin RNAs (shRNAs) induce sequence-specific silencing in mammalian cells. *Genes Dev.* **16,** 948–958.
91. Brummelkamp, T. R., Bernards, R., and Agami, R. (2002) A system for stable expression of short interfering RNAs in mammalian cells. *Science* **296,** 550–553.
92. Paul, C. P., Good, P. D., Winer, I., and Engelke, D. R. (2002) Effective expression of small interfering RNA in human cells. *Nat. Biotechnol.* **20,** 505–508.
93. Miyagishi, M. and Taira, K. (2002) U6 promoter driven siRNAs with four uridine 3' overhangs efficiently suppress targeted gene expression in mammalian cells. *Nat. Biotechnol.* **20,** 497–500.
94. Sui, G., Soohoo, C., Affar, B., et al. (2002) A DNA vector-based RNAi technology to suppress gene expression in mammalian cells. *Proc. Natl. Acad. Sci. USA* **99,** 5515–5520.
95. Kawasaki, H. and Taira, K. (2003) Short hairpin type of dsRNAs that are controlled by tRNA(Val) promoter significantly induce RNAi-mediated gene silencing in the cytoplasm of human cells. *Nucleic Acids Res.* **31,** 700–707.
96. Zeng, Y., Wagner, E. J., and Cullen, B. R. (2002) Both natural and designed microRNAs can inhibit the expression of cognate mRNAs when expressed in human cells. *Mol. Cell.* **9,** 1–20.
97. Xia, H., Mao, Q., Paulson, H. L., and Davidson, B. (2002) siRNA-mediated gene silencing in vitro and in vivo. *Nat. Biotechnol.* **20,** 1006–1010.
98. Barton, G. M. and Medzhitov, R. (2002) Retroviral delivery of small interfering RNA into primary cells. *Proc. Natl. Acad. Sci. USA* **99,** 14,943–14,945.
99. Rubinson, D. A., Dillon, C. P., Dwiatkowski, A. U., et al. (2003) A lentivirus-based system to functionally silence genes in primary mammalian cells, stem cells and transgenic mice by RNA interference. *Nat. Genet.* **33,** 401–406.

100. Tiscornia, G., Singer, O., Ikawa, M., and Verma, I. M. (2003) A general method for gene knockdown in mice by using lentiviral vectors expressing small interfering RNA. *Proc. Natl. Acad. Sci. USA* **100,** 1844–1848.
101. Shen, C., Buck, A. K., Liu, X., Winkler, M., and Reske, S. W. (2003) Gene silencing by adenovirus-delivered siRNA. *FEBS Lett.* **539,** 111–114.
102. Stewart, S. A., Dykxhoorn. D, M., Palliser, D., et al. (2003) Lentivirus-delivered stable gene silencing by RNAi in primary cells. *RNA* **9,** 493–501.
103. Dirac, A. M. and Bernards, R. (2003) Reversal of senescence in mouse fibroblasts through lentiviral suppression of p53. *J. Biol. Chem.* **278,** 11,731–11,734.
104. Brummelkamp, T. R., Bernards, R., and Agami, R. (2002) Stable suppression of tumorigenicity by virus-mediated RNA interference. *Cancer Cell* **2,** 243–247.
105. Hasuwa, H., Kaseda, K., Einarsdottir, T., and Okabe, M. (2002) Small interfering RNA and gene silencing in transgenic mice and rats. *FEBS Lett.* **532,** 227–230.
106. Kunath, T., Gish, G., Lickert, H., Jones, N., Pawson, T., and Rossant, J. (2003) Transgenic RNA interference in ES cell-derived embryos recapitulates a genetic null phenotype. *Nat. Biotechnol.* **21,** 559–561.
107. Reich, S. J., Fosnot, J., Kuroki, A., et al. (2003) Small interfering RNA (siRNA) targeting VEGF effectively inhibits ocular neovascularization in a mouse model. *Mol. Vis.* **9,** 210–216.
108. Verma UN, Surabhi, R. M., Schmalties, A., Becerra, C., and Gaynor, R. B. (2003) Small interfering RNAs directed against beta-catenin inhibits the in vitro and in vivo growth of colon cancer cells. *Clin. Cancer Res.* **9,** 1291–300.
109. Makimura, H., Mizunno, T. M., Mastaitis, J. W., Agami, R., and Mobbs, C. V. (2002) Reducing hypothalamic AGRP by RNA interference increases metabolic rate and decreases body weight without influencing food intake. *BMC Neurosci.* **3,** 18.
110. Lewis, D. L., Hagstrom, J. E., Loomis, A. G., Wolff, J. A., and Herweijer, H. (2002) Efficient delivery of siRNA for inhibition of gene expression in postnatal mice. *Nat. Genet.* **32,** 107–108.

II

Cardiovascular

3

Local Application of Antisense for Prevention of Restenosis

Patrick L. Iversen, Nicholas Kipshidze, Jeffrey W. Moses, and Martin B. Leon

1. Introduction

In 1977, percutaneous transluminal coronary angioplasty (PTCA) was first introduced *(1)*, and it has become an effective treatment for limited coronary artery disease *(2)*. PTCA treatment has since become more extensive and gained favor as an alternative treatment for coronary artery bypass grafting *(3)*. The artery is injured at the site of PTCA, leading to wound healing responses including thrombosis, smooth muscle proliferation and migration, elastic recoil, and vascular remodeling. Each of these responses may contribute to recurrent obstruction or vessel narrowing, referred to as restenosis.

The clinical applicability of antisense technology, however, has been limited due to a relative lack of target specificity, slow uptake across the cell membranes, and rapid intracellular degradation of the oligonucleotide *(4)*. The only randomized study in humans with c-myc antisense demonstrated no reduction in restenosis after stent implantation when arteries were pretreated with the drug *(5)*. Recently introduced, AVI-4126 belongs to a family of molecules known as the phosphorodiamidate morpholino oligomers (PMOs). These oligomers comprise (dimethylamino) phosphinylideneoxy-linked morpholino subunits. The morpholino subunits contain a heterocyclic base recognition moiety of DNA (A, C, G, T) attached to a substituted morpholine ring system. In general, PMOs are capable of binding to RNA in a sequence-specific fashion with sufficient avidity to be useful for inhibition of the translation of mRNA into protein in vivo, a result commonly referred to as an "antisense" effect. Although PMOs share many similarities with other substances capable of pro-

From: *Methods in Molecular Medicine, Vol. 106: Antisense Therapeutics, Second Edition*
Edited by: I. Phillips © Humana Press Inc., Totowa, NJ

ducing antisense effects, such as DNA, RNA, and their analogous oligonucle-otide analogs such as the phosphorothioates, there are important differences. Most important, PMOs are uncharged and resistant to degradation under bio-logical conditions. The combination of efficacy, potency, and lack of nonspe-cific activities of the PMO chemistry compelled us to reexamine the approach to antisense to *c-myc* for the prevention of restenosis following balloon angioplasty.

AVI-4126 is an antisense PMO with sequence complementary to the trans-lation initiation start site of the *c-myc* mRNA. The mechanism of action of AVI-4126 involves interference with ribosomal assembly, thus preventing translation of *c-myc*, and interference with intron 1–exon 2 splicing of the *c-myc* pre-mRNA, preventing appropriate translation of the *c-myc* mRNA. The IC_{50} for inhibition of *c-myc* by AVI-4126 is 0.3 μM in cell culture (*6*). The cellular response to AVI-4126 is diminished cell growth associated with arrest of cells in the G_0/G_1 phase of the cell cycle. Inhibition of *c-myc* would also interfere with expression of downstream genes such as those associated with cellular adhesion, the cell cycle, and connective tissue matrix remodeling.

1.1. Designing Treatment Regimens for Preventing Restenosis

Interpretation of antisense data tends to be more complex than for small molecule inhibitors. This is due to the fact that the antisense mechanism of action involves inhibition of protein synthesis but the rate of protein turnover is generally not influenced. The equation for inhibition of c-myc expression is follows:

$$[MYC]^t = [MYC]_{ss} + \text{induced } [c\text{-}myc \text{ mRNA}] - [MYC]_{turnover} \qquad (1)$$

in which *MYC* is the protein, *c-myc* mRNA is the transcript, and *c-myc* is the gene. The magnitude of injury will determine the magnitude of new transcrip-tion of *c-myc* mRNA, and the amount of *c-myc* mRNA is also a balance between the rate of synthesis and rate of decay.

The rate of *MYC* synthesis depends on the concentration of *c-myc* mRNA. Equation 1 can be simplified to Eq. 2 as follows:

$$[MYC]^t = [MYC]_{ss} + \text{injury-induced } MYC \text{ synthesis} - MYC_{turnover} \qquad (2)$$

Critical questions in the development of a delivery tool for AVI-4126 are as follows: (1) What is the time course for *MYC* synthesis? (2) Can AVI-4126 inhibit MYC expression in the appropriate cells in the vessel wall following appropriate magnitude and duration of injury? (3) Are the amount and duration of AVI-4126 delivered relative to injury to the vessel wall sufficient? The ex-pected result of a study in which the initial synthesis of MYC is not inhibited is shown in **Fig. 1B**.

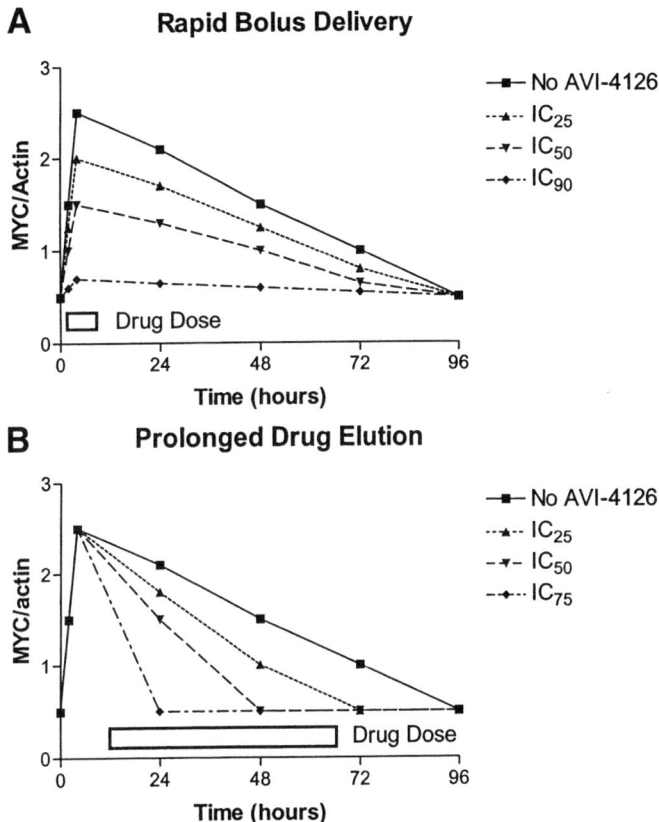

Fig. 1. *MYC* expression as dependent variable and time in hours as independent variable. (**A**) Situation when AVI-4126 is delivered to vessel wall at or near time of vascular injury. MYC concentration is efficiently prevented by inhibiting expression of new protein. (**B**) Situation when AVI-4126 is delivered to vessel wall shortly after induced *MYC* protein synthesis has begun. The *MYC* protein level is elevated prior to AVI-4126 blockade. In this situation, antisense inhibition is reflected only in creating a steeper slope in the decline in *MYC* concentrations in the vessel wall due to inhibition of lingering translation, which occurs approx 3 h after balloon injury.

1.2. MYC Expression Time Course

Considerations important to the evaluation of *MYC* expression and antisense inhibition include the stability of the transcript and the translated protein. The half-life of the transcribed mRNA is between 30 min and 1 h (*7*). The half-life of the *MYC* protein is 20–50 min (*8,9*). *MYC* expression is induced within 2–6 hours of injury (*10*).

Table 1
Comparison of No Injury with 7 D Postinjury[a]

No Injury	7 d Postinjury	p value	Interpretation
0.50 ± 0.11 (3)	0.42 ± 0.06 (12)	0.53; 13 df	No difference

Numbers in parentheses indicate the number of vessels evaluated.
[a]Intensity of MYC/β-actin.

The injury caused by stent placement produces a rapid increase in *MYC* detected by Western blot within 3 h. The peak *MYC* expression is followed by reduced but ongoing transcription and translation of *c-myc* mRNA that occurs simultaneously with the normal degradation of *MYC* protein. The expected result of a polymer-coated stent delivering AVI-4126 into the vascular wall will be to reduce the translation of *c-myc* mRNA. The rate of *MYC* degradation is not influenced, so the result will be observation of an enhanced rate of loss of *MYC*.

Concern for lingering *MYC* expression influences duration for antisense present in the injured vessels. This combined with a report of *MYC* expression at 7 d postinjury prompted further evaluation at 7 d (**Table 1**). Swine coronary vessels were injured by balloon overstretch, and then vessels were recovered 2 h later. The expression of *MYC* was determined by western blot and a ratio of blot intensity for MYC divided by intensity for β-actin, an internal control, to compensate for variability in sample preparation.

These data indicate that elevated *MYC* expression is shorter than 7 d after vessel injury. This is true for bare stents and polymer-coated stents. Therefore, delivery of AVI-4126 does not require prolonged (>4 d) release.

2. Catheter Delivery

Local drug delivery was designed to bring the antisense agent to the coronary artery during the period of time corresponding to peak injury response. The earliest attempts to deliver antisense agents for prevention of restenosis involved a rat carotid artery model using adventitial *(11)* or surgical application *(12)*. The initial clinically applicable devices were catheter-based, providing local delivery as a bolus injection, and then the catheter was withdrawn. The combination of antisense targeting to *c-myc* with a catheter-based delivery to coronary arteries of pigs for prevention of restenosis began with phosphorothioate oligonucleotides *(13)*. The bolus injection of phosphorothioate oligomers produced reduced heart rate, blood pressure (BP), and cardiac output in primate models, which, in some cases, were lethal *(14–18)*. The PMOs have been evaluated for similar effects after iv bolus injections in both primates (GLP studies by Sierra Biomedical) and man (GCP studies at MDS

Table 2
Rapid Bolus Local Delivery in Rabbit Iliac Vessels

	Control	Treated
MLD	0.8 ± 0.2	1.6 ± 0.3^a
Late loss	1.8 ± 0.3	0.9 ± 0.2^a
Lumen (mm^2)	0.62 ± 0.73	1.89 ± 0.35^a
Intima (mm^2)	1.67 ± 0.44	0.82 ± 0.32^a

[a]$p < 0.05$.

Harris). No alterations in heart rate, BP, or cardiac output have been observed. In summary, bolus injections of PMOs by local catheter-based delivery devices are feasible.

2.1. Transport Catheter Studies in Rabbit Iliac Vessels

Twenty-five male New Zealand white atherosclerotic rabbits maintained on a diet of 0.25% cholesterol were anesthetized, a Transport Catheter™ was inserted into the iliac artery, and PTCA was performed (8 atm for 30 s, three times). The endoluminal delivery of saline or 0.5 mg of AVI-4126 to the PTCA site was at 2 atm via the outer balloon for 2 min *(19)*. The area of the intima and media were determined by planimetry (**Table 2**). Quantitative angiography from these animals shows that the maximum lumen diameter (MLD) at the time of harvest (60 d after PTCA) was significantly greater in the antisense-treated group than in the control animals. The morphometric analysis confirms the angiography in demonstrating significantly greater lumen area than in the control The intimal area was also significantly smaller in the AVI-4126 treated animals. We also observed positive remodeling of the vessel. Vessel area was significantly greater ($p < 0.05$) in the treated animals.

2.2. Infiltrator Catheter Studies in Swine

We evaluated the long-term influence of intramural delivery of advanced c-myc antisense on neointimal hyperplasia following stenting in a pig model *(20)*. In acute experiments, different doses (from 500 µg to 5 mg) of Resten-NG ($n = 11$) or saline ($n = 14$) were delivered prior to the stent implantation site with the Infiltrator™ delivery system. Animals were sacrificed at 2, 6, and 18 h after interventions, and excised vessels were analyzed for c-myc expression by Western blot. In chronic experiments ($n = 20$), saline or 1, 5, and 5 mg of AVI-4126 was delivered in the same fashion, and animals were sacrificed at 28 d following intervention.

Table 3
Rapid Bolus Local Delivery in Swine Coronary Vessels

	Control	Treated
Lumen area (mm^2)	3.26 ± 1.57	5.62 ± 1.40[a]
Intimal area (mm^2)	3.88 ± 1.04	1.95 ± 0.91[a]
IA/IS[b]	4.08 ± 0.76	2.13 ± 0.55[a]

[a]$p < 0.05$.
[b]IA/IS = intimal area (IA) divided by injury score (IS).

Western blot analysis demonstrated inhibition of *c-myc* expression and was dose dependent. Morphometry showed that the intimal area was significantly reduced relative to the control (**Table 3**). There was a statistically significant reduction of intimal areas in the 5- and 10-mg groups (2.01 ± 0.66 and 1.95 ± 0.91, respectively; $p < 0.001$) but no significant reduction in the 1-mg group (2.81 ± 0.56; $p > 0.5$) in comparison with control. This study demonstrated that intramural delivery of advanced *c-myc* neutrally charged antisense morpholino compound completely inhibits *c-myc* expression and dramatically reduces neointimal formation in a dose-dependent fashion in a porcine coronary stent restenosis model while allowing for complete vascular healing.

3. Coated Stents

3.1. Phosphorylcholine Matrix for Drug Delivery

Phosphorylcholine stents were loaded with AVI-4126 using soak-trap (ST) and dry-trap (DT) methods. Twelve pigs underwent AVI-4126 phosphorylcholine coronary stent implantation (three stents/animal). Two to 6 h postprocedure, three pigs were sacrificed, and stented segments were analyzed by Western blot for c-myc expression. In chronic experiments, nine pigs (27 stent sites) were sacrificed at 28 d following intervention and vessels were perfusion fixed. High-performance liquid chromatography (HPLC) analysis of plasma showed a minimal presence of the antisense oligomer. Western blot analysis involved determination of both *MYC* and β-actin (an internal control protein) band intensities. The ratio of *MYC* to β-actin was 48% lower in the AVI-4126-treated vessels than in the untreated control vessels with stent implantation. The concentration of AVI-4126 in those vessels was 52 nM, as determined by HPLC. Quantitative histological morphometry (**Table 4**) showed that the neointimal area was significantly reduced (by 40%) in the ST group compared with control (2.2 ± 0.7 vs 3.7 ± 0.7 mm^2, respectively; $p = 0.0077$). Immunostaining and electron microscopy demonstrated complete endothelialization, without fibrin deposition, thrombosis, or necrosis, in all groups.

Table 4
Drug Eluting Stent in Swine Coronary Vessels

	Control	Treated
MLD	2.3 ± 0.6	3.3 ± 0.6^a
Late loss	1.1 ± 0.2	0.4 ± 0.5^a
Lumen area (mm^2)	3.4 ± 0.9	5.2 ± 0.9^a
Intima area (mm^2)	3.9 ± 0.8	2.3 ± 1.1^a

[a]$p < 0.05$.

Control arteries exhibited a substantial neointima consisting mostly of stellate and spindle-shaped cells, in a loose extracellular matrix. The neointima from treated arteries with antisense-loaded stent implantation was significantly smaller in size. Most important, there was no difference in appearance of reendothelialization. Transmission electron microscopy revealed a virtually normal appearance of endothelium. A semiquantitative histological grading system demonstrated similar smooth muscle cell (SMC) colonization in all groups and minimal residual fibrin deposition for the ST-eluting stents. However, DT and control phosphorylcholine stents had higher intimal fibrin scores.

We also observed less inflammation after implantation of the antisense-loaded stent. In general, the neointima of the ST- and DT-coated stents consisted of SMCs, matrix proteoglycans, and minimal focal regions of residual fibrin adjacent to the stent struts. Focal medial necrosis or intimal hemorrhage was an infrequent observation within any of the control or drug-coated stents. Its favorable influence on hyperplasia (reduction of intima by 40%) in the absence of endothelial toxicity may represent an advantage over more destructive methods such as brachytherapy *(21)* or cytotoxic inhibitors *(22)*. Indeed, recently local antiproliferative strategies including pharmacological stent coatings (e.g., paclitaxel, rapamycin) have demonstrated inhibition of SMC proliferation in vitro, reduced neointimal thickening in animal models of restenosis, and produced promising results in pilot human studies *(23)*. However, questions remain about the reendothelialization process after stent implantation with certain cytotoxic compounds that could put patients at risk for late stent thrombosis and cause late complications *(23)*. In contrast with other chemotherapeutics (paclitaxel, actinomycin D) Resten-NG inhibits the cell cycle in the G_1 phase. Compounds that inhibit the cell cycle in the early phase are often less toxic. Therefore, Resten-NG as well as Rapamycin fits this description.

Table 5
Rate of MYC Loss in Injured Vessels

	MYC:β-actin and GAPDH ratios		
Group	3-h vessels	24-h vessels	dx/dt
Control	1.51 ± 0.31 (12)	1.24 ± 0.79 (12)	0.27
Low-dose stent	1.68 ± 0.59 (6)	0.82 ± 0.45 (6)	0.86
High-dose stent	2.61 ± 1.77 (5)	1.15 ± 0.47 (6)	1.46

Numbers in parentheses indicate the number of vessels evaluated.

3.2. Experimental Matrices in Coated Stents

Some polymer coatings induce *MYC* to greater levels perhaps as a result of greater vessel wall injury. The bare stent vessel represents the best measure of the rate of the concentrations of *MYC* returning to steady state, which is approx 4 d.

Table 5 shows tabular data describing the rate of *MYC* loss in the injured vessels. The column referred to as *dx/dt* represents the difference in *MYC*:internal control ratio (both β-actin and GAPDH were utilized as internal controls) from the 3-h point to the 24-h point. This rate of *MYC* loss appears to be dose dependent.

If we assume that the injury does not alter steady-state *MYC* expression, then the time to recovery in the control group can be determined by $(1.51 - 0.5)/(0.27/21) = 78.3$ h. If synthesis of *MYC* were inhibited by 100%, then the *MYC* half-life would bring *MYC* to steady state in 5 half-lives, or approx 2.5 h. In the case of the low-dose stent, we observe $(1.68 - 0.5)/(0.86/21) = 28.8$ h, which represents a 63% reduction in the time to *MYC* steady state (IC_{63}). The high-dose stent would be $(2.61 - 0.5)/(1.46/21) = 30.4$ h or a 61.2% inhibition (IC_{61}). The amount of AVI-4126 in the vessel wall was determined in these studies, and the dose dependence of the rate of loss of *MYC* vs the vessel concentrations measured at 24 h is presented in **Table 6**. The studies demonstrate a remarkable agreement of inhibition of *MYC* with a resident amount of AVI-4126 in the vessel wall.

Table 6 reveals an excellent agreement between the published IC_{50} of 300 n*M* for Resten-NG in inhibiting *MYC* (*6*) and the observed (IC_{61}) at the 415 n*M* in the high-dose group. Inhibition by the low-dose stent was essentially equal to the high-dose stent, and the concentration at 24 h was is somewhat less but still in reasonable agreement given the difference between cell culture and in vivo blood vessels.

Table 6
Concentration of Resten-NG in Vessel Walls (nM)

Group	24-h vessels
Control	63.4 ± 92.8 (12)
Low-dose stent	233 ± 55.1 (6)
High-dose stent	415 ± 204 (6)

Numbers in parentheses indicate the number of vessels evaluated.

The studies did not include a polymer-coated stent or scrambled PMO polymer-coated stent as controls. The polymer coating the stent tends to induce greater MYC expression observed at the 3-h time point over the bare stent. Analysis of the rate of loss of *MYC* tends to minimize the shortfall in direct comparisons of observed *MYC* level at a given time point.

It can be concluded that AVI-4126 inhibited *MYC* expression in the current polymer-coated stent format in proportion to the amount of AVI-4126 that remained resident in the vessel wall. The degree of inhibition is quantitatively in good agreement with earlier studies investigating *MYC* inhibition. The polymer-coated stent delivered micromolar concentrations into the vessel wall within 3 h of placement. The delivery would need to be faster in order to prevent initial expression of *MYC* demonstrated by the substantial *MYC* detected at 3 h. The polymer tended to add to vessel injury, as measured by elevated *MYC* 3 h after placement. Ultimate success will require polymers capable of rapid elution of AVI-4126 with minimal capacity to inflame or otherwise additionally injure the vessel wall.

4. Perflourocarbon Gas Microbubble Carriers for Site-Specific Drug Delivery

Perfluorobutane gas microbubbles with a coating of dextrose and albumin efficiently bind antisense oligomers *(24)*. These 0.3- to 10-μm particles bind to sites of vascular injury. Furthermore, the perfluorobutane gas is an effective cell membrane fluidizer. The potential advantages of microbubble carrier delivery include minimal additional vessel injury from delivery, no resident polymer to degrade leading to eventual inflammation, rapid bolus delivery, and highly feasible repeated delivery. In addition, the potential for perfluorocarbon gas microbubble carriers (PGMCs) to deliver to vessel regions both proximal and distal to stents in vessels suggests that this mode of delivery will be excellent as an adjuvant to a variety of catheter and coated-stent delivery techniques.

Table 7
PGMC Delivery in Swine Carotid Vessels

	Control	Treated
Maximal intimal thickness (mm^2)	0.34 ± 0.06	0.20 ± 0.06[a]
Lumen area (mm^2)	12.4 ± 6.6	21.2 ± 2.9[a]
Area of stenosis (%)	19 ± 8	8 ± 2[a]

[a]$p < 0.05$.

4.1. PGMC Delivery Assisted by Diagnostic Ultrasound

Based on the aforementioned advantages, 21 pigs received AVI-4217 (a pig version of AVI-4126) bound to PGMC, AVI-4217 alone, or no antisense treatment after carotid balloon injury *(25)*. The vessels were evaluated 30 d postinjury. The results are presented in **Table 7**.

4.2. PGMC for Site-Specific Delivery of AVI-4217 in Porcine Coronary Vessels

The aforementioned results are impressive, so additional studies that did not involve ultrasound for site-specific delivery of the AVI-4217 were conducted in coronary vessels with stents. Seven pigs underwent stent implantation (three stents/animal). Four pigs received an iv injection of PGMC and 1 mg of AVI-4217 and two served as controls. Four hours postprocedure, three pigs were sacrificed and stented segments analyzed by HPLC and Western blot. In chronic experiments, four pigs were sacrificed at 28 d.

HPLC analysis of plasma samples of the treated animals showed a minimal detected concentration of AVI-4217, but analysis of treated vessels demonstrated easily detected AVI-4217. Western blot analysis of the stented vessels demonstrated modest inhibition of *c-myc* with no alteration in expression of p21 or p27. Morphometry showed that the neointimal area was significantly reduced (**Table 8**). These data represent a limited number of vessels and studies are under way to expand the use of PGMC for site-specific delivery of antisense agents.

5. Clinical Studies with AVI-4126

5.1. Phase I Clinical Studies

A phase I study was conducted to evaluate the safety and pharmacokinetic properties of AVI-4126 at five dose levels (1, 3, 10, 30, and 90 mg) administered intravenously. Six subjects were tested at each dose level. Safety laboratory assessments (chemistry, hematology and urinalysis) were performed at

Table 8
PGMC Delivery in Swine Coronary Vessels

	Control	Treated
Intimal area (mm^2)	4.8 ± 1.7	2.8 ± 2.0[a]
Lumen area (mm^2)	3.3 ± 0.7	6.1 ± 3.2
Area of stenosis (%)	57.8 ± 13.2	33.3 ± 24.6
Inflammatory score	0.7 ± 0.5	0.2 ± 0.3
Intimal vascularity	0.4 ± 0.5	0.2 ± 0.3
Intimal smooth muscle	3.0 ± 0.0	3.0 ± 0.0
Adventitial fibrosis	1.2 ± 0.8	0.7 ± 0.6

[a]$p < 0.05$

baseline, and at 24 h, 48 h, 72 h, 1 wk, and 2 wk postdose. Adverse experiences were collected on an ongoing basis from the time of dosing to discharge from the study at follow-up wk 2.

The most frequent adverse events reported included lower-extremity aches and headache. The majority of adverse events were graded mild in intensity and were generally self-limiting. Serum complement C3a was measured. Four subjects had elevated C3a greater than twice the upper limit of normal (normal = 0–400 ng/mL), including two subjects in the 3-mg dose cohort, one subject in the 10-mg dose cohort, and one subject in the 90-mg dose cohort. Three of the four elevations occurred at 24 h postdose and one occurred at 0.5 h postdose. In light of pharmacokinetic studies of the investigational compound, elevations of C3a at 24 h postdose are unlikely related to administration of AVI-4126 but, rather, spurious assay results. Further, there was no concurrent clinical symptomatology accompanying the elevated C3a, which is expected with elevated split complement levels.

5.2. Phase II Clinical Studies: AVAIL

Dr. M. Leon is the principal investigator of the AVAIL phase II randomized, evaluator-blinded three-arm study. It will be conducted at up to five investigational sites. The patients will be scheduled for angiography and angioplasty following confirmed stenosis or in-stent restenosis. Patients will receive study treatment (3 mg of AVI-4126, 10 mg of AVI-4126, or the Infiltrator catheter alone) following angiographic procedures. Patients will return for physical examinations and safety evaluations at 1 mo, 3 mo, and 6 mo, post–study treatment. The final angiographic and IVUS measurement outcomes will occur at the 6-mo visit unless ischemic changes warrant earlier angiographic evaluation. So far 42 patients have enrolled, some with 6-mo follow-up. There have been no serious AVI-4126-related adverse events.

6. Conclusion

The most robust of the observations to date include the fact that AVI-4126 is safe and effective in the vascular application in multiple species and conducted by multiple investigators. Three different methods for local delivery have been described, each with advantages and limitations. These observations fall short of proof that AVI-4126 will be effective in the treatment of human restenosis. Efficacy in animal models is encouraging. Furthermore, clinical trials with AVI-4126 indicate that the agent is very safe. The last remaining question is, Will AVI-4126 find a place in the future therapeutic regimen for the prevention of restenosis?

Acknowledgments

We wish to thank the exceptional effort of the AVI BioPharma chemistry group, including Dwight Weller, Doreen Weller, and Mary Martin, for the preparation of high purity and well defined oligonucleotides. We also wish to thank Derek Knapp for expert technical assistance, particularly with the Western blot analysis; and the analytical group at AVI BioPharma, including Murali Reddy and Tom Whitehead, for the high precision analysis of blood and tissue for AVI-4126. Finally, we wish to thank Dr. Thomas Porter and his laboratory at the University of Nebraska Medical Center for their efforts regarding PGMCs.

References

1. Gruntzig, A. R., Senning, A., and Siegenthaler, W. E. (1977) Nonoperative dilation of coronary artery stenosis: percutaneous transluminal coronary angioplasty. *N. Engl. J. Med.* **301,** 61–68.
2. Parisi, A. F., Folland, E. D., and Hartigan, P. (1992) A comparison of angioplasty with medical therapy in the treatment of single-vessel coronary artery disease: veterans Affairs ACME Investigators. *N. Engl. J. Med.* **326,** 10–16.
3. King, S. B., Lembo, N. J., and Weintraub, W. S. (1994) A randomized trial comparing coronary angioplasty with coronary bypass surgery. Emory Angioplasty versus Surgery Trial (EAST). *N. Engl. J. Med.* **331,** 1044–1050.
4. Zalewski, A., Shi, Y., Mannion, J. D., et al. (1998) Synthetic DNA-based compounds for the prevention of coronary restenosis: current status and future challenges. In *Clinical Trials of Genetic Therapy with Antisense DNA and DNA Vectors* (Wickstrom, E., ed.), Marcel Dekker New York, pp. 363–393.
5. Serruys, P., Kutryk, M. J. B., Bruining, N., et al. (1998) Antisense oligonucleotide against c-myc administered with the Transport delivery catheter for the prevention of in-stent restenosis: results of the randomized ITALICS trial. *Suppl. Circ. Sci. Sessions Abstr.* **98(17),** I-1–I-1016.
6. Hudziak, R. M., Summerton, J., Weller, D. D., and Iversen, P. L. (2000) Antiproliferative effects of steric blocking phosphorodiamidate Morpholino

antisense agents directed against c-myc. *Antisense Nucleic Acid Drug Devel.* **10,** 163–176.

7. Dani, C., Blanchard, J. M., Piechaczyk, M., El Sabouty, S., Marty, L., and Jeanteur, P. (1984) Extreme instability of myc mRNA in normal and transformed human cells. *Proc. Natl. Acad. Sci. USA* **81,** 7046–7050.

8. Hann, S. R. and Eisenman, R. N. (1984) Proteins encoded by the human c-myc oncogene: differential expression in neoplastic cells. *Mol. Cell. Biol.* **4,** 2486–2497.

9. Waters, C. M., Littlewood, T. D., Hancock, D. C., Moore, J. P., and Evan, G. I. (1991) c-myc protein expression in untransformed fibroblasts. *Oncogene* **6,** 797–805.

10. Miano, J. M., et al. (1991) Smooth muscle cell immediate early gene and growth factor activation follows vascular injury: a putative mechanism for autocrine growth. *Atheroscl. Thromb.* **13,** 211–219.

11. Simons, M., Edelman, E. R., DeKeyser, J.-L., Langer, R., and Rosenberg, R. D. (1992) Antisense c-myb oligonucleotides inhibit intimal arterial smooth muscle cell accumulation in vivo. *Nature* **359,** 67–80.

12. Morishita, R., Gibbons, G. H., Ellison, K. E., et al. (1993) Single intraluminal delivery of antisense cdc2 kinase and proliferating-cell nuclear antigen oligonucle-otides results in chronic inhibition of neointimal hyperplasia. *Proc. Natl. Acad. Sci. USA* **90,** 8474–8478.

13. Shi, Y., Fard, A., Galeo, A., et al. (1994) Transcatheter delivery of c-myc antisense oligomers reduces neointimal formation in a porcine model of coronary artery balloon injury. *Circulation* **90,** 944–951.

14. Cornish, K. G., Iversen, P. L., Smith, L., Arneson, M., and Bayever, E. (1993) Car-diovascular effects of a phosphorothioate oligonucleotide with sequence antisense to p53 in the conscious rhesus monkey. *Pharmacol. Commun.* **3,** 239–247.

15. Galbraith, W. M., Hobson, W. C., Giclas, P. C., Schechter, P. J., and Agrawal, S. (1994) Complement activation and hemodynamic changes following intravenous administration of phosphorothioate oligonucleotides in the monkey. *Antisense Res. Dev.* **4,** 201–206.

16. Wallace, T. L., Bazemore, S. A., Kornburst, D. J., and Cossum, P. A. (1996) Single-dose hemodynamic toxicity and pharmacokinetics of a partial phosphorothioate anti-HIV oligonucleotide (AR177) after intravenous infusion to cynomologousd monkeys. *J. Pharmacol. Exp. Ther.* **278,** 1306–1312.

17. Henry, S. P., Bolte, H., Auletta, C., and Kornburst, D. J. (1997) Evaluation of the toxicity of ISIS 2302 a phosphorothioate oligonucleotide, in a four week study in cynamologous monkeys. *Toxicology* **120,** 145–155.

18. Iversen, P. L., Cornish, K. G., Iversen, L. J., Mata, J. E., and Bylund, D. B. (1999) Bolus intravenous injection of phosphorothioate oligonucleotides causes hypoten-sion by acting as a1-adrenergic receptor antagonists. *Toxicol. Appl. Pharmacol.* **160,** 289–296.

19. Kipshidze, N., Keane, E., Stein, D., et al. (2001) Local delivery of c-myc neu-trally charged antisense oligonucleotides with transport catheter inhibits myointimal hyperplasia and positively affects vascular remodeling in the rabbit balloon injury model. *Cathet. Cardiovasc. Interventions* **54,** 247–256

20. Kipshidze, N. N., Kim, H.-S., Iversen, P. L., et al. (2002) Intramural delivery of advanced antisense oligonucleotides with infiltrator cathetor inhibits c-myc expression and intimal hyperplasia in the porcine. *J. Am. Coll. Cardiol.* **39(10),** 1686–1691.

21. Sheppard, R. and Eisenberg, M. J. (2001) Intracoronary radiotherapy for restenosis. *N. Engl. J. Med.* **344(4),** 295–297.

22. Herdeg, C., Oberhoff, M., Baumbach, A., et al. (2000) Local paclitaxel delivery for the prevention of restenosis: biological effects and efficacy in vivo. *J. Am. Coll. Cardiol.* **35(7),** 1969–76.

23. Sousa, J. E., Costa, M. A., Abizaid, A., et al. (2001) Lack of neointimal proliferation after implantation of sirolimus-coated stents in human coronary arteries: a quantitative coronary angiography and three-dimensional intravascular ultrasound study. *Circulation* **103(2),** 192–195.

24. Porter, T. R., Iversen, P. L., Li, S., and Xie, F. (1996) Interaction of Diagnostic Ultrasound with Synthetic Oligonucleotide-Labeled Perfluorocarbon-Exposed Sonicated Dextrose Albumin Microbubbles. *J. Ultrasound Med.* **15,** 577–584.

25. Porter, T. R., Hiser, W., Kricsfeld, D., Xie, F., Iversen, P., and Radio, S. (2000) Inhibition of carotid artery neointimal formation with intravenous microbubbles. *Ultrasound Med. Biol.* **27,** 259–265.

4

Antisense Therapeutics for Hypertension

Targeting the Renin–Angiotensin System

M. Ian Phillips and Birgitta Kimura

1. Introduction

Antisense therapeutics can be applied to hypertension. Hypertension is a widespread disease, yet in spite of several excellent drugs for the treatment of hypertension, the number of patients with controlled hypertension is about 27% in the United States. In all other countries studied, even those such as the United Kingdom and Canada, where cost is not a major factor, patient compliance is even worse. Part of the reason for this problem is that current antihypertensive drugs must be taken daily. Some drugs, such as β-blockers, are selective but not highly specific, and side effects are a common reason for patients interrupting their treatment. By not controlling their BP over 24 h, patients experience troughs and peaks in blood pressure (BP). A common regimen of taking a pill in the morning increases the likelihood of stroke or heart attacks at that time, when control is at its weakest.

We have proposed an antisense therapy because antisense effects are long lasting and highly specific, and, therefore, could provide prolonged control of BP without peaks and troughs or side effects. By taking advantage of the targets that the pharmaceutical industry has long used, it is not necessary to analyze more candidate genes. We know that targeting $\beta 1$ receptors, calcium receptors, and components of the renin–angiotensin system (RAS) can reduce BP in hypertensive patients. We have developed antisense inhibition of these target genes for reducing high BP. This chapter focuses on the proof-of-principle studies that have applied antisense inhibition to the RAS.

From: *Methods in Molecular Medicine, Vol. 106: Antisense Therapeutics, Second Edition*
Edited by: I. Phillips © Humana Press Inc., Totowa, NJ

The RAS is important in BP regulation, volume regulation, and vascular tissue growth. Angiotensin II (Ang II), an octapeptide, is the active peptide of the system. It is formed from angiotensin I (Ang I) by angiotensin-converting enzyme (ACE). Ang I is formed from angiotensinogen (AGT) by renin. Ang II is the ligand for Ang II type 1 receptors (AT$_1$R) and Ang II type 2 receptors (AT$_2$R). In addition, Ang II metabolites—Ang III, Ang IV, and Ang 1–7—are active and may have independent receptors. All the components of RAS are present in the brain as well as in the periphery, although renin levels in the brain are very low. Both brain RAS and the blood-borne RAS are important for BP regulation. Brain RAS is also involved in drinking, salt intake, the baroreflex, and hormonal release from the paraventricular nucleus (PVN). The peripheral tissue RAS is involved in cardiac hypertrophy and hyperplasia. AT$_1$Rs have been shown to mediate BP and the growth effects of Ang II *(1–4)*. The role of the AT$_2$R is still uncertain, although it has been implicated in apoptosis and has effects opposing those of the AT$_1$R *(4–7)*. However, mice with AT$_1$Rs but lacking AT$_2$Rs did not develop hypertrophy in response to Ang II infusion *(8)*, suggesting that the relationship between the two receptor types is more complex.

The rate-limiting step in the RAS cascade is the conversion to the decapeptide Ang I. Increases in AGT production have been shown to have an effect on BP in both humans and experimental animals *(9–11)*. Transgenic mice that produce high levels of Ang II are hypertensive *(12–17)*, and hypertensive rat models have increased levels of Ang II *(18–22)*. The spontaneously hypertensive rat (SHR) also has increased density of Ang II receptors *(23,24)*.

Human genetic studies have shown that the AGT gene is linked to hypertension. In French and Utah populations, the AGT 235T variant is more frequent in hypertensive patients than in control subjects *(25–28)*. The ACE gene insertion/deletion variant has also been implicated, but the association of this gene with hypertension may depend both on the ethnic group and whether males or females are studied *(29,30)*. There is also some evidence for the involvement of AT$_1$R gene polymorphism in human hypertension and arterial stiffness *(1,3,31)*.

The pressor effects of circulating Ang II have been known since the 1930s *(32)* and ACE inhibitors are one of the preferred classes of drugs used to treat high BP. Both ACE inhibitors and the newer AT$_1$R antagonists decrease left ventricular hypertrophy as well as hypertension *(2,4,33,34)*. However, these drugs have to be taken daily, and despite being effective in controlling hypertension, only 27% of patients with hypertension take the drugs consistently *(33–35)*. A shocking 73% of patients with hypertension do not comply with their drug treatment *(36)*. Cardiovascular disease is the leading cause of death in the United States and Europe, and the World Health Organization estimates that worldwide 17 million people die of cardiovascular disease every year (www.americanheart.org). Hodgson and Cai *(37)* reported that in the United States the costs of treating hypertension were $108.8 billion in 1998, and the American Heart Association estimated the direct and indirect costs of cardiovascular disease to be $329 billion 2002 *(37)*. Clearly, new treatments for hypertension are needed. We propose that a gene therapy approach would offer several advantages that could increase compliance. One is that the treatment would be long lasting (week, month, or longer). Another is that because of the high specificity of gene targeting there would be few side effects.

One approach is to target the mRNA of components involved in hypertension. Even though hypertension is a multifactorial disease, inhibition of the RAS is a promising strategy, because it is already known that pharmacological depression of the system decreases BP. We propose antisense inhibition of well-documented drug targets such as ACE, AGT, and AT_1R. Preclinically we have tested antisense oligonucleotides (AS-ODNs), plasmids, and viral vectors to decrease the levels of components of the RAS (**Tables 1** and **2**) *(33,34,38,39)*.

2. Antisense Oligonucleotides

AS-ODNs consist of short, 12- to 20-base long DNA sequences complementary to the mRNA producing the protein of interest. They bind to the mRNA and prevent translation of the specific protein encoded in the mRNA. To prevent breakdown in the circulation, the oligonucleotides are phosphorothioated, or otherwise modified, to increase stability. AS-ODNs can be administered by themselves, but delivery in liposomes, in liposomes coupled to Sendai virus, or to carrier molecules increases uptake and prolongs the effect of AS-ODNs *(40–47)*. In vitro experiments have shown that AS-ODNs enter the cell and the nucleus *(48)*.

3. Plasmid Vectors

Full-length antisense mRNA can be manufactured in plasmid vectors under the control of a promotor *(49)*. Theoretically, plasmids should be effective longer than AS-ODNs, but, practically, the difference is negligible. There is also the problem of efficient uptake, but recent studies using liposomes and receptor-mediated uptake have shown adequate effects of antisense mRNA *(50,51)*. Plasmid vectors have the potential to express antisense mRNA in specific cell types if cell-specific promoters or cell-specific delivery systems are used *(52)*. Thus, they would be advantageous when a transient expression is required.

Table 1
Antisense Against Brain RAS Vasoconstrictor Genes as Gene Therapy for Hypertension

Target gene	Construct	Route of delivery	Animal model	Max Δ BP (mmHg)	Duration of effect	Reference
AT_1R	AS-ODN	icv	SHR	–45	7 d	*60,62,63*
AT_1R	AS-ODN	icv	CIH	–35	4 d	*65*
AT_1R	AS-ODN	icv	2K1C chronic	–20	>5 d	*64*
AT_1R	AAV	icv	SHR	–40	>9 wk	*54*
AGT	AS-ODN	icv	SHR	–35	ND[a]	*60,72,73*
AGT	AS-ODN	icv	CIH	–40	2 d	*65*
Renin	AS-ODN	icv	SHR	–20	3 ds	*84*
AGE 2	Decoy ODN	icv	SHR	–30	7 d	*94*

[a]ND, not determined.

54

Table 2
Antisense Against Peripheral RAS Vasoconstrictor Genes as Gene Therapy for Hypertension

Target gene	Construct	Route of delivery	Animal model	Max Δ BP (mmHg)	Duration of effect	Reference
AT$_1$R	AS-ODN	ic	CIH	–35	ND^a	65
AT$_1$R	AS-ODN	iv	2K1C acute	–30	>7 d	68
AT$_1$R	LNSV	ic in 5-d-old rats	SHR	–45	> 120 d	55,70
AT$_1$R	LNSV	ic in 5-d-old rats	60% fructose	–20	>2 wk	95
AGT	AS-ODN	Portal vein	SHR	–20	4 d	79
AGT	AS-ODN	iv	SHR	–30	7 d	46,76–78
AGT	P AS-AGT	iv	SHR	–20	8 d	51
AGT	AAV	ic in 5-d-old rats	SHR	–25	6 mo	53
AGE 2	Decoy ODN	Portal vein	SHR	–20	6 d	96

^{a}ND, not determined.

4. Viral Vectors

Viral vectors containing cDNA in the antisense orientation can potentially integrate into the genome and express antisense mRNA for components of RAS. Our results show long-term attenuation of hypertension and cardiac hypertrophy (**Tables 1–3**) *(33,34,49,53,54)*. The adenoassociated virus (AAV) was used because it is safe, stable, long acting, and appropriate for gene therapy in adult models. Retroviruses preferentially infect dividing cells and integrate randomly into the genome *(33,34,49)*. They have been used in infant models studying the development of hypertension *(55–57)*. However, they are not suitable for adult gene therapy. Lentiviruses (based on human immuneodeficiency virus, simian immunodeficiency virus, or feline immunodeficiency virus) can infect nondividing cells and have a large carrying capacity. However, they integrate randomly, which may disrupt other genes and cause mutagenesis *(49,58,59)*. Adenoviruses have very good uptake and do not integrate into the host genome. They show high levels of short-term expression. Their major disadvantage for a realistic long-term therapy is the immune and inflammatory response that they cause *(33,49,58,59)*. AAV does not cause an immune response. It may integrate into the genome when it is modified and certainly has very long-lasting effects. The wild-type AAV does not cause any known diseases and cannot proliferate without a helper virus. It is likely to be the safest of the viral vectors. The disadvantage of the AAV vector is its small carrying capacity for a vector delivering genes. AAV can only accommodate 4.4 kb, and, thus, the number of promoters, enhancers, and length of AS-mRNA is limited. In addition, the deletion of the *rep* sequence removes the site-specific integration of the vector *(33,34,49,58)*. Nevertheless, one of the advantages of the antisense approach is that it is not necessary to have a full-length DNA, and, therefore, shorter AS-DNA sequences can be used effectively in AAV.

5. RAS Gene Therapy

We began targeting AT_1R and AGT for AS-ODNs in 1993 for hypertension gene therapy *(60)*. In the past 10 yr, all of the components of the RAS have been targeted for gene therapy. Some studies have aimed at understanding the mechanisms of the actions of RAS, and others to decrease hypertension, myocardial dysfunction, and growth effects of the RAS.

6. AT_1 Receptors

Using AS-ODNs in an intact animal showed that AT_1R antisense injected into the brain lateral ventricle of SHRs decreased BP by about 25 mmHg within 24 h and caused a 16–40% decrease in AT_1Rs in the PVN and organum vasculosum laminae terminalis (OVLT) *(60)*. Subsequently, these results have

Table 3
Effects of Antisense to RAS on Growth

Target gene	Construct	Route of delivery	Animal model	Effect studied	Magnitude	Reference
AT1	LNSV	ic in 5-d-old	TGR mRen2	Hypertrophy	90% decrease	71
AGT	AS-ODN	IV	SHR	Hypertrophy	60% decrease	46
AGT	AS-ODN	IV	SHR	Media of aorta	32% decrease	78
AGT	AAV	ic in 5-d-old	SHR	Hypertrophy	About 25% decrease	53
ACE	AS-ODN	Into injured artery	SD balloon catheter	Neointima formation injury	Injured control: 0.24 mm^2; antisense treated: 0.1 mm^2	18

ic = intracardiac.

been corroborated by other studies from our laboratory as well as other research-ers *(61–63)*. AT_1R AS-ODN applied to the central RAS can also attenuate BP in nongenetic models of hypertension. These include the surgical model, two-kidney one-clip (2K1C), and the environmental model, cold-induced hyperten-sion (CIH) *(64,65)* (*see* **Table 1**). In addition to the effect on BP, spontaneous drinking and Ang II– and isoproterenol-induced drinking are decreased by the AT_1R antisense *(65–67)*. Saline (i.e., osmotic)-induced drinking is not affected *(67)*. In early studies, AS-ODNs applied to the peripheral circulation did not elicit a response. However, when they were mixed with liposomes, the uptake into peripheral organs increased, and AT_1AS-ODNs were shown to decrease BP by 25–35 mmHg and AT_1Rs in kidney and arteries in both CIH and 2K1C hypertensive rats *(65,68)* (*see* **Table 2**). AT_1-AS-ODNs administered prior to ischemia-reperfusion also protected against myocardial dysfunction *(69)*. The effects of AT_1AS-ODNs are transient, lasting about 1 wk, with a maximum effect seen after 2 to 3 d.

Viral vectors, on the other hand, enabled very long-term expression of antisense. We have used AAV to deliver AT_1AS mRNA to both the central and the peripheral system of SHRs and obtained attenuation of hypertension by approx 25 mmHg (*see* **Table 1**). The reduction in BP lasts at least 9 wk *(54)*, and we have recorded normalization of BP in double transgenic mice for 6 mo. These mice have a gene for human renin and a gene for human angiotensin. They therefore constantly overexpress Ang II and are hypertensive. AAV delivery of AS to AT_1R dramatically reduced BP within a few days, and the effect persisted for as long as the mice were tested. Our colleagues have used a lentiviral vector (LNSV) to deliver AT_1AS mRNA to 5-d-old SHRs and obtained similar results (*see* **Table 2**) *(55,57,70)*. Indeed, the attenuation of hypertension as well as of hypertrophy could be shown to persist in offspring of the treated rats *(56,71)*.

7. Angiotensinogen

The earliest studies targeting AGT in the brain showed a substantial, up to 40 mmHg, decrease in BP in SHRs and a decrease in hypothalamic AGT *(60,72)*. Subsequent research has confirmed that intracerebroventricular (icv) injections of AGT-AS-ODN decrease BP in SHRs. Injections into the PVN did not affect BP, although it did decrease the release of vasopressin *(73)*. The CIH and 2K1C hypertensive models also respond to AGT-AS-ODN icv treatment with a decrease in BP *(64,65)*; the data are summarized in **Table 1**. The drink-ing response in CIH animals is also attenuated *(65)*. In a normotensive rat the drinking response to renin and isoproterenol is attenuated by icv AGT-AS-ODN injection, whereas the drinking responses to carbachol, Ang II, and water deprivation are unaffected *(74,75)*. Early studies reported no effect of AS-ODN injected into the peripheral circulation. This appears to have been due to failure to reach the target organs in sufficient numbers. We compared the effect of naked AS-ODN and liposome-encapsulated AS-ODN directed against AGT on BP, AGT and Ang II concentration, and hepatic uptake. We found that naked AS-ODN was without effect, whereas with the liposome-encapsulated AS-ODN, BP was lowered in SHRs. This was accompanied by lowered peripheral AGT and Ang II in the liver after injection *(76)*. Similar results were obtained when AS-ODN was coupled to carrier molecules that targeted delivery to the liver (*see* **Table 2**) *(77–79)*. In addition to the effects on protein levels and BP, AGT-AS-ODN attenuated hypertrophy of the heart and the smooth muscle in the aorta (*see* **Table 3**) *(46,78)*.

AGT-AS-ODNs attenuated hypertension for 4 to 5 d *(77,79)*. When a full-length AGT cDNA was inserted into a plasmid under the control of cytomegalovirus (CMV) promotor and injected with liposomes in SHRs, the decrease in BP lasted for 8 d *(51)*. The same construct delivered by AAV to 5-d-old SHR caused a delay in the development of hypertension, attenuated hypertrophy, and reduced the degree of hypertension for at least 6 mo *(53)*.

8. Angiotensin Converting Enzyme

AS-ODNs directed against ACE mRNA have been used to lower the amount of vascular ACE and the formation of neointima after balloon catheter injury (*see* **Table 3**) *(80)*. ACE-AS-ODNs have also been shown to improve cardiac performance after ischemia-reperfusion injury *(81)*.

ACE is present at low to moderate levels in large areas of the brain, and in high levels in the nucleas tractus solitarii (NTS), and is likely to generate Ang II locally *(82)*. Increasing the levels of ACE in the brain by transfection with a plasmid containing the human ACE gene caused an increase in BP, heart rate, and Ang II and ACE levels that lasted 2 wk *(83)*. Unpublished results from our laboratory testing three different sequences of ACE-AS-ODNs showed a decrease in BP in hypertensive SHRs of 15–25 mmHg.

9. Renin

Renin-AS-ODNs decreased BP by about 20 mmHg for 2 d when injected into the lateral ventricle of SHRs. mRNA for renin was also suppressed by the treatment *(84)*. So far, no studies have been done with renin-AS-ODNs systemically. A complete study with β_1-AS-ODNs showed that the release of renin was inhibited and BP decreased for up to 1 mo in SHRs *(43)*.

10. AT$_2$

AT$_2$-AS-ODNs infused into the kidney of uninephrectomized normotensive rats increased BP by about 20 mmHg throughout the infusion period *(85)*.

11. Conclusion

The use of gene therapy to correct genetic abnormalities and to treat diseases is getting closer to being clinically relevant. At least one AS-ODN has been approved by the Food and Drug Administration to treat CMV retinitis *(86,87)*, and AAV-RPE65 has been used to restore sight in a canine model of blindness *(88)*. Adenovirus is being used in phase I and phase II clinical trials in cancer patients *(89–93)*. AS-ODNs and viral vectors have been introduced into both the periphery and the central system of the brain. For therapeutic uses, peripheral administration is more likely to be clinically acceptable. The preclinical studies described above have demonstrated that antisense to AGT, AT_1, and ACE can successfully lower BP in hypertensive rats and attenuate hypertrophy in adult animals when administered systemically. The decrease in BP has been reported to vary between 15 and 40 mmHg *(46,51,53,65,76,79)*. Although these effects would be highly advantageous clinically, it is probably not possible to achieve a "cure" for hypertension with AS-ODNs because the mechanism is a competition between copies of mRNA and the amount of oligonucleotide delivered to cells. Clearly, there is a dose-dependent effect. We found a correlation between the dose of an AGT-AS-producing plasmid and BP, and a large decrease in BP was also found with AGT-AS-ODN using a carrier protein targeting the liver *(46,51)*. Another factor favoring the dose-dependent mechanism is that pharmacological drugs, both ACE inhibitors and AT_1R antagonists, are able to normalize BP alone, although some new approaches have tried combinations.

The next step in research is to solve the problems in increasing the uptake of AS-ODNs and viral vectors, and in the case of viral vectors, to ascertain that they do not cause adverse effects over a very long time, such as immune responses or tumorigenesis. Nevertheless, antisense gene therapy has the potential to provide extended protection against hypertension, cardiovascular disease, and a multitude of other diseases. Here we have reviewed its use on the RAS as a target system in hypertension and cardiovascular disease, but clearly any target with a known DNA sequence is a target for antisense inhibition.

Reference

1. Benetos, A., Gautier, S., Ricard, S., et al. (1996) Influence of angiotensin-converting enzyme and angiotensin II type 1 receptor gene polymorphisms on aortic stiffness in normotensive and hypertensive patients. *Circulation* **94,** 698–703.
2. Chung, O. and Unger, T. (1999) Angiotensin II receptor blockade and end-organ protection. *Am. J. Hypertens.* **12,** S150–S156.
3. Kurland, L., Melhus, H., Karlsson, J., et al. (2002) Polymorphisms in the angiotensinogen and angiotensin II type 1 receptor gene are related to change in left ventricular mass during antihypertensive treatment: results from the Swedish

Irbesartan Left Ventricular Hypertrophy Investigation versus Atenolol (SILVHIA) trial. *J. Hypertens.* **20**, 657–663.

4. Unger, T. (2002) The role of the renin-angiotensin system in the development of cardiovascular disease. *Am. J. Cardiol.* **89**, 3A–9A.

5. De Paepe, B., Verstraeten, V. M., De Potter, C. R., et al. (2002) Increased angiotensin II type-2 receptor density in hyperplasia, DCIS and invasive carcinoma of the breast is paralleled with increased iNOS expression. *Histochem. Cell Biol.* **117**, 13–19.

6. De Paepe, B., Verstraeten, V. L., De Potter, C. R., et al. (2001) Growth stimulatory angiotensin II type-1 receptor is upregulated in breast hyperplasia and in situ carcinoma but not in invasive carcinoma. *Histochem. Cell. Biol.* **116**, 247–254.

7. Schmieder, R. E., Erdmann, J., Delles, C., et al. (2001) Effect of the angiotensin II type 2-receptor gene (+1675 G/A) on left ventricular structure in humans. *J. Am. Coll. Cardiol.* **37**, 175–182.

8. Ichihara, S., Senbonmatsu, T., Price, E., Jr., et al. (2001) Angiotensin II type 2 receptor is essential for left ventricular hypertrophy and cardiac fibrosis in chronic angiotensin II-induced hypertension. *Circulation* **104**, 346–351.

9. Bloem, L. J., Foroud, T. M., Ambrosius, W. T., et al. (1997) Association of the angiotensinogen gene to serum angiotensinogen in blacks and whites. *Hypertension* **29**, 1078–1082.

10. Kim, H. S., Krege, J. H., Kluckman, K. D., et al. (1995) Genetic control of blood pressure and the angiotensinogen locus. *Proc. Natl. Acad. Sci. USA* **92**, 2735–2739.

11. Walker, W. G., Whelton, P. K., Saito, H., et al. (1979) Relation between blood pressure and renin, renin substrate, angiotensin II, aldosterone and urinary sodium and potassium in 574 ambulatory subjects. *Hypertension* **1**, 287–291.

12. Fukamizu, A., Sugimura, K., Takimoto, E., et al. (1993) Chimeric renin-angiotensin system demonstrates sustained increase in blood pressure of transgenic mice carrying both human renin and human angiotensinogen genes. *J. Biol. Chem.* **268**, 11,617–11,621.

13. Davisson, R. L., Ding, Y., Stec, D. E., et al. (1999) Novel mechanism of hypertension revealed by cell-specific targeting of human angiotensinogen in transgenic mice. *Physiol. Genomics* **1**, 3–9.

14. Merrill, D. C., Thompson, M. W., Carney, C. L., et al. (1996) Chronic hypertension and altered baroreflex responses in transgenic mice containing the human renin and human angiotensinogen genes. *J. Clin. Invest.* **97**, 1047–1055.

15. Morimoto, S., Cassell, M. D., Beltz, T. G., et al. (2001) Elevated blood pressure in transgenic mice with brain-specific expression of human angiotensinogen driven by the glial fibrillary acidic protein promoter. *Circ. Res.* **89**, 365–372.

16. Ohkubo, H., Kawakami, H., Kakehi, H., et al. (1990) Generation of transgenic mice with elevated blood pressure by introduction of the rat renin and angiotensinogen genes. *PNAS* **87**, 5153–5157.

17. Stec, D. E., Keen, H. L., and Sigmund, C. D. (2002) Lower blood pressure in floxed angiotensinogen mice after adenoviral delivery of cre-recombinase. *Hypertension* **39**, 629–633.

18. Morishita, R., Higaki, J., Miyazaki, M., et al. (1992) Possible role of the vascular renin-angiotensin system in hypertension and vascular hypertrophy. *Hypertension* **19,** II62–II67.

19. Morton, J. J. and Wallace, E. C. (1983) The importance of the renin-angiotensin system in the development and maintenance of hypertension in the two-kidney one-clip hypertensive rat. *Clin. Sci. (Lond.)* **64,** 359–370.

20. Phillips, M. I. and Kimura, B. K. (1986) Levels of brain angiotensin in the spontaneously hypertensive rat and treatment with ramiprilat. *J. Hypertens. Suppl.* **4,** S391–S394.

21. Phillips, M. I. and Kimura, B. (1988) Brain angiotensin in the developing spontaneously hypertensive rat. *J. Hypertens.* **6,** 607–612.

22. Navar, L. G., Von Thun, A. M., Zou, L., et al. (1995) Enhancement of intrarenal angiotensin II levels in 2 kidney 1 clip and angiotensin II induced hypertension. *Blood Pressure Suppl.* **2,** 88–92.

23. Brown, L., Passmore, M., Duce, B., et al. (1997) Angiotensin receptors in cardiac and renal hypertrophy in rats. *J. Mol. Cell. Cardiol.* **29,** 2925–2929.

24. Gutkind, J. S., Kurihara, M., and Saavedra, J. M. (1988) Increased angiotensin II receptors in brain nuclei of DOCA-salt hypertensive rats. *Am. J. Physiol.* **255,** H646–H650.

25. Atwood, L. D., Kammerer, C. M., Samollow, P. B., et al. (1997) Linkage of essential hypertension to the angiotensinogen locus in Mexican Americans. *Hypertension* **30,** 326–330.

26. Corvol, P. and Jeunemaitre, X. (1997) Molecular genetics of human hypertension: role of angiotensinogen. *Endocr. Rev.* **18,** 662–677.

27. Jain, S., Tang, X., Chittampalli, S. N., et al. (2002) Angiotensinogen gene polymorphism at -217 affects basal promoter activity and is associated with hypertension in African-Americans. *J. Biol. Chem.* **277,** 36,889–36,896.

28. Niu, T., Xu, X., Rogus, J., et al. (1998) Angiotensinogen gene and hypertension in Chinese. *J. Clin. Invest.* **101,** 188–194.

29. Agerholm-Larsen, B., Nordestgaard, B. G., and Tybjarg-Hansen, A. (2000) ACE gene polymorphism in cardiovascular disease: meta-analyses of small and large studies in whites. *Arterioscl. Thrombo. Vasc. Biol.* **20,** 484–492.

30. O'Donnell, C. J., Lindpaintner, K., Larson, M. G., et al. (1998) Evidence for association and genetic linkage of the angiotensin-converting enzyme locus with hypertension and blood pressure in men but not women in the Framingham Heart Study. *Circulation* **97,** 1766–1772.

31. Bonnardeaux, A., Davies, E., Jeunemaitre, X., et al. (1994) Angiotensin II type 1 receptor gene polymorphisms in human essential hypertension. *Hypertension* **24,** 63–69.

32. Phillips, M. I. and Schmidt-Ott, K. M. (1999) The discovery of renin 100 years ago. *News Physiol. Sci.* **14,** 271–274.

33. Phillips, M. I. (2001) Gene therapy for hypertension: sense and antisense strategies. *Expert. Opin. Biol. Ther.* **1,** 655–662.

34. Phillips, M. I. (2001) Gene therapy for hypertension: the preclinical data. *Hypertension* **38,** 543–548.
35. Phillips, M. I. (2000) Somatic gene therapy for hypertension. *Braz. J. Med. Biol. Res.* **33,** 715–721.
36. Kaplan, N. M. (1998) *Clinical Hypertension.* Williams & Wilkins, Baltimore, 1998.
37. Hodgson, T. A. and Cai, L. (2001) Medical care expenditures for hypertension, its complications, and its comorbidities. *Med. Care* **39,** 599–615.
38. Kagiyama, S., Kagiyama, T., and Phillips, M. I. (2001) Antisense oligonucleotides strategy in the treatment of hypertension. *Curr. Opin. Mol. Ther.* **3,** 258–264.
39. Phillips, M. I., Wielbo, D., and Gyurko, R. (1994) Antisense inhibition of hypertension: a new strategy for renin-angiotensin candidate genes. *Kidney Int.* **46,** 1554–1556.
40. Dzau, V. J., Mann, M. J., Morishita, R., et al. (1996) Fusigenic viral liposome for gene therapy in cardiovascular diseases. *Proc. Natl. Acad. Sci. USA* **93,** 11,421–11,425.
41. Fillion, P., Desjardins, A., Sayasith, K., et al. (2001) Encapsulation of DNA in negatively charged liposomes and inhibition of bacterial gene expression with fluid liposome-encapsulated antisense oligonucleotides. *Biochim. Biophys. Acta* **1515,** 44–54.
42. Hughes, J. A., Bennett, C. F., Cook, P. D., et al. (1994) Lipid membrane permeability of 2'-modified derivatives of phosphorothioate oligonucleotides. *J. Pharm. Sci.* **83,** 597–600.
43. Zhang, Y. C., Bui, J. D., Shen, L., et al. (2000) Antisense inhibition of beta(1)-adrenergic receptor mRNA in a single dose produces a profound and prolonged reduction in high blood pressure in spontaneously hypertensive rats. *Circulation* **101,** 682–688.
44. Zhang, Y. M., Rusckowski, M., Liu, N., et al. (2001) Cationic liposomes enhance cellular/nuclear localization of 99mTc-antisense oligonucleotides in target tumor cells. *Cancer Biother. Radiopharm.* **16,** 411–419.
45. Morishita, R., Gibbons, G. H., Ellison, K. E., et al. (1993) Single intraluminal delivery of antisense cdc2 kinase and proliferating-cell nuclear antigen oligonucleotides results in chronic inhibition of neointimal hyperplasia. *Proc. Natl. Acad. Sci. USA* **90,** 8474–8478.
46. Makino, N., Sugano, M., Ohtsuka, S., et al. (1999) Chronic antisense therapy for angiotensinogen on cardiac hypertrophy in spontaneously hypertensive rats. *Cardiovasc. Res.* **44,** 543 548.
47. Clare, Z. Y., Kimura, B., Shen, L., et al. (2000) New beta-blocker: prolonged reduction in high blood pressure with beta(1) antisense oligodeoxynucleotides. *Hypertension* **35,** 219–224.
48. Li, B., Hughes, J. A., and Phillips, M. I. (1997) Uptake and efflux of intact antisense phosphorothioate deoxyoligonucleotide directed against angiotensin receptors in bovine adrenal cells. *Neurochem. Int.* **31,** 393–403.

49. Mohuczy, D. and Phillips, M. I. (2000) Designing antisense to inhibit the renin-angiotensin system. *Mol. Cell. Biochem.* **212,** 145–153.
50. Merdan, T., Kopecek, J., and Kissel, T. (2002) Prospects for cationic polymers in gene and oligonucleotide therapy against cancer. *Adv. Drug Deliv. Rev.* **54,** 715–758.
51. Tang, X., Mohuczy, D., Zhang, Y. C., et al. (1999) Intravenous angiotensinogen antisense in AAV-based vector decreases hypertension. *Am. J. Physiol.* **277,** H2392–H2399.
52. Zhang, Y., Jeong, L. H., Boado, R. J., et al. (2002) Receptor-mediated delivery of an antisense gene to human brain cancer cells. *J. Gene Med.* **4,** 183–194.
53. Kimura, B., Mohuczy, D., Tang, X., et al. (2001) Attenuation of hypertension and heart hypertrophy by adeno-associated virus delivering angiotensinogen antisense. *Hypertension* **37,** 376–380.
54. Phillips, M. I., Mohuczy-Dominiak, D., Coffey, M., et al. (1997) Prolonged reduction of high blood pressure with an in vivo, nonpathogenic, adeno-associated viral vector delivery of AT1-R mRNA antisense. *Hypertension* **29,** 374–380.
55. Lu, D., Raizada, M. K., Iyer, S., et al. (1997) Losartan versus gene therapy: chronic control of high blood pressure in spontaneously hypertensive rats. *Hypertension* **30,** 363–370.
56. Metcalfe, B. L., Raizada, M., and Katovich, M. J. (2002) Genetic targeting of the renin-angiotensin system for long-term control of hypertension. *Curr. Hypertens. Rep.* **4,** 25–31.
57. Wang, H., Lu, D., Reaves, P. Y., et al. (2000) Retrovirally mediated delivery of angiotensin II type 1 receptor antisense in vitro and in vivo. *Methods Enzymol.* **314,** 581–590.
58. Hauswirth, W. W. and McInnes, R. R. (1998) Retinal gene therapy 1998: summary of a workshop. *Mol. Vis.* **4,** 11.
59. Sinnayah, P., Lindley, T. E., Staber, P. D., et al. (2002) Selective gene transfer to key cardiovascular regions of the brain: comparison of two viral vector systems. *Hypertension* **39,** 603–608.
60. Gyurko, R., Wielbo, D., and Phillips, M. I. (1993) Antisense inhibition of AT1 receptor mRNA and angiotensinogen mRNA in the brain of spontaneously hypertensive rats reduces hypertension of neurogenic origin. *Regul. Pept.* **49,** 167–174.
61. Ambuhl, P., Gyurko, R., and Phillips, M. I. (1995) A decrease in angiotensin receptor binding in rat brain nuclei by antisense oligonucleotides to the angiotensin AT1 receptor. *Regul. Pept.* **59,** 171–182.
62. Gyurko, R., Tran, D., and Phillips, M. I. (1997) Time course of inhibition of hypertension by antisense oligonucleotides targeted to AT1 angiotensin receptor mRNA in spontaneously hypertensive rats. *Am. J. Hypertens.* **10,** 56S–62S.
63. Piegari, E., Galderisi, U., Berrino, L., et al. (2000) In vivo effects of partial phosphorothioated AT1 receptor antisense oligonucleotides in spontaneously hypertensive and normotensive rats. *Life Sci.* **66,** 2091–2099.
64. Kagiyama, S., Varela, A., Phillips, M. I., et al. (2001) Antisense inhibition of brain renin-angiotensin system decreased blood pressure in chronic 2-kidney, 1 clip hypertensive rats. *Hypertension* **37,** 371–375.

65. Peng, J. F., Kimura, B., Fregly, M. J., et al. (1998) Reduction of cold-induced hypertension by antisense oligodeoxynucleotides to angiotensinogen mRNA and AT1-receptor mRNA in brain and blood. *Hypertension* **31,** 1317–1323.

66. Meng, H., Wielbo, D., Gyurko, R., et al. (1994) Antisense oligonucleotide to AT1 receptor mRNA inhibits central angiotensin induced thirst and vasopressin. *Regul. Pept.* **54,** 543–551.

67. Sakai, R. R., Ma, L. Y., He, P. F., et al. (1995) Intracerebroventricular administration of angiotensin type 1 (AT1) receptor antisense oligonucleotides attenuate thirst in the rat. *Regul. Pept.* **59,** 183–192.

68. Galli, S. M. and Phillips, M. I. (2001) Angiotensin II AT(1A) receptor antisense lowers blood pressure in acute 2-kidney, 1-clip hypertension. *Hypertension* **38,** 674–678.

69. Yang, B., Li, D., Phillips, M. I., et al. (1998) Myocardial angiotensin II receptor expression and ischemia-reperfusion injury. *Vasc. Med.* **3,** 121–130.

70. Iyer, S. N., Lu, D., Katovich, M. J., et al. (1996) Chronic control of high blood pressure in the spontaneously hypertensive rat by delivery of angiotensin type 1 receptor antisense. *Proc. Natl. Acad. Sci. USA* **93,** 9960–9965.

71. Pachori, A. S., Numan, M. T., Ferrario, C. M., et al. (2002) Blood pressure-independent attenuation of cardiac hypertrophy by AT(1)R-AS gene therapy. *Hypertension* **39,** 969–975.

72. Wielbo, D., Sernia, C., Gyurko, R., et al. (1995) Antisense inhibition of hypertension in the spontaneously hypertensive rat. *Hypertension* **25,** 314–319.

73. Kagiyama, S., Tsuchihashi, T., Abe, I., et al. (1999) Antisense inhibition of angiotensinogen attenuates vasopressin release in the paraventricular hypothalamic nucleus of spontaneously hypertensive rats. *Brain Res.* **829,** 120–124.

74. Sinnayah, P., Kachab, E., Haralambidis, J., et al. (1997) Effects of angiotensinogen antisense oligonucleotides on fluid intake in response to different dipsogenic stimuli in the rat. *Brain Res. Mol. Brain Res.* **50,** 43–50.

75. Sinnayah, P., McKinley, M. J., and Coghlan, J. P. (1997) Angiotensinogen antisense oligonucleotides and fluid intake. *Clin. Exp. Hypertens.* **19,** 993–1007.

76. Wielbo, D., Simon, A., Phillips, M. I., et al. (1996) Inhibition of hypertension by peripheral administration of antisense oligodeoxynucleotides. *Hypertension* **28,** 147–151.

77. Makino, N., Sugano, M., Ohtsuka, S., et al. (1998) Intravenous injection with antisense oligodeoxynucleotides against angiotensinogen decreases blood pressure in spontaneously hypertensive rats. *Hypertension* **31,** 1166–1170.

78. Sugano, M., Tsuchida, K., Sawada, S., et al. (2000) Reduction of plasma angiotensin II to normal levels by antisense oligodeoxynucleotides against liver angiotensinogen cannot completely attenuate vascular remodeling in spontaneously hypertensive rats. *J. Hypertens.* **18,** 725–731.

79. Tomita, N., Morishita, R., Higaki, J., et al. (1995) Transient decrease in high blood pressure by in vivo transfer of antisense oligodeoxynucleotides against rat angiotensinogen. *Hypertension* **26,** 131–136.

80. Morishita, R., Gibbons, G. H., Tomita, N., et al. (2000) Antisense oligodeoxynucleotide inhibition of vascular angiotensin-converting enzyme expression attenuates

neointimal formation: evidence for tissue angiotensin-converting enzyme function. *Arterioscl. Thromb. Vasc. Biol.* **20,** 915–922.

81. Chen, H., Mohuczy, D., Li, D., et al. (2001) Protection against ischemia/reperfusion injury and myocardial dysfunction by antisense-oligodeoxynucleotide directed at angiotensin-converting enzyme mRNA. *Gene Ther.* **8,** 804–810.

82. Phillips, M. I. and Kimura, B. (1999) Central nervous system and angiotensin in the development of hypertension. In: *Development of the Hypertensive Phenotype: Basic and Clinical Studies.* (McCarty, R., Blizard, D. A., Chevalier, R. L., eds.), Elsevier Science B.V., pp. 383–411.

83. Nakamura, S., Moriguchi, A., Morishita, R., et al. (1999) Activation of the brain angiotensin system by in vivo human angiotensin-converting enzyme gene transfer in rats. *Hypertension* **34,** 302–308.

84. Kubo, T., Ikezawa, A., Kambe, T., et al. (2001) Renin antisense injected intraventricularly decreases blood pressure in spontaneously hypertensive rats. *Brain Res. Bull.* **56,** 23–28.

85. Moore, A. F., Heiderstadt, N. T., Huang, E., et al. (2001) Selective inhibition of the renal angiotensin type 2 receptor increases blood pressure in conscious rats. *Hypertension* **37,** 1285–1291.

86. Orr, R. M. (2001) Technology evaluation: fomivirsen, Isis Pharmaceuticals Inc/CIBA vision. *Curr. Opin. Mol. Ther.* **3,** 288–294.

87. de Smet, M. D., Meenken, C. J., and van den Horn, G. J. (1999) Fomivirsen—a phosphorothioate oligonucleotide for the treatment of CMV retinitis. *Ocul. Immunol. Inflamm.* **7,** 189–198.

88. Acland, G. M., Aguirre, G. D., Ray, J., et al. (2001) Gene therapy restores vision in a canine model of childhood blindness. *Nat. Genet.* **28,** 92–95.

89. Freytag, S. O., Khil, M., Stricker, H., et al. (2002) Phase I study of replication-competent adenovirus-mediated double suicide gene therapy for the treatment of locally recurrent prostate cancer. *Cancer Res.* **62,** 4968–4976.

90. Reid, T., Galanis, E., Abbruzzese, J., et al. (2001) Intra-arterial administration of a replication-selective adenovirus (dl1520) in patients with colorectal carcinoma metastatic to the liver: a phase I trial. *Gene Ther.* **8,** 1618–1626.

91. Harvey, B. G., Maroni, J., O'Donoghue, K. A., et al. (2002) Safety of local delivery of low- and intermediate-dose adenovirus gene transfer vectors to individuals with a spectrum of morbid conditions. *Hum. Gene Ther.* **13,** 15–63.

92. The, B. S., Aguilar-Cordova, E., Kernen, K., et al. (2001) Phase I/II trial evaluating combined radiotherapy and in situ gene therapy with or without hormonal therapy in the treatment of prostate cancer—a preliminary report. *Int. J. Radiat. Oncol. Biol. Phys.* **51,** 605–613.

93. Lamont, J. P., Nemunaitis, J., Kuhn, J. A., et al. (2000) A prospective phase II trial of ONYX-015 adenovirus and chemotherapy in recurrent squamous cell carcinoma of the head and neck (the Baylor experience). *Ann. Surg. Oncol.* **7,** 588–592.

94. Nishii, T., Moriguchi, A., Morishita, R., et al. (1999) Angiotensinogen gene-activating elements regulate blood pressure in the brain. *Circ. Res.* **85,** 257–263.

95. Katovich, M. J., Reaves, P. Y., Francis, S. C., et al. (2001) Gene therapy attenuates the elevated blood pressure and glucose intolerance in an insulin-resistant model of hypertension. *J. Hypertens.* **19,** 1553–1558.
96. Morishita, R., Higaki, J., Tomita, N., et al. (1996) Role of transcriptional cis-elements, angiotensinogen gene-activating elements, of angiotensinogen gene in blood pressure regulation. *Hypertension* **27,** 502–507.

5

Antisense Strategies for Treatment of Heart Failure

Sian E. Harding, Federica del Monte, and Roger J. Hajjar

1. Introduction

In the mammalian heart, intracellular Ca^{2+} movements are tightly regulated at various levels within the cell (1). The sarcoplasmic reticulum (SR) plays an important role in orchestrating the movement of calcium during each contraction and relaxation. Excitation leads to the opening of voltage-gated L-type Ca^{2+} channels, allowing the entry of a small amount of Ca^{2+} into the cell. Through a coupling mechanism between the L-type Ca^{2+} channel and the SR release channels (ryanodine receptors), a larger amount of Ca^{2+} is released, activating the myofilaments leading to contraction. During relaxation, Ca^{2+} is reaccumulated back into the SR by the SR Ca^{2+} adenosine triphosphatase (ATPase) pump (SERCA2a) and extruded extracellularly by the sarcolemmal Na^+/Ca^{2+} exchanger. The contribution of each of these mechanisms for lowering cytosolic Ca^{2+} varies with species. In humans, approx 70% of the Ca^{2+} is removed by SERCA2a and approx 30% by the Na^+/Ca^{2+} exchanger (1). The Ca^{2+} pumping activity of SERCA2a is influenced by phospholamban. In the unphosphorylated state, phospholamban inhibits the Ca^{2+}-ATPase, whereas phosphorylation of phospholamban by cyclic adenosine monophosphate (cAMP)-dependent protein kinase and by Ca^{2+}-calmodulin-dependent protein kinase reverses this inhibition (2). Ca^{2+} transients recorded from failing human myocardial cells or trabeculae reveal a significantly prolonged Ca^{2+} transient with an elevated end-diastolic intracellular Ca^{2+} (3,4). A decrease in SR Ca^{2+} ATPase activity and Ca^{2+} uptake have been shown to be responsible for abnormal Ca^{2+} homeostasis in both experimental and human failing hearts (5,6). Associated with a defective Ca^{2+} uptake, there is a decrease in the relative ratio

From: *Methods in Molecular Medicine, Vol. 106: Antisense Therapeutics, Second Edition*
Edited by: I. Phillips © Humana Press Inc., Totowa, NJ

of SERCA2a/phospholamban in these failing hearts. Using transgenic and gene transfer approaches, increasing levels of phospholamban relative to SERCA2a in isolated cardiac myocytes significantly altered intracellular Ca^{2+} handling by prolonging the relaxation phase of the Ca^{2+} transient, decreasing Ca^{2+} release, and increasing resting Ca^{2+} *(7,8)*.

A number of studies have attempted to improve contractility by decreasing the phospholamban/SERCA2a ratio. Minamisawa et al. *(9)* showed that phospholamban-deficient mice were characterized by an increase in SR Ca^{2+} uptake and enhanced contractile performance. Similarly, mice overexpressing cardiac SERCA2a displayed an increased contraction and accelerated relaxation of the heart *(10)*. Using adenoviral gene transfer, we have shown that increasing SERCA2a levels relative to phospholamban leads to an increase in SR Ca^{2+} ATPase activity, a faster relaxation phase, an increase in the amount of Ca^{2+} released, and a decrease in diastolic Ca^{2+} *(11–13)*. These results supported the hypothesis that altering the relative ratio of phospholamban to SERCA2a can account for the abnormalities in calcium handling observed in failing ventricular myocardium. Another approach to restore the phospholamban/SERCA2a ratio to normal in failing hearts would be to decrease levels of phospholamban using antisense strategies. Indeed, a number of groups have used phospholamban antisense or dominant-negative constructs to decrease expression of this phosphoprotein and enhance SR Ca^{2+} ATPase activity *(9,14)*. In this chapter, we review the experience of antisense strategies in animal models of heart failure and isolated human cardiomyocytes.

2. Antisense Strategies in Animal Myocytes: SERCA2a Overexpression Compared to Phospholamban Downregulation in Rat and Rabbit Myocytes

Adenoviral gene transfer into isolated myocytes has allowed the investigation of changes in phospholamban in species other than mouse, as well as direct comparison of phospholamban depletion with SERCA2a overexpression in the same preparation. Rat is similar to mouse in the predominance of SERCA2a (92%) compared to other mechanisms of calcium removal, while in rabbit heart (like human) SERCA2a controls approx 70%, and there is a much larger role for the Na^+/Ca^{2+} exchanger to remove excess calcium from the cell *(15)*. Initial results with an antisense strategy for phospholamban in adult myocytes were disappointing *(14)*, but later work showed that even partial (approx 50%) reduction of phospholamban could have effects almost equivalent to SERCA2a overexpression (approx fivefold) on contraction amplitude (**Fig. 1**) and only slightly less on relaxation *(16,17)*. This correlates well with results from transgenic animals, in which phospholamban knockout can be even more effective than SERCA2a. In mice overexpressing SERCA2a, the maxi-

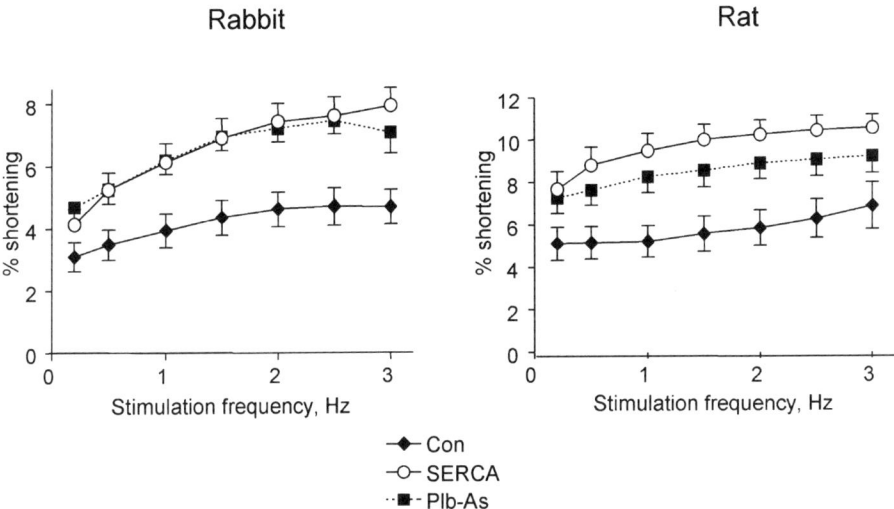

Fig. 1. Frequency-dependent increase in contraction amplitude of rat or rabbit myocytes cultured for 48 h without infection. Rabbit: uninfected controls (Con) (*n* = 14 cells), infected with Ad.SERCA2a.GFP (SERCA) (*n* = 17) or Ad.PlbAs.GFP (Plb-As) (*n* = 17). Both SERCA and Plb-As curves were significantly different from Con (*p* < 0.001 in each case) but were not significantly different from each other. Rat: Con (*n* = 9 cells), SERCA (*n* = 13) or Plb-As (*n* = 11). Both SERCA and Plb-As curves were significantly different from Con (*p* < 0.01 for SERCA; *p* < 0.02 for Plb-As) but were not significantly different from each other. (Adapted from **ref. *13***.)

mum velocity of myocyte relengthening was accelerated by a modest 40% *(18)* compared with 340% in cells from PLB-KO mice *(19)*. In that and another study *(20)*, robust overexpression of SERCA2a mRNA translated into increases in protein of only 20–50%. An upper limit on SERCA2a incorporation has been suggested previously in a quantitative study using adenoviral vectors in chick embryo myocytes *(21)*. This may be particularly relevant to the mouse heart, where the activity of SERCA2a (relative to other Ca^{2+} removal systems) is already high, possibly leaving little scope for increased gain via overexpression of SERCA2a.

Increases in contraction amplitude were similar in either rat or rabbit ventricular myocytes and, like human (*see* **Subheading 3.**), showed a frequency dependence (**Fig. 1**). As might be predicted from the mouse models, adenovirally mediated acceleration of relaxation was less evident in rat than in rabbit, but this was true for both phospholamban-depleted and SERCA2a-overexpressing cells *(17)*. In both rat and rabbit myocytes, the maximum effect of phospholamban antisense/SERCA2a was similar to that of catecholamines, which is consistent with phospholamban phosphorylation as the main mechanism for β-adrenoceptor-mediated lusitropy. Interestingly, however, frequency-dependent acceleration of relaxation, which is not related to phospholamban, was additive with that by SERCA2a overexpression or phospholamban depletion *(17)*.

3. Antisense Strategies in Human Myocytes

Myocardial cells from failing human hearts are characterized by abnormal calcium handling, a negative force-frequency relationship, and decreased SR Ca^{2+} ATPase (SERCA2a) activity. In a study by del Monte et al. *(22)*, ventricular myocytes isolated from nine patients with end-stage heart and 18 donor nonfailing hearts were infected with adenoviruses encoding for either the antisense sequence of phospholamban (Ad.asPL), SERCA2a gene (Ad.SERCA2a), or the reporter genes β-galactosidase and green fluorescent protein (Ad.βgal-GFP). Adenoviral gene transfer with Ad.asPL decreased phospholamban expression over 48 h, increasing the velocity of both contraction and relaxation. Compared to cardiomyocytes infected with Ad.asPL, human myocytes infected with Ad.βgal-GFP had enhanced contraction velocity and relaxation velocity, as shown in **Fig. 2**. The improvement in contraction and relaxation velocities was comparable to that of cardiomyocytes infected with Ad.SERCA2a. Failing human cardiomyocytes had decreased contraction and Ca^{2+} release with increasing frequency (0.1–2 Hz). Ablation of phospholamban restored the frequency response in the failing cardiomyocytes to normal: increasing frequency resulted in enhanced SR Ca^{2+} release and contraction. These results show that gene transfer of asPL can improve the contractile function in failing human myocardium. Targeting phospholamban may provide therapeutic benefits in human heart failure. The study by del Monte et al. *(22)* revealed important findings in terms of targeting calcium cycling proteins in failing human cardiomyocytes: (1) decreasing phospholamban expression restores contractility in failing ventricular cells of different etiologies, and (2) ablation of phospholamban results in similar improvement in contractility to SERCA2a overexpression. These findings also extend previous results that overexpression of SERCA2a improves contractile function in human failing cardiac myocytes *(11)*. Even though the two strategies of decreasing phospholamban or overexpression of SERCA2a improved contractile function in the failing human cardiomyocytes to the same extent, there were certain differences. SR Ca^{2+} ATPase activity increased to a greater degree in cardiomyocytes overexpressing SERCA2a. In failing cardiomyocytes, relieving inhibition to SERCA2a pumps that may be impaired due to oxidative stresses may not restore ATPase activity to normal. The results demonstrate that targeting calcium regulation by ablation of phospholamban improves con-

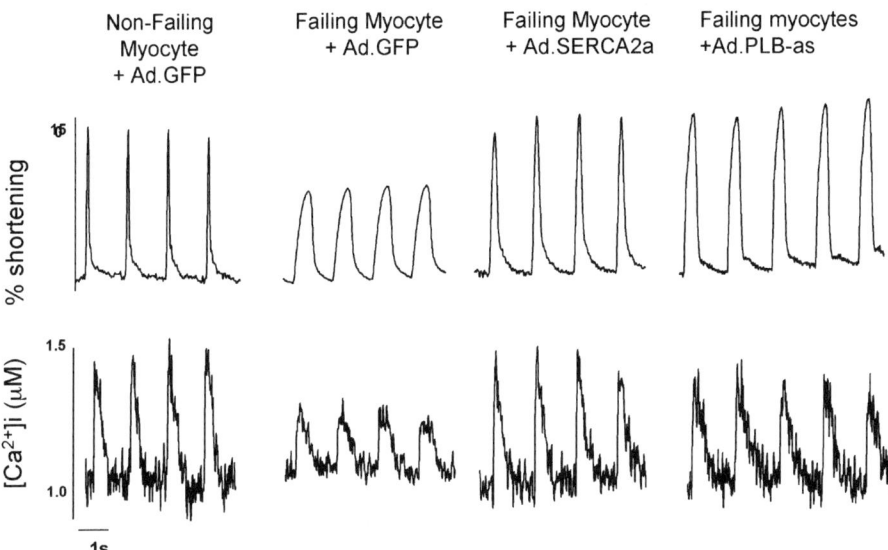

Fig. 2. Recordings from cardiomyocytes isolated from a donor nonfailing heart and from a failing heart infected either with an adenovirus expressing green fluorescent protein, Ad.βgal-GFP, or with Ad.asPL, stimulated at 1 Hz at 37°C. The failing cell had a characteristic decrease in contraction and prolonged relaxation along with a prolonged Ca^{2+} transient. Ablation of phospholamban in the failing cardiomyocyte normalized these parameters.

tractile function in failing human cardiomyocytes. del Monte et al. *(22)* study validates the feasibility of cardiac gene transfer in failing hearts as a therapeutic modality.

There are clearly major disadvantages of using human cardiomyocytes, such as problems with the patient population, necrotic tissue, and relative purification. We have, however, obtained consistent good yields with our technique of isolation. Furthermore, we have been able to maintain these cells in cultures for 6 d while they are expressing the transgene. Even though antisense strategies have been used extensively with oligonucleotides, there is less experience with adenoviral gene delivery of the antisense transgene. One of the concerns of long-term expression of antisense phospholamban is the specificity of the target and whether expression of other proteins will be affected.

3.1. Interaction Between Phospholamban Depletion and β-Adrenoceptor Stimulation

In PLB-KO mice, the effect of β-adrenoceptor stimulation appears reduced, but this is largely because of the tonic effect of phospholamban depletion on basal contraction and relaxation *(23)*. A residual effect of catecholamines on relaxation is observed, and this is probably due to phosphorylation of troponin I, because crossing of PLB-KO mice with a strain having mutant or slow skeletal troponin abolishes the lusitropic effect of β-adrenoceptor stimulation completely *(24,25)*. In rabbit myocytes transfected with phospholamban antisense, it is possible to eliminate the effect of tonic stimulation of contraction by selecting a frequency (*see* **Fig. 1**) at which basal amplitude is unchanged. At this frequency, overexpression of SERCA2a or phospholamban downregulation does not alter the contractile response to isoproterenol, with both maximal effect and concentration dependence unchanged *(16)*. This suggests that cAMP-dependent phosphorylation of the L-type Ca^{2+} channel predominates in mediating the positive inotropic effect of catecholamines. Relaxation, however, is accelerated by SERCA2a and phospholamban antisense adenoviruses at all stimulation frequencies, so that isoproterenol is acting against this background. Isoproterenol had additive effects on both viruses, but the additional effect was modest. For example, SERCA2a overexpression and maximum concentrations of isoproterenol decreased time-to-50% relaxation (R50) by 52 and 54%, respectively, compared to basal, untreated myocytes, and the combination of the two brought the overall decrease to 66% *(16)*. Similarly, the combination of phospholamban antisense and maximum isoproterenol brought the R50 values down by 64% *(16)*. Interestingly, the effect of a pronounced SERCA2a overexpression plus catecholamines was not different for a partial phospholamban depletion plus catecholamines. This again speaks of some intrinsic limitation that does not permit the overexpression of SERCA2a to be effective beyond a certain level.

An additional important observation was that the stimulation of SERCA2a activity decreased the threshold for the lusitropic effect of catecholamines. In untreated myocytes, the first lusitropic concentration was similar to the first inotropic one. However, the sensitivity of R50 to isoproterenol increased by approximately half a log unit after SERCA2a overexpression and by more than one log unit for phospholamban downregulation. This implies that in phospholamban antisense-treated myocytes, the acceleration of relaxation occurred at concentrations 10-fold lower than those that increased L-type Ca^{2+} channel activity. Because increased Ca^{2+} channel activity is related to arrhythmia generation (*see* **Subheading 3.2.**), this dissociation of the two would be predicted to be beneficial particularly in failing human heart, where myocyte relaxation is slowed *(26)*. Note, however, that the failing human heart is especially sensitive to the lusitropic effect of catecholamines, even though the contractile response is reduced or absent *(27)*.

3.2. SERCA2a Activity and Arrhythmogenesis

The choice of species is particularly important for investigation of the potential arrhythmogenic effects of stimulation of SERCA2a activity through phospholamban depletion. Mouse myocardium is heavily dependent on SR Ca^{2+}, and it is well known that mouse myocytes show a high degree of spontaneous Ca^{2+} release from the SR. Human or rabbit myocytes can maintain an unchanged contraction amplitude when SR function is completely inhibited pharmacologically if the stimulation frequency is not too high *(28,29)*, whereas contraction in mouse myocytes is abolished by the same maneuver. Additionally, the action potential of the mouse is extremely short compared to human (or rabbit) and lacks the plateau related to L-type Ca^{2+} channel activation. Probably for these reasons, among others, little note has been made of arrhythmias in the PLB-KO mouse. Alterations in the SR Ca^{2+} release channel and the β-adrenoceptor number in this model also complicate interpretation of phenomena *(24,30)*.

Investigations have been done on rabbit myocytes with adenovirus using either phospholamban antisense or SERCA2a overexpression to increase SERCA2a activity. Ca^{2+} levels in the SR are raised by 50–80% *(31,32)*, and this in itself might be expected to increase the incidence of spontaneous Ca^{2+} release, but no such increase was observed. The greater survival of the myocytes in culture and their ability to withstand high stimulation frequencies without diastolic contracture suggested that the beneficial effect of lowering diastolic Ca^{2+} outweighed the influence of SR load *(31)*.

Even if not arrhythmogenic *per se*, increased SR Ca^{2+} might be predicted to potentiate the arrhythmic effects of catecholamines. This is of particular concern because the final target of the gene transfer is failing human heart, which is prone to arrhythmias because of both changes in the myocyte (e.g., prolonged duration of action potential) and in the myocardium (e.g., areas of necrosis) and is under constant sympathetic tone. Several distinct classes of catecholamine-dependent arrhythmias can be detected in the contracting isolated myocyte, and these are thought to have parallels in the myocardium in vivo. First, β-adrenoceptor agonists produce or accentuate early aftercontractions/afterdepolarizations. These are observed at isoproterenol concentrations below maximum and do not necessarily disrupt stimulated contractions. The extra contraction is close to the main beat and can initially be observed as a second, long phase of relaxation; the associated afterdepolarization occurs before repolarization, often during the plateau phase of the action potential. Reactivation of the L-type Ca^{2+} channel during the plateau is a likely mechanism, although some reverse-mode Na^+/Ca^{2+} exchange might be involved *(33)*. Prolongation of the action potential potentiates aftercontractions, and this accounts for an increased incidence of torsades de pointes (their in vivo correlate *[34]*) in heart failure or long QT syndrome.

Strikingly, SERCA2a overexpression or phospholamban depletion not only did not exacerbate early aftercontractions, but actually reduced their incidence in rabbit or human myocytes *(16,22,31)* (**Fig. 3**). This suggests that the effect of β-adrenoceptor stimulation to increase the open probability of the L-type Ca^{2+} channel is central to the development of catecholamine-induced aftercontractions. Part of the mechanism for the decrease in aftercontractions in myocytes overexpressing SERCA2a may have been a reduction in the duration of action potential *(32)*.

In contrast to the aftercontractions, at high isoproterenol concentrations, contraction in the myocytes is frequently disrupted by arrhythmias, indicating a calcium overload state, and these arrhythmias are characterized by disorganized contraction, loss of synchronization with the stimulation pulse, and waves of contraction likely resulting from spontaneous calcium release from the SR. These arrhythmias have more in common with delayed afterdepolarizations in their occurrence and calcium dependence *(33)*. However, even these arrhythmias were slightly reduced in phospholamban-deficient or SERCA2a-overexpressing rat myocytes *(16)*. This indicates that the higher SR calcium load has not made the cells more sensitive to the arrhythmic effects of β-adrenoceptor stimulation.

4. Phospholamban Downregulation and Heart Failure

Although antisense *per se* has not been used to downregulate phospholamban in animal models of heart failure, a similar strategy has been pursued with dominant-negative constructs, with some success. A pseudophosphorylated mutant of phospholamban was introduced using viral transfer to a hamster model of cardiomyopathy, and progression of heart failure was slowed *(35)*. The same method has been used to transfect rats with myocardial infarction, and again the deterioration of ventricular structure and function was prevented *(36)*. This accords well with the ability of phospholamban knockout mice to rescue various models of heart failure *(9,37,38)* although the benefit is not universal *(39)*. SERCA2a overexpression was also able to reduce mortality and preserve cardiac function in rats with aortic banding *(40,41)* and even to reverse effects due to aging *(42)*. One concern about this strategy of increasing SR Ca^{2+} ATPase by diminishing phospholamban inhibition is the anticipated energy cost for increasing adenosine triphosphate (ATP) hydrolysis. Our group has recently shown that overexpression of SERCA2a in a rat model of heart failure enhances contractility without energetic compromise. In fact, in this

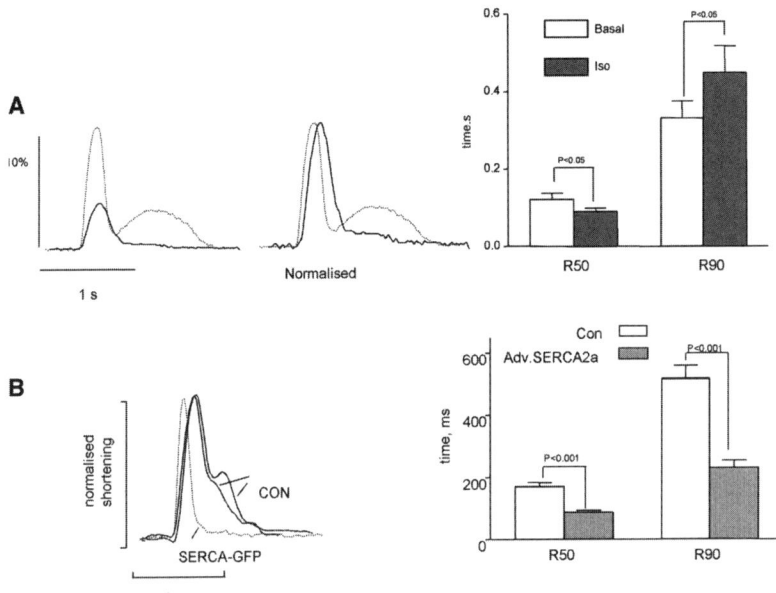

Fig. 3. (**A**) Effects of β-adrenoceptor stimulation on cultured rabbit myocytes: (**left**) sample trace for isoproterenol (iso, 10 nmol/L) on contraction of a 48-h cultured adult rabbit myocyte as (1) % shortening or (2) normalized to show speed of contraction and relaxation; (**right**) averaged R50 (early) and time-to-90% (R90) (late) relaxation ($n = 17$ preparations). (**B**) Effects of SERCA2a overexpression on cultured rabbit myocytes: (**left**) sample contractions of uninfected myocytes (CON) and an Adv.SERCA2a-GFP-overexpressing myocyte (SERCA-GFP) at 48 h; (**right**) averaged R50 and R90. (□) untreated myocytes ($n = 105$ cells/seven preparations); (■) Ad.SERCA2a.GFP-treated myocytes ($n = 99$ cells/seven preparations). (Adapted from **ref. 21**.)

model of heart failure, overexpression of SERCA2a restored the balance between ATP and creatine phosphate.

Two naturally occurring mutations in the human phospholamban gene have been identified, with implications for the strategy of phospholamban downregulation. In the first example, the inheritance of the phospholamban mutation encoding Arg9Cys was linked to the dominant inheritance of dilated cardiomyopathy in a large American family *(43)*. The effects of the phospholamban Arg9Cys mutation were characterized by expression in heterologous cell culture, by the creation of a transgenic mouse, and by analysis of cardiac tissue obtained from an explanted heart. In all cases, the level of phospholamban phosphorylation was reduced markedly. The key effect of the mutation was enhancement of the affinity of Arg9Cys mutant phospholamban for protein kinase A (PKA). In attempting to phosphorylate mutant phospholamban, PKA becomes trapped in a stabilized mutant phospholamban-PKA complex and can no longer dissociate to phosphorylate wild-type phospholamban molecules. Affected individuals must go through life with chronically inhibited SERCA2a and can never draw on their full cardiac reserve. The association of decreased SERCA2a activity and heart failure correlates well with the many animal models.

A second human phospholamban mutation, Leu39stop, was discovered in two large Greek families *(44)*. The heterozygous inheritance of the Leu39stop mutation in one family led to left ventricular hypertrophy in one-third of the older affected family members, without diminished contractile performance. However, the inheritance of two copies of the mutant phospholamban gene led to dilated cardiomyopathy and heart failure in two teenage siblings. In heterologous expression studies, the Leu39stop mutant protein was unstable or misrouted to other membranes, and no protein was detected in the endoplasmic reticulum of these cells or in a cardiac explant from one of the affected individuals. As a result, there was no effect of the mutant protein, in either the homozygous or heterozygous state, on the Ca^{2+} affinity of SERCA2a. Accordingly, these two homozygous mutant individuals can be considered to be equivalent to a phospholamban-null genotype with a phenotype of dilated cardiomyopathy. Thus, in contrast to the benefits of phospholamban ablation in mouse, humans that lack phospholamban develop lethal cardiomyopathy. A caveat in these studies is that the number of affected individuals is very low and the LOD score for linkage of the mutation to the disease is low. The reason for the discrepancy between the cardiac phenotypes in mice and humans has not been uncovered yet. For this reason, gene transfer experiments with antisense phospholamban in large animal models will be needed to give information on contractility and survival in a preclinical model of heart failure in both the short term and long term.

5. Conclusion

From animal experiments, the spectrum of changes observed following adenovirally mediated gene transfer of phospholamban antisense would be predicted to be favorable for the failing human heart since it includes reversal of the depressed contractile response to increasing frequency *(45)*, increased ability of catecholamines to accelerate the slowed relaxation *(26)*, and reduction of catecholamine-induced aftercontractions *(34)*. Interest in stimulation of SERCA2a activity as a possible strategy for gene therapy of the failing myocardium has been generated by the beneficial effects in animal models of cardiomyopathy and in myocytes from diseased human hearts *(9,11,22,38, 41,42,46–48)*. Reduction of the inhibitory action of phospholamban is an attractive option because of the small size of the cDNA inserts needed for antisense/dominant-negative phospholamban constructs, making it suitable for use with the newer generation of adenoassociated viruses. Infection with adenoassociated virus is less immunogenic than with adenovirus and has proved to be longer lasting in the myocardium *(49)*. Our demonstration that partial phospholamban depletion can be as effective as SERCA2a overexpression supports the therapeutic use of this strategy.

Acknowledgments

This work was supported in part by grants from the National Institutes of Health (HL 57623) and the British Heart Foundation, and by a Doris Duke Charitable Foundation Clinician Scientist Award.

References

1. Bers, D. M. (2001) Cardiac excitation-contraction coupling [review]. *Nature* **415,** 198–205.
2. MacLennan, D. H. and Kranias, E. G. (2003) Phospholamban: a crucial regulator of cardiac contractility [review]. *Nat. Rev. Mol. Cell Biol.* **4,** 566–577.
3. Gwathmey, J. K. and Hajjar, R. J. (1990) Intracellular calcium related to force development in twitch contraction of mammalian myocardium. *Cell Calcium* **11,** 531–538.
4. Beuckelmann, D. J., Nabauer, M., and Erdmann, E. (1992) Intracellular calcium handling in isolated ventricular myocytes from patients with terminal heart failure. *Circulation* **85,** 1046–1055.
5. Kiss, E., Ball, N. A., Kranias, E. G., and Walsh, R. A. (1995) Differential changes in cardiac phospholamban and sarcoplasmic reticular Ca(2+)-ATPase protein levels: effects on Ca2+ transport and mechanics in compensated pressure-overload hypertrophy and congestive heart failure. *Circ. Res.* **77,** 759–764.
6. Hasenfuss, G. (1998) Alterations of calcium-regulatory proteins in heart failure. *Cardiovasc. Res.* **37,** 279–289.
7. Kadambi, V. J., Ponniah, S., Harrer, J. M., et al. (1996) Cardiac-specific overexpression of phospholamban alters calcium kinetics and resultant cardiomyocyte mechanics in transgenic mice. *J. Clin. Invest.* **97,** 533–539.
8. Davia, K., Hajjar, R. J., Terracciano, C. M. N., et al. (1999) Functional alterations in adult rat myocytes after overexpression of phospholamban using adenovirus. *Physiol. Genomics* **1,** 41–50, 1999.
9. Minamisawa, S., Hoshijima, M., Chu, G., et al. (1999) Chronic phospholamban-sarcoplasmic reticulum calcium ATPase interaction is the critical calcium cycling defect in dilated cardiomyopathy. *Cell* 99, 313–322.
10. Hashimoto, K., Perez, N. G., Kusuoka, H., Baker, D. L., Periasamy, M., and Marban, E. (2000) Frequency-dependent changes in calcium cycling and contractile activation in SERCA2a transgenic mice. *Basic Res. Cardiol.* **95,** 144–151.
11. del Monte, F., Harding, S. E., Schmidt, U., et al. (1999) Restoration of contractile function in isolated cardiomyocytes from failing human hearts by gene transfer of SERCA2a. *Circulation* **100,** 2308–2311.

12. Hajjar, R. J., Schmidt, U., Kang, J. X., Matsuki, T., and Rosenzweig, A. (1997) Adenoviral gene transfer of phospholamban in isolated rat cardiomyocytes: rescue effects by concomitant gene transfer of sarcoplasmic reticulum Ca ATPase. *Circ. Res.* **81,** 145–153.

13. Hajjar, R. J., Kang, J. X., Gwathmey, J. K., and Rosenzweig, A. (1997) Physiological effects of adenoviral gene transfer of sarcoplasmic reticulum calcium ATPase in isolated rat myocytes. *Circulation* **95,** 423–429.

14. He, H., Meyer, M., Martin, J. L., et al. (1999) Effects of mutant and antisense RNA of phospholamban on SR Ca(2+)-ATPase activity and cardiac myocyte contractility. *Circulation* **100,** 974–980.

15. Bers, D. M. (2000) Calcium fluxes involved in control of cardiac myocyte contraction. *Circ. Res.* **87,** 275–281.

16. Chaudhri, B., del Monte, F., Hajjar, R. J., and Harding, S. E. (2002) Interaction between increased SERCA2a activity and b-adrenoceptor stimulation in adult rabbit myocytes. *Am. J. Physiol.* **283,** H2450–H2457.

17. Chaudhri, B., del Monte, F., Hajjar, R. J., and Harding, S. E. (2003) Contractile effects of adenovirally-mediated increases in SERCA2a activity: a comparison between adult rat and rabbit ventricular myocytes. *Mol. Cell. Biochem.* **251,** 103–109.

18. He, H., Giordano, F. J., Hilal Dandan, R., et al. (1997) Overexpression of the rat sarcoplasmic reticulum Ca2+ ATPase gene in the heart of transgenic mice accelerates calcium transients and cardiac relaxation. *J. Clin. Invest.* **100,** 380–389.

19. Wolska, B. M., Stojanovic, M. O., Luo, W., Kranias, E. G., and Solaro, R. J. (1996) Effect of ablation of phospholamban on dynamics of cardiac myocyte contraction and intracellular Ca^{2+}. *Am. J. Physiol.* **271,** C391–C397.

20. Baker, D. L., Hashimoto, K., Grupp, I. L., et al. (1998) Targeted overexpression of the sarcoplasmic reticulum Ca^{2+}- ATPase increases cardiac contractility in transgenic mouse hearts. *Circ. Res.* **83,** 1205–1214.

21. Sumbilla, C., Cavagna, M., Zhong, L., Ma, H., Lewis, D., Farrance, I., and Inesi, G. (1999) Comparison of SERCA1 and SERCA2a expressed in COS-1 cells and cardiac myocytes. *Am. J. Physiol.* **277,** H2381–H2391.

22. del Monte, F., Harding, S. E., Dec, G. W., Gwathmey, J. K., and Hajjar, R. J. (2002) Targeting phospholamban in human heart failure by gene transfer. *Circulation* **105,** 904–907.

23. Pan, B. S., Hannon, J. D., Wiedmann, R., et al. (1999) Effects of isoproterenol on twitch contraction of wild type and phospholamban-deficient murine ventricular myocardium. *J. Mol. Cell. Cardiol.* **31,** 159–166.

24. Wolska, B. M., Arteaga, G. M., Pena, J. R., et al. (2002) Expression of slow skeletal troponin I in hearts of phospholamban knockout mice alters the relaxant effect of beta-adrenergic stimulation. *Circ. Res.* **90,** 882–888.

25. Pi, Y., Kemnitz, K. R., Zhang, D., Kranias, E. G., and Walker, J. W. (2002) Phosphorylation of troponin I controls cardiac twitch dynamics: evidence from phosphorylation site mutants expressed on a troponin I-null background in mice. *Circ. Res.* **90,** 649–656.

26. del Monte, F., O'Gara, P., Poole-Wilson, P. A., Yacoub, M. H., and Harding, S. E. (1995) Cell geometry and contractile abnormalities of myocytes from failing human left ventricle. *Cardiovasc. Res.* **30,** 281–290.

27. Harding, S. E., Brown, L. A., del Monte, F., et al. (1996) Acceleration of contraction by b-adrenoceptor stimulation is greater in ventricular myocytes from failing than non-failing human hearts. *Basic Res. Cardiol.* **91(Suppl 2),** 53–56.

28. Davia, K., Davies, C. H., and Harding, S. E. (1997) Effects of inhibition of sarcoplasmic reticulum calcium uptake on contraction of myocytes from failing human ventricle. *Cardiovasc. Res.* **33,** 88–97.

29. Ranu, H. K., Terracciano, C. M., Davia, K., et al. (2002) Effects of Na(+)/Ca(2+)-exchanger overexpression on excitation-contraction coupling in adult rabbit ventricular myocytes. *J. Mol. Cell. Cardiol.* **34,** 389–400.

30. Chu, G., Ferguson, D. G., Edes, I., Kiss, E., Sato, Y., and Kranias, E. G. (1998) Phospholamban ablation and compensatory responses in the mammalian heart. *Ann. NY Acad. Sci.* **853,** 49–62.

31. Davia, K., Bernobich, E., Ranu, H. K., et al. (2001) SERCA2a overexpression decreases the incidence of aftercontractions in adult rabbit ventricular myocytes. *J. Mol. Cell. Cardiol.* **33,** 1005–1015.

32. Terracciano, C. M. N., Hajjar, R. J., and Harding, S. E. (2002) Overexpression of SERCA2a accelerates repolarisation in rabbit ventricular myocytes. *Cell Calcium* **31,** 299–305.

33. Bers, D. M. (2002) Calcium and cardiac rhythms: physiological and pathophysiological. *Circ. Res.* **90,** 14–17.

34. Priori, S. G., Napolitano, R., and Schwartz, P. J. (1991) Electrophysiologic mechanisms involved in the development of torsades de pointes. *Cardiovasc. Drugs Ther.* **5,** 203–212.

35. Hoshijima, M., Ikeda, Y., Iwanaga, Y., et al. (2002) Chronic suppression of heart-failure progression by a pseudophosphorylated mutant of phospholamban via in vivo cardiac rAAV gene delivery. *Nat. Med.* **8,** 864–871.

36. Iwanaga, Y., Hoshijima, M., Gu, Y., et al. (2004) Chronic phospholamban inhibition prevents progressive cardiac dysfunction and pathological remodeling after infarction in rats. *J. Clin. Invest.* **113,** 727–736.

37. Sato, Y., Kiriazis, H., Yatani, A., et al. (2001) Rescue of contractile parameters and myocyte hypertrophy in calsequestrin overexpressing myocardium by phospholamban ablation. *J. Biol. Chem.* **276,** 9392–9399.

38. Freeman, K., Lerman, I., Kranias, E. G., et al. (2001) Alterations in cardiac adrenergic signaling and calcium cycling differentlally affect the progression of cardiomyopathy. *J. Clin. Invest.* **107,** 967–974.

39. Song, Q., Schmidt, A. G., Hahn, H. S., et al. (2003) Rescue of cardiomyocyte dysfunction by phospholamban ablation does not prevent ventricular failure in genetic hypertrophy. *J. Clin. Invest.* **111,** 859–867.

40. del Monte, F., Williams, E., Lebeche, D., et al. (2001) Improvement in survival and cardiac metabolism after gene transfer of sarcoplasmic reticulum Ca(2+)-ATPase in a rat model of heart failure. *Circulation* **104,** 1424–1429.

41. Miyamoto, M. I., del Monte, F., Schmidt, U., et al. (2000) Adenoviral gene transfer of SERCA2a improves left-ventricular function in aortic-banded rats in transition to heart failure. *Proc. Natl. Acad. Sci. USA* **97,** 793–798.
42. Schmidt, U., del Monte, F., Miyamoto, M. I., Matsui, T., Gwathmey, J. K., Rosenzweig, A., and Hajjar, R. J. (2000) Restoration of diastolic function in senescent rat hearts through adenoviral gene transfer of sarcoplasmic reticulum Ca(2+)-ATPase. *Circulation* **101,** 790–796.
43. Schmitt, J. P., Kamisago, M., Asahi, M., et al. (2003) Dilated cardiomyopathy and heart failure caused by a mutation in phospholamban. *Science* **299,** 1410–1413.
44. Haghighi, K., Kolokathis, F., Pater, L., et al. (2003) Human phospholamban null results in lethal dilated cardiomyopathy revealing a critical difference between mouse and human. *J. Clin. Invest.* **111,** 869–876.
45. Davies, C. H., Davia, K., Bennett, J. G., Pepper, J. R., Poole-Wilson, P. A., and Harding, S. E. (1995) Reduced contraction and altered frequency response of isolated ventricular myocytes from patients with heart failure. *Circulation* **92,** 2540–2549.
46. Trost, S U., Belke, D. D., Bluhm, W. F., Meyer, M., Swanson, E., and Dillmann, W. H. (2002) Overexpression of the sarcoplasmic reticulum Ca(2+)-ATPase improves myocardial contractility in diabetic cardiomyopathy. *Diabetes* **51,** 1166–1171.
47. Ito, K., Yan, X., Feng, X., Manning, W. J., Dillmann, W. H., and Lorell, B. H. (2001) Transgenic expression of sarcoplasmic reticulum Ca(2+) ATPase modifies the transition from hypertrophy to early heart failure. *Circ. Res.* **89,** 422–429.
48. Slack, J. P., Grupp, I. L., Dash, R., et al. (2001) The enhanced contractility of the phospholamban-deficient mouse heart persists with aging. *J. Mol. Cell. Cardiol.* **33,** 1031–1040.
49. del Monte, F., Harding, S. E., and Hajjar, R. J. (2000) Manipulation of SERCA2a in the heart by gene transfer. In: *Molecular Approaches to the Therapy of Heart Failure* (Hasensfus, G. and Marban, E., eds.).

III

Cancer

6

Clinical Studies of Antisense Oligonucleotides for Cancer Therapy

Rosanne M. Orr and F. Andrew Dorr

1. Introduction
1.1. New Drugs for Cancer Therapy

Until now, the clinical demise of cancer has relied on surgical resection and the inhibition of tumor cell proliferation using ionizing radiation or chemotherapeutic drugs designed to perturb DNA synthesis or the mitotic event. The development of cytotoxic agents has resulted in improvements in the treatment of leukemia, lymphoma, testicular cancer, and many other solid tumor types *(1)*. Hormone-based drugs have also been useful for breast and prostate cancers *(2)*. Although much success has been achieved, cytotoxic modalities walk the therapeutic tightrope of toxicities to normal tissues vs cancer cells, and drug resistance is generally present *de novo* or develops with treatment. Over the last decade or so, more attention has been focused on different therapeutic approaches. These include the development of monoclonal antibodies (MAbs) to specifically target cancer cells, and small molecule-inhibitors of cell-signaling pathways that have been linked to oncogenesis or maintenance of the malignant phenotype. For example, the former approach has seen the development and licensing of Herceptin® (trastuzumab; Genentech/Roche), a humanized MAb that targets erbB2/HER2, a receptor tyrosine kinase that is overexpressed in some 30% of breast cancers and has shown promising clinical activity when used in combination with other drugs for the treatment of metastatic breast cancer *(3–5)*. Promising small-molecule inhibitors of cell-signaling pathways include Gleevec™ (STI571, imitanib mesylate; Novartis) and Iressa™ (ZD1839, gefitinib; AstraZeneca). Gleevec is a small-molecule

From: *Methods in Molecular Medicine, Vol. 106: Antisense Therapeutics, Second Edition*
Edited by: I. Phillips © Humana Press Inc., Totowa, NJ

inhibitor (a phenylaminopyridine) of the tyrosine kinase encoded by the *abl* gene *(6)*. In chronic myelogenous leukemia (CML), a chromosomal translocation forms the Philadelphia chromosome that expresses a *bcr/abl* gene resulting in the expression of a constitutively active Abl tyrosine kinase. Gleevec has shown excellent clinical activity against CML and Philadelphia chromosome–positive acute lymphoblastic leukemia (ALL) *(7)*. Gleevec is not uniquely selective for inhibition of Abl tyrosine kinase and has additional inhibitory activities against the c-Kit receptor tyrosine kinase and the platelet-derived growth factor receptor tyrosine kinase *(6)*. Resistance to Gleevec characterized by overexpression or mutations in the *bcr/abl* gene is now emerging *(8)*. Iressa™ is another small-molecule (an aniloquinazoline) tyrosine kinase inhibitor that targets the epidermal growth factor receptor (EGFR) that is overexpressed in several cancers *(9)*. Iressa has shown clinical responses at well-tolerated doses in phase I and is currently in phase II clinical studies as a monotherapy and phase III combination trials against non–small cell lung cancer (NSCLC) *(10)*.

1.2. Antisense Alternatives

An alternative strategy in this era of molecular therapeutics is the antisense approach in which antisense oligonucleotides (AS-ODNs) hybridize to complementary sequences of target RNA, resulting in the prevention of message translation into protein product. This is generally achieved by activation of the enzyme RNase H that recognizes the RNA:DNA duplex and cleaves the target RNA or by steric blocking at the ribosome (reviewed by Crooke *[11]*). Thus, in theory, antisense technology has the potential to selectively correct aberrant gene expression that may be related to many diseases *(12)*. Chemical modifications of AS-ODNs are required to limit degradation by serum exonucleases and intracellular exo- and endonucleases. Phosphorothioates, possessing sulfur substitutions at nonbridging oxygens on the phosphate backbone, were the first generation of AS- ODNs to enter clinical studies. Vitravene® (fomivirsen; Isis/Ciba Vision) was the first phosphorathioate oligonucleotide (P-ODN) to be licensed in 1998. Vitravene is a 21mer P-ODN that targets the major immediate-early gene of cytomegalovirus (CMV) and is administered by intravitreal injection to patients with acquired immunodeficiency syndrome for the treatment of CMV-induced retinitis *(13)*. However, systemic administration of antisense phosphorothioates in preclinical models and clinical trials can result in toxic side effects that are unrelated to target inhibition. In this chapter, we review AS-ODNs that have entered clinical trials for cancer therapy. The mRNA targets of the AS-ODNs are given in **Table 1** and stages of clinical development in **Table 2**.

Table 1
mRNA Targets for AS-ODNs in Clinical Trials

Oligonucleotide	Target	RNA-binding region	Length	Sequence[a]	Chemistry[b]
OL(1)p53	*p53*	Exon 10	20mer	CCCTGCTCCCCCCTGGCTCC	PS
INX-3001	*c-myb*	Codons 2–9	24mer	TATGCTGTGCCGGGGTCTTCGGGC	PS
Genasense	*bcl-2*	Codons 1–6	18mer	TCTCCCAGCGTGCGCCAT	PS
ISIS 5132	*c-raf*	3'-UTR	20mer	TCCCGCCTGTGACATGCATT	PS
Affinitak	*PKC-* α	3'-UTR	20mer	GTTCTCGCTGGTGAGTTTCA	PS
ISIS 2503	Ha-*ras*	Translation initiation region	20mer	TCCGTCATCGCTCCTCAGGG	PS
GTI-2040	R2 subunit of RR	Coding region	20mer	Not disclosed	PS
GTI-2501	R1 subunit of RR	Coding region	20mer	Not disclosed	PS
GEM®231	*PKA-R1* α	Codons 8–13	18mer	*GCGU*GCCTCCTCAC*UGGC*	MBO
MG98	DNA MeTase	3'-UTR	20mer	Not disclosed	MBO
AP 12009	*TGF-* β2	Not disclosed	18mer	Not disclosed	PS
OGX-011	Clusterin	Translation initiation region	21mer	*CAG*CAGCAGATGCTTCATC*AT*	2'-MOE

[a]Italicized bases = 2-O-methyl substitutions and underlined bases = 2'-methoxyethyl substitutions. [b]PS, phosphorothioate; MBO, mixed backbone oligonucleotide; 2'-MOE, 2'-methoxyethyl substitutions.

Table 2
Clinical Development of AS-ODNs

Oligonucleotide	Phase I	Phase II	Phase III	Study type
OL(1)p53	✓			Monotherapy
INX-3001	✓			Monotherapy
Genasense	✓	✓	✓	Combination
ISIS 5132	✓	✓		Monotherapy
Affinitak	✓	✓	✓	Combination
ISIS 2503	✓	✓		Combination
GTI-2040	✓	✓		Combination
GTI-2501	✓			Monotherapy
GEM®231	✓	✓		Combination
MG98	✓	✓		Monotherapy
AP 12009	✓	✓		Monotherapy
OGX-011	✓			Monotherapy

2. p53

The *p53* tumor suppressor gene is responsive to elements of cellular stress, and the p53 protein is induced following DNA damage leading to the transcriptional activation of genes involved in cell-cycle checkpoints, DNA repair, and apoptosis (reviewed by Amundson et al. *[14]*). Checkpoint controls at G_1 and G_2 phases of the cell cycle allow cells to repair the damage to prevent the onset of misrepresentative DNA synthesis or to undergo apoptosis. Both wild-type and mutant p53 levels have been detected in acute myelogenous leukemia (AML) and myelodysplastic syndrome *(15)*. In preclinical studies of a 20mer antisense phosphorothioate (OL[1]p53) targeting exon 10 of *p53*, cytotoxicity to freshly isolated AML cells was consistently shown whereas a control oligonucleotide had no effects *(16)*. These results prompted a dose-finding phase I study of OL(1)p53 in patients with hematological malignancies. Sixteen patients with refractory AML (*n* = 6) or advanced myelodysplastic syndrome (*n* = 10) received OL(1)p53 (synthesized by Lynx) by continuous iv infusion at dosages of 0.05–0.25 mg/kg/h for 10 d *(17)*. No significant toxicities were observed at these levels, and 36% of OL(1)p53 was recovered intact in the urine. Although no clinical responses were achieved, this early clinical study demonstrated a favorable pharmacokinetic profile of a P-ODN administered by systemic infusion. It was hypothesized that transient depletion of p53 in cells experiencing DNA damage following exposure to an anthracycline, e.g., would lead to subsequent overexpression of p53 resulting in apoptosis. This proof of principle has recently been published in a report examining the effects of a combination of OL(1)p53 and idarubicin (4-demethoxy-daunorubicin; Pharmacia) on the viability of WMN lymphoma cells expressing wild-type p53 *(18)*. At concentrations that had minimal toxicities as single agents, the combination was lethal to this lymphoma cell line by 72 h but spared normal hematopoietic progenitor cells. This opens up a new avenue for lymphoma therapy with the caveat that the cells must express wild-type p53.

3. *c-myb*

The *c-myb* protooncogene encodes a transcription factor that is downregulated during differentiation of hematopoietic cells and commonly upregulated in leukemias (reviewed by Weston *[19]*). The implication that Myb expression is in some way linked to the pathogenesis and maintenance of leukemias led to extensive preclinical and pilot clinical studies of AS-ODNs targeting *c-myb* (reviewed by Gewirtz *[20]* and Orr *[21]*). Early in vitro studies using unmodified AS-ODNs targeting codons 2–7 and 6–11 of *c-myb* demonstrated inhibition of proliferation of a T-cell leukemia line and cloning efficiency of primary AML and CML cultures, whereas the sense counterparts were inactive. In five primary CML cultures, from patients in blast crisis, Bcr/Abl expression was depleted. The concentrations of AS-ODNs used were nontoxic to normal bone marrow progenitor cells *(22)*. An extension of these studies showed responses of CML cells harvested from patients in the chronic phase of this disease, suggesting that AS-ODNs targeting *c-myb* might be employed for ex vivo bone marrow purging and systemic treatment of CML *(23)*.

A 24mer P-ODN targeting codons 2–9 of *c-myb*, administered by continuous infusion, prolonged the survival of severe combined immunodeficient (SCID) mice bearing the K562 erythroleukemia (Philadelphia chromosome positive) *(24)*. In a clinical study, this P-ODN was administered to 18 CML patients (13 in blast crisis) by systemic infusion at dosages of 0.3–2 mg/kg/d for 7 d. No drug-related toxicities were observed, some disease stabilization was noted, and one patient in blast crisis appeared to revert to the chronic phase of the disease and survived another 14 mo *(20)*.

A recently published study of this P ODN, used for ex vivo bone marrow purging in allograft-ineligible CML patients, suggested that the short half-life of both the RNA and protein (<1 h) made *c-myb* a suitable target for inhibition in short-term bone marrow cultures *(25)*. In this pilot study, involving 25 patients, CD34$^+$ cells were purged with this P-ODN at approx 18 μM for 24 h ($n = 19$) or 72 h ($n = 5$). Patients (20 in chronic phase and 5 in accelerated phase) received busulphan (4 mg/kg/d × 4 d) and cyclophosphamide (60 mg/kg/d × 2 d, 21 patients in Philadelphia) or busulphan alone (16 mg/kg/d × 4 d, 4 patients in

London, UK) prior to bone marrow reengraftment. Following purging, *c-myb* RNA levels were reduced in approx 50% of bone marrow cultures. After 100 d, 14 patients who did not require backup infusions of unmanipulated bone marrow were evaluable for cytogenetics. Of these 14 patients, 6 had a cytogenetic response, as measured by a reduction in the percentage of Bcr/Abl-positive cells. In this study, the 72-h purging regimen resulted in prolonged marrow dysplasia. Lynx supplied this P-ODN for the clinical studies and Inex subsequently acquired the rights to this compound, now designated INX-3001. Clinical studies of INX-3001 were restarted in 2002 in a physician-sponsored trial at the University of Pennsylvania under the direction of A. M. Gewirtz, with the National Cancer Institute (Bethesda, MD) supplying 1 kg of INX-3001 (A. M. Gewirtz, personal communication).

There have been two small clinical studies of ex vivo marrow purging in CML using AS-ODNs targeted to *bcr/abl*. In the first study, the bone marrow from five patients expressing the B_2A_2 break-point junction and three expressing the B_3A_2 were purged with the complementary 26mer P-ODNs (Lynx) for 24 or 72 h at 150 μg/mL (approx 20μM) *(26)*. In contrast to c-Myb, the Bcr/Abl protein has a half-life of more than 48 h *(27)*. Patients, all with advanced CML, were conditioned with busulphan (4 mg/kg/d × 4 d) and etoposide (20 mg/kg/d × 2 d). A karyotypic response was reported in two cases, and three patients remained in the chronic second phase of the disease in the follow-up period, suggesting the feasibility of oligonucleotide bone marrow purging and autograft. In the second study, CD34+ cells from three patients, two expressing the B_3A_2 and one expressing the B_2A_2 break-point junction, were permeabilized with streptolysin O (up to 10 U/10^6 cells) for 10 min in serum-free medium in the presence of the respective complementary 18mer chimeric methylphosphonate/phosphodiester oligonucleotides *(28)*. Cell membrane pores were sealed by the addition of a human albumin solution, and incubations continued for a further 60 min followed by cryopreservation prior to transplant. Although engraftment in busulphan-conditioned patients (16 mg/kg total dose over 4 d) was slow, this study demonstrated that cell permeabilization, to facilitate intracellular oligonucleotide delivery, was not significantly toxic. We have not included details of the *bcr/abl* targeted oligonucleotides in **Table 1** because no complete phase I studies have been reported and the patient numbers involved were small.

4. *bcl-2*

The majority of follicular lymphomas and some high-grade lymphomas possess a t(14;18) chromosomal translocation in which the immunoglobulin G heavy chain region on chromosome 14 becomes juxtaposed to the *bcl-2* gene on chromosome 18, leading to deregulation of *bcl-2* expression (*[29]*, and

reviewed by Cotter *[30]*). The Bcl-2 protein is a member of a family of proteins that reside within the mitochondrial membrane and are key regulators of
programmed cell death, morphologically characterized as apoptosis. Bcl-2 is
an antiapoptotic protein *(31)*, and overexpression may be linked to tumorigenesis and chemoresistance. In a bioassay approach, 20mer phosphorothioate-
capped phosphodiesters targeting the initiation codon of *bcl-2* were incubated
with DoHH2 lymphoma cells. Viable cells were then injected intravenously
into SCID mice and the antisense-treated cells failed to engraft, whereas all
mice receiving control oligonucleotides developed lymphomas *(32)*. Following engraftment of untreated DoHH2 cells, further studies showed that in vivo
sc infusion of 5 mg/kg/d × 21 d of an 18mer P-ODN targeting the first six
codons of the open reading frame of *bcl-2* resulted in complete eradication of
this tumor *(33)*. This P-ODN (Genasense™, oblimersen sodium, G3139;
Genta) entered a phase I study against non-Hodgkin's lymphoma (NHL) using
a protocol of continuous sc infusion. In this dose-finding study, Genasense was
administered over 14 d to patients with Bcl-2-positive relapsed NHL at doses
from 4.6 to 195.8 mg/m^2/d *(34)*. All patients experienced inflammation at the
infusion site, and dose-limiting toxicities (DLTs), associated with oglionucleotide plasma levels of >4 µg/mL were thrombocytopenia, hypotension, fever, and asthenia. The maximum tolerated dose (MTD) was 147.2 mg/m^2/d
(4.1 mg/kg/d). There were one complete and two minor responses, nine cases
of stable disease and nine cases of progressive disease. The patient who
achieved a complete response after receiving 2 mg/kg/d remains in long-term
remission. A reduction in Bcl-2 protein was measured in 7 of 16 evaluable
patients in lymph node biopsies, peripheral blood mononuclear cells (PBMCs),
or bone marrow. A phase II study of Genasense in patients with mantle cell
lymphoma receiving 3 mg/kg/d for 7 d over a 21-d cycle has produced encouraging early results, with disease stabilization noted in 2 of 8 patients with chemotherapy naïve disease after six cycles of treatment *(35)*.

The implication that Bcl-2 plays a major role as an antiapoptotic protein has
led to many preclinical and clinical studies of Genasense, used either as a single
agent or in combination therapy, against many different hematological malignancies and solid tumors (reviewed by Klasa et al. *[36]*). Of particular interest
is the report of a xenograft model using Gleevec-resistant TF-1 leukemia cells
transformed with Bcr/Abl in which Genasense treatment (7 mg/kg/d × 14 d)
led to increased longevity over the control group of mice. In addition, cells
harvested from mice that had received Genasense treatment for 7 d displayed
an increased sensitivity to Gleevec and conventional cytotoxic agents *(37)*,
thus demonstrating the utility of the antisense approach in this new era of
molecular medicine. In a phase I study of Genasense (4–7 mg/kg/d × 10 d by
continuous iv infusion) in combination with fludarabine, cytarabine, and

filgrastim (FLAG) administered on d 5–10 at incremental doses, in the treatment of relapsed AML or ALL, responses were noted in 9 of 20 evaluable patients. Of the six patients who received Genasense™ at 7 mg/kg/d with full-dose FLAG, three achieved a complete remission and one remains in remission after 15 mo *(36,38)*. There were no toxicities that could be attributed to Genasense treatment, and an MTD was not defined. Further combination studies of Genasense in relapsed AML have been initiated. For an excellent review of preclinical and ongoing clinical studies of Genasense, we recommend that the reader see **ref. *36***. To conserve space in this chapter, we highlight the remaining clinical experiences with Genasense against chronic lymphocytic leukemia (CLL); multiple myeloma; malignant melanoma; prostate, breast, and colorectal cancers; NSCLC; and small cell lung cancer (SCLC) *(36)*.

In a nonrandomized clinical study of Genasense for patients with relapsed/refractory CLL, fever, hypotension, hypoglycemia, and back pain were experienced by patients receiving a continuous iv infusion of 5–7 mg/kg/d for 5–7 d every 3 wk. A lower dose of 3 mg/kg/d was well tolerated as a monotherapy and in combination with fludarabine and cyclophosphamide. There is an indication that tolerance to Genasense may be disease specific. Genasense continues to be evaluated in randomized trials in combination with fludarabine and cyclophosphamide for second-line treatment of CLL. A randomized trial has also been initiated for Genasense, with or without dexamethasone, in patients with relapsed/refractory multiple myeloma. Genasense (3–7 mg/kg/d × 9 d by continuous iv infusion) was evaluated in a phase I study of 17 patients with metastatic colorectal cancer, who were also receiving irinotecan (280–350 mg/m^2 intravenously on d 6). Cycles were repeated every 3 wk and one patient achieved a partial response and stable disease was noted in three patients. At 5 mg/kg/d of Genasense and 350 mg/m^2 of irinotecan, diarrhea, nausea/vomiting, and neutropenia were dose limiting whereas lower doses of irinotecan were well tolerated, even when combined with higher doses of Genasense *(36)*. In a phase I study of Genasense and weekly docetaxel in eight patients with advanced breast cancer and other solid tumors, patients received Genasense (1–4 mg/kg/d) as a continuous iv infusion for 3 wk together with docetaxel (35 mg/kg intravenously on d 8, 15, and 22) with the cycles repeated every 4 wk. Transient toxicities included thrombocytopenia, fatigue, mucositis, neutropenia, and transaminitis *(39)*. At a dose of 3 mg/kg/d of Genasense, reduced Bcl-2 levels in PBMCs and tumor responses were observed in two patients with breast cancer. This study was extended to include a further nine patients who received Genasense (5–8 mg/kg/d on d 1–5, 12, 13, 19, and 20) and docetaxel (35 mg/m^2 on d 6, 14, and 21) on a 4-wk cycle. The overall results of the study showed two patients with partial responses and four patients with disease stabilization.

Aberrant Bcl-2 expression has been linked to a poor prognosis in malignant melanoma *(40)*. A nonrandomized study was initiated to evaluate Genasense in combination with dacarbazine in 24 patients with malignant melanoma, including patients who were resistant to dacarbazine as a first-line therapy *(41,42)*. The first cohort of patients received Genasense as a continuous iv infusion (0.6–6.5 mg/kg/d × 14 d) with dacarbazine at 200 mg/m^2 (daily × 5 starting at d 5). The second cohort received Genasense by twice daily sc injection (5.3–7.7 mg/kg/d × 14 d) with dacarbazine at 800 mg/m^2 on d 5, and the third cohort received Genasense by continuous iv infusion (5–12 mg/kg/d × 5 d) with dacarbazine at 1000 mg/m^2 on d 6. Treatment cycles were repeated every 4 wk. Transient but manageable toxicities were fever, fatigue, lymphopenia, and transaminitis, but thrombocytopenia was dose limiting at a dose of 12 mg/kg/d of Genasense in combination with dacarbazine at 1000 mg/m^2. Of 14 evaluable patients, there was 1 complete response, 2 partial responses, and 3 minor responses leading to an overall median survival of >17 mo. Serial cutaneous biopsies revealed inhibition of Bcl-2 protein expression in 10 of 12 patients whose plasma concentrations of Genasense were >1µg/mL. A multicenter randomized study of dacarbazine with or without Genasense is currently under way for the treatment of stage III/IV malignant melanoma. A marketing application has recently been filed with the US Food and Drug Administration based on this trial, but final results of the study have not been presented publicly.

In prostate cancer, in which Bcl-2 overexpression has been implicated in the development of androgen independence, Genasense has undergone clinical evaluation as a single agent or in combination with paclitaxel, mitoxantrone, or docetaxel *(36,43)*. A phase II study of Genasense at a dosage of 7 mg/kg/d through d 1–8 in combination with docetaxel (75 mg/m^2 intravenously on d 6) with cycles of therapy repeated every 21 d was initiated for 31 men with metastatic hormone-refractory prostate cancer. Common adverse effects of the treatment schedule were neutropenia, fatigue, and fever. A >50% reduction in circulating prostate-specific antigen was noted in 15 of 31 patients, and 4 of 15 patients with measurable disease experienced a partial response *(43)*. Antitumor responses support the concept of Genasense treatment of androgen-independent prostate cancer.

Patients with SCLC often respond initially to conventional chemotherapy but invariably relapse with chemoresistant disease. In chemorefractory SCLC, a phase I/II study has evaluated Genasense, administered by continuous iv infusion (3 mg/kg/d × 8 d), in combination with paclitaxel (175 mg/m^2 on d 6) *(44)*. Stable disease was observed in 2 of 12 patients and one patient remained stable for >30 wk. Hematological toxicities in the first few patients were attributed to paclitaxel and resulted in a dose reduction of paclitaxel to

150 mg/m^2. Other clinical studies have been initiated for the treatment of patients with extensive untreated disease using Genasense in combination with carboplatin and etoposide *(36)*.

Finally, Genasense is undergoing clinical evaluation in phase III randomized multicenter trials in patients with advanced stage IIIB/IV NSCLC who have received previous chemotherapy. This study aims to compare responses and survival and safety profiles of Genasense in combination with docetaxel vs docetaxel alone *(36)*.

The above clinical studies demonstrate that Genasense exhibits a good safety profile even when administered with conventional cytotoxics. There are also suggestions that lower tolerance to Genasense therapy may be linked to specific disease states and that therapeutic benefit may be governed by maintenance of circulating intact drug. Currently, Genasense™ is being evaluated in randomized phase III trials in CLL, malignant melanoma, and multiple myeloma in combination with standard treatments. In addition, under its own sponsorship or in collaboration with the National Cancer Institute, Genta has other ongoing nonrandomized clinical trials in renal, pancreatic, and pediatric solid tumors.

5. c-*raf*-1

The Raf family of proteins (A-Raf, B-Raf, and Raf-1) consists of a series of serine/threonine kinases involved in signal transduction pathways *(45)*. Raf-1, encoded by c-*raf*-1, plays an integral role in the Ras/mitogen-activated protein kinase (MAPK) signaling cascade initiated by growth factor/receptor interactions at the cell surface leading to the activation of oncogenic Ras. Activated Ras recruits Raf-1 to the cell membrane, and following activation by phosphorylation, Raf-1 activates MEK1 (MAPK kinase) and MEK2, which, in turn, activate extracellular signal-related kinase 1 (ERK1) and ERK2. ERK1 and ERK2 translocate to the nucleus, where they are responsible for the activation of specific transcription factors associated with cell survival and proliferation *(46)*. In addition, Raf-1 may be activated by Bcl-2 on the outer mitochondrial membrane and has been linked to suppression of apoptosis *(47)*, indicating that Raf-1 has other cellular roles that are independent of the Ras/MAPK signaling pathway. Isis has developed a 20mer P-ODN, ISIS 5132, that targets the 3'-untranslated region (UTR) of c-*raf*-1. In vitro studies of ISIS 5132 using A549 lung cells showed inhibition of c-*raf*-1 mRNA and Raf-1 protein expression concurrent with cytotoxicity (IC$_{50}$ concentrations of about 100 n*M*) *(48)*. In vivo antitumor activity of ISIS 5132 (6–25 mg/kg/d) was demonstrated against several human tumor xenografts leading to the initiation of phase I clinical studies *(49)*.

Phase I studies of ISIS 5132 from three different centers have been published. In the first trial *(50)*, 31 patients (29 evaluable) with various solid tumor types were treated with ISIS 5132 at doses ranging from 0.5 to 6 mg/kg *(50)*. ISIS 5132 was administered by 2-h infusions three times a wk for three wk followed by one wk of rest between cycles. Some patients experienced anemia, fever, fatigue, or slight elevations in activated partial thromboplastin time (aPPT) during infusions, but no DLTs were recorded. Transient elevations of activated complement (C3a) appeared to be dose related, and the trial was closed when plasma levels of intact ISIS 5132 approached levels associated with complement activation in primates. Inhibition of c-*raf*-1 expression in PBMCs was measured in 13 of 14 patients receiving 2.5 mg/kg or above, and 1 patient with colorectal cancer and another with renal carcinoma experienced disease stabilization for several months *(51)*. In the second phase I study, 34 patients with refractory solid tumors received ISIS 5132 (0.5–5 mg/kg/d × 3 wk) as a continuous iv infusion with 1 wk of rest between cycles *(52)*. As with the other study, similar mild toxicities but no DLTs were observed up to 4 mg/kg/d. Fever associated with hypotension was observed in one patient receiving 5 mg/kg/d after the first 24 h of infusion. Prolonged stabilization of disease was experienced by two patients with renal cell carcinoma and pancreatic cancer.

In the third phase I study, ISIS 5132 was administered to 22 patients with advanced refractory solid tumors by 24-h infusions (6–30 mg/kg) three times a wk followed by 1 wk of rest between cycles *(53)*. Mild toxicities among 20 of 22 evaluable patients were fever, fatigue, nausea/vomiting, and thrombocytopenia. Laboratory investigations showed increased levels of serum tumor necrosis factor-α (TNF-α), aPTT, and complement activation (C3a and Bb). At 30 mg/kg, two of four patients experienced DLTs. Acute hemolytic anemia developed in one patient and acute renal failure accompanied by generalized edema in the other after the first infusion of ISIS 5132. The MTD for ISIS 5132 in this infusion schedule was defined as 24 mg/kg. At this level, plasma steady-state oligonucleotide concentrations were approx 2 μM intact ISIS 5132 (55.3% of total measured oglionucleotide). In this study, c-*raf*-1 mRNA measurements in PBMCs revealed no inhibition at any dose level in the time points studied. In addition, there were no objective responses although 5 of 13 evaluable patients (after three cycles) had stable disease. The results of the three phase I studies suggest that the efficacy and toxicity of AS-ODN therapy may be dependent on administration schedules.

A phase I trial of ISIS 5132 with 5-fluoruracil (5-FU) and leucovorin (LV) was conducted to assess the feasibility of combination studies *(54)*. ISIS 5132 (1–3 mg/kg/d) was administered by continuous iv infusion to 14 patients with refractory cancers. 5-FU (425 mg/m^2) and LV (20 mg/m^2) were administered as an iv bolus on d 1–5 of a 4-wk cycle. DLTs were recorded as neutropenia, thrombocytopenia, and mucositis at 3 mg/kg/d of ISIS 5132 whereas 2 mg/kg/ d was well tolerated. Disease stabilization through six cycles of treatment was observed in one patient with renal cell, one patient with colorectal, and one patient with pancreatic carcinoma.

ISIS 5132 entered phase II studies, and 22 patients with recurrent ovarian cancer received 4 mg/kg/d for 21 d by continuous infusion but no objective responses were reported *(55)*. In a second phase II in patients with locally advanced colorectal cancer or with metastatic disease, ISIS 5132 was administered at 2 mg/kg/d by continuous iv infusion for 21 d with cycles repeating every 28 d *(56)*. Although stable disease was noted in 5 of 15 patients, there were no measurable objective responses. Isis has terminated further development of ISIS 5132 against cancer in order to concentrate on other AS-ODNs with greater potential therapeutic benefit.

6. Protein Kinase C-α

Protein kinase C-α (PKC-α) is a member of a family of phospholipid-dependent serine/threonine kinases of which there are many different isozymes *(57)*. These isozymes have been placed into three classifications dependent on activation requirements and biochemical properties. Group A consists of PKC-α, -βI, -βII, and -γ isoforms activated by 1,2-diacylglycerol, and inappropriate expression of PKC-α has been linked specifically to increased proliferation, anchorage-independent growth, and cell transformation *(58,59)*. Although PKC inhibitors, such as staurosporine analogs, have been studied extensively in preclinical models and, to a lesser extent, in the clinic, these compounds lack isoform specificity and some possess inhibitory properties against other cellular protein kinases. Isis/Eli Lilly have developed a 20mer P-ODN (ISIS 3521, LY900003, Affinitak™) that targets the 3'-UTR of *PKC*-α. In cell culture studies, Affinitak was shown to inhibit mRNA and protein expression in A549 lung and T24 bladder carcinoma cells with resultant inhibition of proliferation at 100–200 n*M* concentrations (reviewed by Holmlund et al. *[60]*). Both cell lines express high levels of PKC-α and are sensitive to staurosporine. In vivo antitumor activity of Affinitak was demonstrated in sc and orthotopic models of the U-87 glioblastoma at doses of 20 mg/kg/d for 21d or 42 d, respectively *(61)*. Affinitak was also active against MDA-MB-231 breast and CALU-1 lung tumors *(60)*.

As with ISIS 5132, Affinitak entered phase I clinical studies using three different schedules of administration. In the first study, 36 patients with advanced cancer received Affinitak (0.15–6 mg/kg/d) by thrice-weekly 2-h infusions for 3 wk followed by 1 wk of rest between cycles *(62)*. There were no DLTs recorded in this study although mild to moderate toxicities experienced were nausea, vomiting, fatigue, fever, chills, and thrombocytopenia. There were transient increases in C3a complement levels following infusions, which returned to baseline levels prior to the next scheduled infusion time. Transient elevations of aPTT, which did not appear to be dose related, were also noted in some patients at the end of the infusion. Dose escalations beyond 6 mg/kg were not considered because the plasma levels of Affinitak (four patients had plasma levels of intact oglionucleotide of >30 μg/mL) were approaching those associated with complement activation in primates. Disease stabilization was experienced in 10 of 34 evaluable patients after two cycles of treatment. Two patients with NHL, who received 9 or 17 cycles of treatment, achieved complete responses.

In the second study, Affinitak (0.5–3 mg/kg/d) was administered to 21 patients with advanced cancer by continuous infusion for 21 d followed by 1 wk of rest between cycles *(63)*. DLTs were thrombocytopenia and fatigue at 3 mg/kg/d and the MTD was defined as 2 mg/kg/d. At 2 mg/kg/d, the oglionucleotide steady-state plasma levels were 0.5 μg/mL, of which approximately 50% represented intact ISIS 3521 at the end of the infusion. Objective tumor responses of several months' duration were recorded in three of four patients with ovarian cancer, thus providing a rationale for phase II studies of Affinitak in ovarian cancer and other solid tumors. In the third phase I study of Affinitak as a single agent, 11 patients with refractory cancer received Affinitak (6–24 mg/kg) as a 24-h infusion administered once weekly *(64)*. Mild toxicities experienced were nausea, vomiting, fever, chills, fatigue, headache, myalgias, hemorrhage, and thrombocytopenia. Transient elevations of C3a and Bb complement were correlated with slight increases in aPTT. This schedule was considered acceptable and one patient with colorectal cancer experienced stable disease for >3 mo. To assess possible pharmacokinetic/toxicity interactions of Affinitak with conventional chemotherapy, a phase I study was conducted in combination with 5-FU and LV in patients with advanced cancer *(65)*. Affinitak (1, 1.5, or 2 mg/kg/d × 21 d) was administered as a continuous iv infusion simultaneously with 5-FU (425 mg/m^2/d × 5 d) and LV (20 mg/m^2/d × 5 d) with the cycle repeated every 4 to 5 wk. Using this schedule, no DLTs were observed or changes in the pharmacokinetic parameters of the respective drugs. Objective partial responses (≥50% tumor shrinkage) were recorded in two patients and minor responses in another four patients with colorectal cancer, suggesting that further studies of efficacy might be warranted in a phase II clinical setting.

Following the two complete responses seen in lymphoma *(62)*, Affinitak entered into a phase II study against NHL *(66)*. In this multicenter study, 29 patients with advanced low-grade or follicular lymphoma received Affinitak (2 mg/kg/d) by continuous iv infusion for 3 wk followed by 1 wk of rest between cycles. Thrombocytopenia and fatigue were related to Affinitak administration, and of the 21 patients evaluable for efficacy, 1 patient experienced a 73% reduction in measurable disease and 16 patients had stable disease. The modest antitumor activity in this phase II trial suggested a possible role for Affinitak in combination studies.

In another phase II trial, Affinitak at the aforementioned dose schedule was administered to 15 patients with metastatic breast cancer for two cycles of treatment *(67)*. Toxicities of thrombocytopenia (2 patients) and infection (4 patients) were recorded, but of 13 of 15 evaluable patients no objective antitumor responses were observed. This study indicated that Affinitak as a monotherapy was not beneficial for the treatment of metastatic breast cancer.

Combination studies of Affinitak have been conducted that have progressed from phase I to phase II trials. Affinitak (2 mg/kg/d) was administered by continuous iv infusion for 14 d in combination with cisplatin (80 mg/m^2 on d 1) and gemcitabine (1000 mg/m^2 on d 1 and 8) to seven patients with advanced cancer *(68)*. No pharmacokinetic interactions were recorded, and this schedule was well tolerated with only one patient experiencing a DLT of fatigue. Evidence of antitumor activity was seen in two patients, one with NSCLC (partial response) and the other with pancreatic cancer (minor response). The results of this study led to the initiation of a phase II trial of this drug combination and schedule in patients with NSCLC *(69)*. Of the 31 patients evaluable for antitumor effects, there were 1 complete and 11 partial responses and 17 patients experienced stable disease, indicating promising efficacy of this combination in NSCLC. In another phase I combination study, Affinitak (1, 1.5, and 2 mg/kg/d) was administered by continuous iv infusion for 14 d in combination with carboplatin (area under the curve 5 or 6) and paclitaxel (175 mg/m^2) on d 4 of a 21-d cycle to 17 patients with metastatic disease *(70)*. Omission of Affinitak from the first cycle of treatment allowed for assessment of toxicities with and without Affinitak, which did not differ. Of 16 evaluable patients, 7 partial responses were achieved in 10 patients with NSCLC. These encouraging responses in NSCLC led to a phase II study of this combination and schedule in 53 patients with an advanced stage of the disease *(71)*. Patients received a median of six treatment cycles, and toxicities were neutropenia and thrombocytopenia. Of 48 evaluable NSCLC patients in the phase I and phase II trials, 1 patient experienced a complete response, 21 experienced partial responses, and 11 experienced disease stabilization. The median survival time for the 53 patients was >15.9 mo, compared with approx 8 mo observed in other trials for patients receiving standard chemotherapy alone. Based on these results, 616 patients were recruited into a randomized phase III trial to assess the ability of Affinitak to prolong lives when combined with carboplatin and paclitaxel. The results of this study were disappointing, with a median survival of 9.7 mo for patients treated with chemotherapy alone, compared with 10 mo for patients receiving chemotherapy plus Affinitak.

Further phase II studies of Affinitak (2 mg/kg/d × 14 d), administered by continuous iv infusion, in combination with docetaxel (75 mg/m^2 on d 3) have been reported for the treatment of NSCLC *(72)*. Of 53 patients evaluable for safety, the common experienced toxicities were neutropenia, thrombocytopenia, and fatigue. Of 36 patients evaluable for antitumor response, 5 patients experienced a partial response (2 of whom had previously failed to respond with paclitaxel) and 15 patients experienced disease stabilization.

To summarize, the favored schedule for Affinitak administration is 2 mg/kg/d by continuous iv infusion for 14 d when used in combination with other chemotherapeutic agents. The continued development of Affinitak is questionable following the results of the phase III combination study in NSCLC.

7. Ha-ras

Ras proteins (Ha-ras, Ki-ras, and N-ras) are small cell-signaling proteins that are anchored to the inner cell membrane by farnesylation and are active in a guanosine 5'-triphosphate (GTP)–bound conformation and inactive in a guanosine 5'-diphosphate (GDP)–bound state. As discussed under **Subheading 5.**, Ras proteins initiate the Ras/MAPK signaling cascade leading to nuclear events that favor proliferation *(45)*. Ras proteins also affect other cell-signaling pathways through activation of phosphoinositide-3-kinase *(73)*. *Ras* genes commonly acquire transforming potential through point mutations, resulting in amino acid changes that lock Ras proteins into the GTP-bound active conformation, and such mutations are often found in human tumors *(74)*. Isis has developed a 20mer P-ODN (ISIS 2503) that targets the translation initiation region of Ha-*ras* mRNA. In cell culture studies, ISIS 2503 inhibited the expression of Ha-*ras* RNA and Ras protein with subsequent inhibition of proliferation of T24 bladder cells with an IC$_{50}$ concentration of approx 100 nM *(75)*. In xenograft models, ISIS 2503 displayed antitumor activities against breast, lung, colon, and pancreatic tumors *(60,76)*. Activity against the Mia-PaCa-2 pancreatic and CALU-1 lung tumors, which possess Ki-*ras* mutations, suggested a broad range of utility for ISIS 2503.

In a phase I clinical study, ISIS 2503 (3–18 mg/kg) was administered as a 24-h infusion once weekly for 3 wk followed by 1 wk of rest to 19 patients with advanced solid tumors *(60,77)*. At doses of 6 mg/kg and above, most patients experienced chills and fevers that occurred at the end of the infusion period and subsequently subsided. At 18 mg/kg dose, steady-state plasma concentrations were approx 14 μg/mL with 65% of total drug being intact ISIS 2503 at 24 h. Although no objective responses were recorded, one patient with melanoma and one patient with pancreatic carcinoma experienced stable disease.

In a second phase I study, ISIS 2503 (1–10 mg/kg/d) was administered as a continuous iv infusion for 2 wk followed by 1 wk of rest between cycles to 23 patients with advanced cancer *(78)*. Minor toxicities included fatigue, fever, nausea, vomiting, anemia, and thrombocytopenia. At 10 mg/kg/d, circulating oglionucleotide levels of >8 μg/mL were achieved, of which nearly 70% was intact ISIS 2503. Although no objective responses were recorded, stable disease was observed in one patient with liposarcoma treated for 10 cycles (1 mg/kg/d), one patient with mesothelioma for 6 cycles (2 mg/kg/d), one patient with colon carcinoma for 8 cycles (4.5 mg/kg/d), and one patient with pancreatic carcinoma for 9 cycles (8 mg/kg/d). A phase I combination study of ISIS 2503 in combination with gemcitabine has been reported for patients with advanced cancer *(79)*. ISIS 2503 was administered at 4 or 6 mg/kg/d as a continuous infusion for 14 d with gemcitabine at 1000 mg/m^2 on d 1 and 8 with cycles repeated every 3 wk. Non-DLTs were neutropenia, thrombocytopenia, anorexia, nausea, fatigue, and fever. Of the 27 patients enrolled in the study, disease stabilization was observed in five patients receiving more than six cycles of treatment and a partial response was noted in 1 patient with metastatic breast cancer.

Interim results of a phase II combination study of ISIS 2503 (6 mg/kg/d) administered by continuous iv infusion for 14 d of a 21-d cycle with gemcitabine (1000 mg/m^2 on d 1 and 8) have been reported for 48 patients with pancreatic carcinoma *(80)*. Mild common toxicities were fatigue, neutropenia, and thrombocytopenia. A complete response and six partial responses were recorded in patients evaluable for a 6-mo follow-up. The patient survival data during this time exceeded the requirements for a positive clinical result. The activity of ISIS 2503, targeting Ha-*ras*, in pancreatic cancer is very interesting because 90% of pancreatic tumors express a mutation in Ki-*ras* *(81)*.

8. Ribonucleotide Reductase

Ribonucleotide reductase (RR) is the enzyme responsible for the production of 2'-deoxynucleotides from ribonucleotides and consists of two subunits. The R1 subunit binds nucleoside diphosphate substrates and also possesses two allosteric nucleoside triphosphate effector sites that govern both overall activity and substrate specificity to ensure a balanced supply of 2'-deoxynucleotides for DNA synthesis. The R2 subunit contains an iron center that generates a tyrosyl free radical that is necessary for the catalytic reaction *(82)*. In addition, the R2 subunit cooperates with a number of oncogenes and therefore is an attractive target for inhibition in cancer therapy *(83)*.

Lorus has developed 20mer P-ODNs to each of these RR subunits. GTI-2040 selectively inhibited the production of the R2 subunit at 200 n*M* concentrations, resulting in cytotoxicity to a number of human cell lines in vitro (reviewed by Orr *[84]*). In xenograft studies of GTI-2040 (0.5–30 mg/kg/d intravenously every other day for up to 44 d), antitumor activity was observed in colon, pancreatic, liver, lung, breast, ovarian, brain, lymphoma, and skin cancers *(84,85)*. In addition, tumor regression was noted when GTI-2040 was administered in combination with other chemotherapeutic agents *(84,86)*.

In a phase I study, GTI-2040 (18.5–222 mg/m^2/d) was administered by continuous iv infusion for 3 wk followed by 1 wk of rest between cycles to 27 patients with advanced cancer *(84,86)*. Mild toxicities included anorexia, nausea, hypotension, chills, and fever. One patient experienced dose-limiting fatigue, and DLTs of diarrhea and hepatotoxicity were experienced at the highest dose of 222 mg/m^2/d. For phase II studies, 185 mg/m^2/d (5 mg/kg/d) was recommended. At this level, plasma concentrations of GTI-2040 of approx 1 µg/mL were deemed sufficient to achieve target suppression in clinical studies.

Currently, GTI-2040 is in phase II clinical trials in combination with capecitabine for the treatment of renal cell carcinoma. Interim data have shown disease stabilization and tumor responses in some of the 21 evaluable patients.

The sister compound to GTI-2040, GTI-2501, targets the R1 subunit of RR and has displayed target specificity in vitro and in vivo. In human tumor models, GTI-2501 exhibited marked antitumor activity against colon, pancreatic, lung, breast, ovarian, skin, prostate, brain, and renal cancers *(87,88)*. In addition, GTI-2501 prevented metastatic tumor spread and was effective in combination with other chemotherapeutic agents, producing results ranging from pronounced tumor growth delay to complete tumor regressions *(88)*. GTI-2501 has completed phase I studies in patients with advanced cancer and is currently in phase II clinical trials against hormone-refractory prostate cancer.

9. Protein Kinase A

The cyclic adenosine monophosphate–dependent protein kinase A (PKA) exists as two isoforms, PKA-I and PKA-II. Although both proteins share a common catalytic subunit, each contains a different regulatory component, RI and RII. PKA-II expression is found in normal tissues and is linked with growth arrest and differentiation, whereas PKA-I expression plays a role in cell proliferation and neoplastic transformation *(89)*. PKA-I is involved with signal transduction pathways initiated at the cell membrane, such as EGF binding *(90)*, and overexpression has been correlated with unfavorable features in several cancers, such as ovarian *(91)*. Hybridon has developed an 18mer AS-ODN (GEM®231) targeting the RIα subunit of PKA. GEM®231 is a second-generation MBO P-ODN of RNA/DNA composition in which the four flanking nucleosides at each end of the oglionucleotide contain 2'-*O*-methyl substitutions. In cell-culture studies, GEM®231 was cytotoxic to ZR-75-1 breast cancer cells at micromolar concentrations. Cooperative growth inhibition with apoptosis was observed using suboptimal concentrations of GEM®231, C225 (an MAb) and docetaxel *(92)*. GEM®231 treatment of PC-3M androgen-independent prostate cells inhibited the expression of RIα mRNA and protein concomitant with inhibition of growth and the onset of apoptosis at 200 nM concentrations. In addition, in PC-3M xenograft studies, GEM®231 treatment (approx 4 mg/kg/d) produced a significant tumor growth delay *(93)*. GEM®231 has been shown to have more acceptable pharmacological properties than P-ODNs without 2'-O-methyl substitutions *(94)* and also to have activity by oral administration in preclinical studies *(95)*.

In a phase I clinical study, GEM®231 (20–360 mg/m^2, 2.5–9 mg/kg) was administered to 14 patients with refractory solid tumors by 2-h infusions twice weekly *(96)*. DLTs of transient elevations of aPTT and reversible transaminitis were experienced at 360 mg/m^2, and one patient with colorectal cancer had disease stabilization. Although a dose of 240 mg/m^2 was considered safe for 4 wk, alternative schedules of GEM®231 have been explored. In another phase I study, GEM®231 (80–180 mg/m^2) was administered by continuous iv infusion for 3 d (1 patient) or 5 d (13 patients) *(97)*. No objective antitumor activity was observed, and minor toxicities were elevations of aPTT and fatigue, with the DLT at the highest dose being transaminitis. To date, GEM®231 has completed phase I/II combination studies with paclitaxel and docetaxel with data analysis awaited *(98)*. Studies with irinotecan are in progress, and GEM®231 is being evaluated as a monotherapy in phase II trials.

10. DNA Methyltransferase 1

DNA methyltransferase 1 (DNA MeTase) is the major DNA-methylating enzyme and methylates cytosine bases at a CpG dinucleotide site. Unmethylated CpG regions are associated with transcriptionally active DNA, whereas hypermethylated DNA is linked with gene silencing. In tumor cells, the patterns of DNA methylation are altered and hypermethylation of tumor suppressor genes that affect cell-cycle regulation is a common event *(99)*. Researchers at MethylGene screened a series of second-generation mixed backbone oligonucleotides, with the same chemical modifications as GEM®231 to identify AS-ODNs with inhibitory activities to DNA MeTase mRNA expression. MG88 (targeting the 5' region) and MG98 (targeting the 3'-UTR) were found to be the most potent AS-ODNs with IC$_{50}$ concentrations of 40 and 45 nM, respectively *(100)*. Following the observations of tumor growth delays or regressions in xenografted tumors together with an acceptable safety profile in primates *(101)*, MG98 was entered into phase I studies. In the first phase I study, MG98 (40–240 mg/m^2/d) was administered to 14 patients as a continuous iv infusion for 3 wk with 1 wk of rest between cycles *(102)*. DLTs resulting from this schedule were defined as transaminitis, fatigue, and thrombocytopenia. A transient decrease in MeTase was observed in 7 of 10 patients evaluated. In the second phase I study, MG98 (40–480 mg/m^2) was administered to 19 patients with solid tumors by twice weekly 2-h infusions for 3 wk followed by 1 wk of rest between cycles *(103)*. An increase in aPTT was seen shortly after the infusions, similar to the observations of others using 2-h infusion schedules of AS-ODNs. Other mild toxicities experienced were fatigue, fever, nausea, anorexia, anemia, transaminitis, and thrombocytopenia, while the DLTs were defined as rigors and fatigue (two of three patients) at the highest dose of 480 mg/m². Inhibition of MeTase mRNA was recorded in two of nine patients evaluated, and there was one partial response of more than six cycles of treatment duration in one patient with renal cell carcinoma. Using this 2-h infusion schedule, MG98 has entered into phase II studies for metastatic head and neck cancer at 240 mg/m², and for advanced renal cell carcinoma at 360 mg/m², and MethylGene and MGI Pharama have initiated a third trial against myelodysplastic syndrome/AML.

11. Other Studies

AP 12009, a P-ODN targeting transforming growth factor β-2 (TGF-β2) mRNA has demonstrated preclinical efficacy against glioma and an acceptable safety profile when administered by intrathecal bolus or continuous intracerebral infusions to rabbits or cynomolgus monkeys in the micrograms-per-kilograms range (3–120 µg/kg) *(104)*. AP 12009 entered phase I clinical studies in patients with malignant glioma and was administered by high-flow intratumoral microperfusion *(105)*. Good safety and tolerability profiles were shown in preliminary dose escalations with first efficacy indications assessed by magnetic resonance imaging and survival. Antisense Pharma GmbH has received orphan drug status for AP 12009 in the treatment of malignant glioma and AP 12009 is currently in phase II studies.

Overexpression of clusterin, also called testosterone-repressed prostate message-2, has been associated with a hormone-independent and chemorefractory phenotype in prostate cancer xenograft models *(106)*. In preclinical studies, a second-generation P-ODN with 2'-methoxyethyl flanking regions (OGX-1, ISIS 112989) targeting the mRNA of clusterin was shown to enhance the antitumor activity of paclitaxel against the PC-3 prostate tumor in vitro and in vivo *(107)*. In addition, the in vivo tissue half-life of OGX-1 was significantly longer than its P-ODN counterpart, allowing for weekly rather than daily administration to maintain tissue levels. OncoGenex and Isis have initiated two phase I studies, the first in patients with prostate cancer and the second in patients with solid tumors.

12. Conclusion

There is now a wealth of information on phase I trials of P-ODNs in the field of oncology. Using different schedules, safety and tolerability profiles have been demonstrated with MTDs defined. Pharmacokinetic and pharmacodynamic profiles are encouraging, with plasma levels of intact P-ODNs reaching concentrations predicted to be of therapeutic benefit to patients. Second-generation chemistries are being developed, and GEM®231 (Hybridon) has completed phase II combination studies in cancer. Isis has explored AS-ODNs with 2'-methoxyethyl-modified flanking regions, and ISIS 104838, targeting TNF-α, is currently in phase II studies for the treatment of rheumatoid arthritis and psoriasis, and OGX-1 is in phase I studies for the treatment of cancer. Fine-tuning of antisense chemistries and delivery systems may result in improved therapies for the future with inclusion of AS-ODNs into new and conventional anticancer drug regimens.

Acknowledgments

We thank Patrick Iversen (AVI Biopharma), Alan Gewirtz (University of Pennsylvania), Stanley Frankel (Genta), Russell Martin (Hybridon), and Aiping Young (Lorus) for helpful discussions.

References

1. Sikora, K., Advani, S., Koroltchouk, V., et al. (1999) Essential drugs for cancer therapy: a World Health Organisation consultation. *Ann. Oncol.* **10,** 385–390.
2. De Vita, V. T., Hellman S., and Rosenberg, S. A., eds. (2001) *Cancer: Principles and Practice of Ooncology,* 6th ed. Lippincott, Philadelphia.
3. Goldenberg, M. M. (1999) Trastuzumab, a recombinant DNA-derived humanised monoclonal antibody, a novel agent for the treatment of metastatic breast cancer. *Clin. Ther.* **21,** 309–318.
4. Shak, S. (1999) Overview of the trastuzumab (Herceptin) anti-HER2 monoclonal antibody clinical programme in HER2 overexpressing metastatic breast cancer. Herceptin multinational investigator group. *Semin. Oncol.* **26,** 71–77.
5. Slamon, D., Leyland-Jones, B., Skak, S., et al. (2001) Use of chemotherapy plus a monoclonal antibody against HER2 for metastatic breast cancer that overexpresses HER2. *N. Engl. J. Med.* **344,** 783–792.
6. O'Dwyer, M. E. and Druker, B. J. (2001) The role of the tyrosine kinase inhibitor STI571 in the treatment of cancer. *Curr. Cancer Drug Targets* **1,** 49–57.
7. Druker, B. J., Sawyers, C. L., Kantarjian, H., et al. (2001) Activity of a specific inhibitor of the BCR-ABL tyrosine kinase in the blast crisis of chronic myeloid leukemia and acute lymphoblastic leukemia with the Philadelphia chromosome. *N. Engl. J. Med.* **344,** 1038–1042.
8. Gorre, M. E., Mohammed, M., Ellwood, K., et al. (2001) Clinical resistance to STI-571 cancer therapy caused by BCR-ABL mutation or amplification. *Science* **293,** 876–880.
9. Baselga, J. and Averbuch, S. D. (2000) ZD1839 ("Iressa"), as an anticancer agent. *Drugs* **60,** 33–40.
10. Ronson, M., Hammond, L. A., Ferry, D., et al. (2002) ZD1839, a selective oral epidermal growth factor receptor-tyrosine kinase inhibitor, is well tolerated and

active in patients with solid, malignant tumors: Results of a phase I trial. *J. Clin. Oncol.* **20,** 2240–2250.

11. Crooke, S. T. (2000) Progress in antisense technology: the end of the beginning, in *Methods in Enzymology* (Phillips, M. I., ed.), Academic, San Diego, pp. 3–45.

12. Orr, R. M. and O'Neill, C. F. (2000) Patent review: therapeutic applications for antisense oligonucleotides 1999–2000. *Curr. Opin. Mol. Ther.* **2,** 325–331.

13. Orr, R. M. (2001) Technology evaluation: Fomivirsen, Isis Pharmaceuticals Inc/ CIBA Vision. *Curr. Opin. Mol. Ther.* **3,** 288–294.

14. Amundson, S. A., Myers, T. G., and Fornace, A. J. Jr. (1998) Roles for p53 in growth arrest and apoptosis: putting on the brakes after genotoxic stress. *Oncogene* **17,** 3287–3299.

15. Imamura, J., Miyoshi, I., and Koeffler, H. P. (1994) p53 in hematologic malignancies. *Blood* **84,** 2412–2421.

16. Bayever, E., Haines, K. M., Iverson, P. L., et al. (1994) Selective cytotoxicity to human leukemic myeloblasts produced by oligodeoxyribonucleotide phosphorothioates complementary to p53 nucleotide sequences. *Leuk. Lymphoma* **12,** 223–231.

17. Bishop, M. R., Iversen, P. L., Bayever, E., et al. (1996) Phase I trial of an antisense oligonucleotide OL(1)p53 in hematologic malignancies. *J. Clin. Oncol.* **14,** 1320–1326.

18. Sharp, J. G., Bishop, M. R., Copple, B., et al. (2001) Oligonucleotide enhanced cytotoxicity of Idarubicin for lymphoma cells. *Leuk. Lymphoma* **42,** 417–427.

19. Weston, K. (1999) Reassessing the role of C-MYB in tumorigenesis. *Oncogene* **18,** 3034–3038.

20. Gewirtz, A. M. (1997) Developing oligonucleotide therapeutics for human leukemia. *Anti-Cancer Drug Des.* **12,** 341–358.

21. Orr, R. M. (1999) Technology evaluation: leukemia therapy, University of Pennsylvania. *Curr. Opin. Mol. Ther.* **1,** 399–403.

22. Calabretta, B., Sims, R. B., Valtieri, M., et al. (1991) Normal and leukemic hematopoietic cells manifest differential sensitivity to inhibitory effects of c-myb AS oligodeoxynucleotides: an *in vitro* study relevant to bone marrow purging. *Proc. Natl. Acad. Sci. USA* **88,** 2351–2355.

23. Ratajczak, M. Z., Hijiya, N., Catani, L., et al. (1992) Acute- and chronic-phase chronic myelogenous leukemia colony-forming units are highly sensitive to the growth inhibitory effects of *c-myb* AS oligodeoxynucleotides. *Blood* **79,** 1956–1961.

24. Ratajczak, M. Z., Kant, J. A., Luger, S. M., et al. (1992) *In vivo* treatment of human leukemia in a scid mouse model with *c-myb* AS oligodeoxynucleotides. *Proc. Natl. Acad. Sci. USA* **89,** 11823–11827.

25. Luger, S. M., O'Brien, S. G., Ratajczak, J., et al. (2002) Oligodeoxynucleotide-mediated inhibition of *c-myb* gene expression in autografted bone marrow: a pilot study. *Blood* **99,** 1150–1158.

26. De Fabritiis, P., Petti, M. C., Montefusco, E., et al. (1998) BCR-ABL antisense oligodeoxynucleotide in vitro purging and autologous bone marrow transplantation for patients with chronic myelogenous leukemia in advanced phase. *Blood* **91,** 3156–3162.

27. Dhut, S., Chaplin, T., and Young, B. (1990) BCR-ABL and BCR proteins: Biochemical characterization and localization. *Leukemia* **4,** 745–750.
28. Clarke, R. E., Grzybowski, J., Broughton, C. M., et al. (1999) Clinical use of streptolysin-O to facilitate antisense oligodeoxynucleotide delivery for purging autografts in chronic myeloid leukaemia. *Bone Marrow Transplant.* **23,** 1303–1308.
29. Reed, J. C., Tsujimoto, Y., Epstein, S., et al. (1989) Regulation of *bcl-2* gene expression in lymphoid cell lines containing t(14:18) or normal #18 chromosomes. *Oncogene Res.* **4,** 271–282.
30. Cotter, F. E. (1993) Molecular pathology of lymphoma, in *The Molecular Pathology of Cancer* (Lemoine, N. R. and Wright, N. A., eds.), Cold Spring Harbor Laboratory, Cold Spring Harbor, NY, pp. 157–174.
31. Hockenberry, D., Nunez, G., Milliman, C., Schreiber, R. D., and Korsmeyer, S. J. (1990) Bcl-2 is an inner mitochondrial membrane protein that blocks programmed cell death. *Nature (Lond.)* **348,** 334–336.
32. Cotter, F. E., Johnson, P., Hall, P., et al. (1994) Antisense oligonucleotides suppress B-cell lymphoma growth in a scid-hu mouse model. *Oncogene* **9,** 3049–3055.
33. Cotter, F. E., Corbo, M., Raynaud, F., et al. (1996) *Bcl-2* antisense therapy in lymphoma: *in vitro* and *in vivo* mechanisms, efficacy, pharmacokinetic and toxicity studies. *Ann. Oncol.* **7,** 3 (abstract).
34. Waters, J. S., Webb, A., Cunningham, D., et al. (2000) Phase I clinical and pharmacokinetic study of Bcl-2 antisense oligonucleotide therapy in patients with non-Hodgkin's lymphoma. *J. Clin. Oncol.* **18,** 1812–1823.
35. Leonard, J. P., Coleman, M., Vose, J., et al. (2003) Phase II study of oblimersen sodium (G3139) alone and with R-CHOP in mantle cell lymphoma (MCL). *Proc. Am. Soc. Clin. Oncol.* **22,** 566 (abstract).
36. Klasa, R. J., Gillum, A. M., Klem, R. E., and Frankel, S. R. (2002) Oblimersen Bcl-2 antisense: facilitating apoptosis in anticancer treatment. *Antisense Nucleic Acid Drug Dev.,* **12,** 193–213.
37. Tauchi, T., Nakajima, A., Sumi, M., Shimamoto, T., Sashida, G., and Ohyashiki, K. (2002) G3139 (Bcl-2 antisense oligonucleotide) is active against Gleevec-resistant BCR-ABL-positive cells. *Proc. Am. Assoc. Cancer Res.* **43,** 4702 (abstract).
38. Marcucci, G., Byrd, J. C., Dai, G., et al. (2003) Phase I and pharmacodynamic studies of G3139, a Bcl-2 antisense oligonucleotide, in combination with chemotherapy in refractory or relapsed leukemia. *Blood* **101,** 425–432.
39. Chen, H. X., Marshall, J. L., Trocky, N., et al. (2000) A phase I study of BCL-2 antisense G3139 (GENTA) and weekly docetaxel in patients with advanced breast cancer and other solid tumors. *Proc. Am. Soc. Clin. Oncol.* **19,** 692 (abstract).
40. Grover, R. and Wilson, G. D. (1996) Bcl-2 expression in malignant melanoma and its prognostic significance. *Eur. J. Surg. Oncol.* **22,** 347–349.
41. Jansen, B., Wacheck, V., Heere-Ress, E., et al. (2000) Chemosensitisation of malignant melanoma by BCL-2 antisense therapy. *Lancet* **356,** 1728–1733.
42. Jansen, B., Wacheck, V., Heere-Ress, E., et al. (2001) Clinical, pharmacologic, and pharmacodynamic study of Genasense (G3139, Bcl-2 antisense oligonucleotide) and dacarbazine (DTIC) in patients with malignant melanoma. *Proc. Am. Soc. Clin. Oncol.* **20,** 1426 (abstract).

43. Chi, K.N., Murray, R.N., Gleave, M.E., et al. (2003) A phase II study of oblimersen sodium (G3139) and docetaxel (D) in patients (pts) with metastatic hormone-refractory prostate cancer (HRPC). *Proc. Am. Soc. Clin. Oncol.* **22**, 393 (abstract).

44. Rudin, C. M., Otterson, G. A., Mauer, A. M., et al. (2002) A pilot trial of G3139, a bcl-2 antisense oligonucleotide, and paclitaxel in patients with chemorefractory small-cell lung cancer. *Ann. Oncol.* **13**, 539–545.

45. Daum, G., Eisenmann-Tappe, I., Fries, H. W., Troppmair, J., and Rapp, U. R. (1994) The ins and outs of Raf kinases. *Trends Biol. Sci.* **19**, 474–480.

46. McCormick, F. (1995) Ras-related proteins in signal transduction and growth control. *Mol. Reprod. Dev.* **42**, 500–506.

47. Wang, H.-G., Rapp, U. R., and Reed, J. C. (1996) Bcl-2 targets the protein kinase Raf-1 to mitochondria. *Cell* **87**, 629–638.

48. Monia, B. P., Johnston, J. F., Geiger, T., Muller, M., and Fabbro, D. (1996) Antitumor activity of a phosphorothioate antisense oligodeoxynucleotide targeted against c-raf kinase. *Nature Med.* **2**, 668–675.

49. ISIS 5132 investigator's brochure. (1998) Isis Pharmaceuticals Inc., Carlsbad, CA.

50. Stevenson, J. P., Yao, K.-S., Gallagher, M., et al. (1999) Phase I clinical/pharmacokinetic and pharmacodynamic trial of the c-*raf*-1 antisense oligonucleotide ISIS 5132 (CGP 69846A). *J. Clin. Oncol.* **17**, 2227–2236.

51. O'Dwyer, P. J., Stevenson, J. P., Gallagher, M., et al. (1999) c-raf-1 depletion and tumor responses in patients treated with the c-raf-1 antisense oligodeoxynucleotide ISIS 5132 (CGP 69846A). *Clin. Cancer Res.* **5**, 3977–3982.

52. Cunningham, C. C., Holmlund, J. T., Schiller, J. H., et al. (2000) A phase I trial of c-raf kinase antisense oligonucleotide ISIS 5132 administered as a continuous intravenous infusion in patients with advanced cancer. *Clin. Cancer Res.* **6**, 1626–1631.

53. Rudin, C. M., Holmlund, J., Fleming, G. F., et al. (2001) Phase I trial of ISIS 5132, an antisense oligonucleotide inhibitor of c-raf-1, administered by 2-hour weekly infusion to patients with advanced cancer. *Clin. Cancer Res.* **7**, 1214–1220.

54. Stevenson, J. P., Gallagher, M., Ryan, W. F., et al. (1999) Phase I trial of the c-Raf-1 antisense oligonucleotide (ODN) ISIS 5132 administered as a 21-day continuous IV infusion in combination with 5-fluorouracil (5-FU) and leucovorin (LV) as a daily x 5 IV bolus. *Clin. Cancer Res.* **5**, 579 (abstract).

55. Oza, A.M., Elit, L., Swenerton, K., et al. (2003) Phase II study of CGP 69846A (ISIS 5132) in recurrent epithelial ovarian cancer: an NCIC clinical trials study group (NCIC IND. 116). *Gynecol. Oncol.* **89**, 129–133.

56. Cripps, M. C., Figueredo, A. T., Oza, A. M., et al. (2002) Phase II randomized study of ISIS 3521 and ISIS 5132 in patients with locally advanced or metastatic colorectal cancer: a National Cancer Institute of Canada clinical trials group study. *Clin. Cancer Res.* **8**, 2188–2192.

57. Nishizuka, Y. (1992) Intracellular signalling by hydrolysis of phospholipids and activation of protein kinase C. *Science* **258**, 607–614.

58. Blobe, G. C., Obeid, L. M., and Hannun, Y. A. (1994) Regulation of protein kinase C and role in cancer therapy. *Cancer Metastasis Rev.* **13**, 411–431.

59. Ways, D. K., Kukoly, C. A., deVente, J., et al. (1995) MCF-7 breast cancer cells transfected with protein kinase C-α exhibit altered expression of other protein kinase C isoforms and display a more aggressive phenotype. *J. Clin. Invest.* **95,** 1906–1915.

60. Holmlund, J. T., Monia, B. P., Kwoh, J., and Dorr, F. A. (1999) Toward antisense oligonucleotide therapy for cancer: ISIS compounds in clinical development. *Curr. Opin. Mol. Ther.* **1,** 372–385.

61. Yazaki, T., Ahmad, S., Chahlavi, A., et al. (1996) Treatment of glioblastoma U-87 by systemic administration of an antisense protein kinase C-α phosphorothioate oligodeoxynucleotide. *Mol. Pharmacol.* **50,** 236–242.

62. Nemunaitis, J., Holmlund, J. T., Kraynak, M., et al. (1999) Phase I evaluation of ISIS 3521, an antisense oligodeoxynucleotide to protein kinase c-alpha, in patients with advanced cancer. *J. Clin. Oncol.* **17,** 3586–3595.

63. Yuen, A. R., Halsey, J., Fisher, G. A., et al. (1999) Phase I study of an antisense oligonucleotide to protein kinase C-α (ISIS 3521/CGP 64128A) in patients with cancer. *Clin. Cancer Res.* **5,** 3357–3363.

64. Advani, R., Fisher, G. A., Grant, P., et al. (1999) A phase I trial of an antisense oligonucleotide targeted to protein kinase C-α (ISIS 3521/ISI641A) delivered as a 24-hour continuous infusion (CI). *Proc. Am. Soc. Clin. Oncol.* **18,** 609 (abstract).

65. Mani, S., Rudin, C. M., Kunkel, K., et al. (2002) Phase I clinical and pharmacokinetic study of protein kinase C-α antisense oligonucleotide ISIS 3521 administered in combination with 5-fluorouracil and leucovorin in patients with advanced cancer. *Clin. Cancer Res.* **8,** 1042–1048.

66. Emmanouilides, C. E., Saleh, A., Laufman, L., et al. (2002) Phase II trial of the efficacy and safety of ISIS 3521/LY900003, an antisense inhibitor of PKC-alpha, in patients with low-grade, non-Hodgkin's lymphoma. *Proc. Am. Soc. Clin. Oncol.* **21,** 1124 (abstract).

67. Gradishar, W. J., O'Neill, A., Cobleigh, M., Goldstein, L. J., and Davidson, N. E. (2001) A phase II trial with antisense oligonucleotide ISIS 3521/Cgp 64128a in patients (Pts) with metastatic breast cancer (MBC): ECOG trial 3197. *Proc. Am. Soc. Clin. Oncol.* **20,** 171 (abstract).

68. Villalona-Calero, M. A., Figueroa, J., Nadella, P., et al. (2001) Phase I and pharmacokinetic (PK) study of the protein kinase C alpha (PKC-α) inhibitor ISIS-3521 in combination with cisplatin and gemcitabine in patients with solid malignancies. *Clin. Cancer Res.* **7,** 132 (abstract).

69. Ritch, P. S., Belt, R., George, S., et al. (2002) Phase I/II trial of ISIS 3521/LY900003, an antisense inhibitor of PKC-alpha with cisplatin and gemcitabine in advanced non-small cell lung cancer. *Proc. Am. Soc. Clin. Oncol.* **21,** 1233 (abstract).

70. Yuen, A., Sikic, B. I., Advani, R., et al. (1999) A phase I trial of ISIS 3521 (ISI641A), an antisense inhibitor of protein kinase C alpha, combined with carboplatin and paclitaxel in patients with cancer. *Clin. Cancer Res.* **5,** 580 (abstract).

71. Yuen, A., Halsey, J., Fisher, G., et al. (2001) Phase II trial of ISIS 3521, an antisense inhibitor of PKC-α, with carboplatin and paclitaxel in non-small cell

lung cancer: Updated survival and time to progression data. *Clin. Cancer Res.* **7,** 140 (abstract).

72. Moore, M. R., Saleh, M., Jones, C. M., et al. (2002) Phase II trial of ISIS 3521/ LY900003, an antisense inhibitor of PKC-alpha, with docetaxel in non-small cell lung cancer (NSCLC). *Proc.Am. Soc. Clin. Oncol.* **21,** 1186 (abstract).

73. Yan, J., Roy, S., Apolloni, A., Lane, A., and Hancock, J. F. (1998) Ras isoforms vary in their ability to activate Raf-1 and phosphoinositide-3-kinase. *J. Biol. Chem.* **273,** 24,052–24,056.

74. Bos, J.L. (1989) *Ras* oncogenes in human cancer: a review. *Cancer Res.* **49,** 4682– 4689.

75. Chen, G., Oh, S., Monia, B. P., and Stacey, D. W. (1996) Antisense oligonucleotides demonstrate a dominant role of c-Ki-RAS proteins in regulating the proliferation of diploid human fibroblasts. *J. Biol. Chem.* **271,** 28,259–28,265.

76. Cowsert, L. M. (1997) *In vitro* and *in vivo* activity of antisense inhibitors of *ras*: potential for clinical development. *Anti-Cancer Drug Des.* **12,** 359–371.

77. Gordon, M. S., Sandler, A. B., Holmlund, J. T., et al. (1999) A phase I trial of ISIS 2503, an antisense inhibitor of H-ras, administered by a 24-hour (hr) weekly infusion to patients (pts) with advanced cancer. *Proc. Am. Soc. Clin. Oncol.* **18,** 604 (abstract).

78. Cunningham, C. C., Holmlund, J. T., Geary, R. S., et al. (2001) A phase I trial of H-ras antisense oligonucleotide ISIS 2503 administered as a continuous intravenous infusion in patients with advanced carcinoma. *Cancer* **92,** 1265–1271.

79. Adjei, A. A., Dy, G. K., Erlichman, C., et al. (2003) A phase I trial of ISIS 2503, an antisense inhibitor of H-ras, in combination with gemcitabine in patients with advanced cancer. *Clin. Cancer Res.* **9,** 115–123..

80. Burch, P. A., Alberts, S. R., Schroeder, M. T., et al. (2003) Gemcitabine and ISIS 2503 for patients with pancreatic adenocarcinoma (ACA): a North Central Cancer Treatment Group (NCCTG) phase II study. *Proc. Am. Soc. Clin. Oncol.* **22,** 259.

81. Friess, H., Kleef, J., Korc, M., and Buchler, M. W. (1999) Molecular aspects of pancreatic cancer and future perspectives. *Dig. Surg.* **16,** 281–290.

82. Reichard, P. (1993) From RNA to DNA, why so many ribonucleotide reductases? *Science* **260,** 1773–1777.

83. Fan, H. Z., Villegas, C., Huang, A., and Wright, J. A. (1998) The mammalian ribonucleotide reductase R2 component cooperates with a variety of oncogenes in mechanisms of cellular transformation. *Cancer Res.* **58,** 1650–1653.

84. Orr, R. M. (2001) GTI-2040 Lorus Therapeutics. *Curr. Opin. Invest. Drugs* **2,** 1462–1466.

85. Wright, J. A., Feng, N. P., Jin, H. N., Wang, M., Lee, Y., and Young, A. (2001) GTI-2040, an outstanding antisense antitumor agent that targets the R2 component of human ribonucleotide reductase: from the laboratory to the clinic. *Proc. Am. Assoc. Cancer Res.* **42,** 4559 (abstract).

86. Janisch, L. A., Schilsky, R. L., Vogelzang, N. J., et al. (2001) Phase I study of GTI-2040 by continuous intravenous infusion (CVI) in patients with advanced cancer. *Proc. Am. Soc. Clin. Oncol.* **20,** 469 (abstract).

87. Tu, G. and Tu, X. (2001) GTI-2501 Lorus Therapeutics. *Curr. Opin. Invest. Drugs* **2,** 1467–1470.
88. Wright, J. A., Feng, N. P., Jin, H. N., Wang, M., Lee, Y., and Young, A. (2001) GTI-2501, an outstanding antitumor agent that targets the R1 component of human ribonucleotide reductase. *Proc. Am. Assoc. Cancer Res.* **42,** 4560 (abstract).
89. Cho-Chung, Y. S. and Clair, T. (1993) The regulatory subunit of cAMP-dependent protein kinase as a target for chemotherapy of cancer and other cellular dysfunctional-related diseases. *Pharmacol. Ther.* **60,** 265–288.
90. Tortora, G., Damiano, V., Bianco, C., et al. (1997) The RIα subunit of protein kinase A (PKA) binds to Grb2 and allows PKA interaction with the activated EGF-receptor. *Oncogene* **14,** 923–928.
91. McDaid, H. M., Cairns, M. T., Atkinson, R. J., et al. (1999) Increased expression of the RIα subunit of the c-AMP-dependent protein kinase A is associated with advanced stage ovarian cancer. *Br. J. Cancer* **79,** 933–939.
92. Tortora, G., Caputo, R., Pomatico, G., et al. (1999) Cooperative inhibitory effect of novel mixed backbone oligonucleotide targeting protein kinase A in combination with docetaxel and anti-epidermal growth factor-receptor antibody on human breast cancer cell growth. *Clin. Cancer Res.* **5,** 875–881.
93. Cho, Y. S., Kim, M.-Y., Tan, L., Srivastava, R., Agrawal, S., and Cho-Chung, Y. S. (2002) Protein kinase A R1α antisense inhibition of PC3M prostate cancer cell growth: *Bcl-2* hyperphosphorylation, Bax up-regulation and Bad hypophosphorylation. *Clin. Cancer Res.* **8,** 607–614.
94. Agrawal, S., Jiang, Z., Zhao, Q., et al. (1997) Mixed-backbone oligonucleotides as second generation antisense oligonucleotides: *in vitro* and *in vivo* studies. *Proc. Natl. Acad. Sci. USA* **94,** 2620–2625.
95. Tortora, G., Bianco, R., Damiano, V., et al. (2000) Oral antisense that targets protein kinase A cooperates with taxol and inhibits tumor growth, angiogenesis, and growth factor production. *Clin. Cancer Res.* **6,** 2506–2512.
96. Chen, H. X., Marshall, J. L., Ness, E., et al. (2000) A safety and pharmacokinetice study of a mixed-backbone oligonucleotide (GEM231) targeting the type I protein kinase A by two-hour infusions in patients with refractory solid tumors. *Clin. Cancer Res.* **6,** 1259–1266.
97. Goel, S., Desai, K., Bulgara, A., et al. (2003) A safety study of a mixed-backbone oligonucleotide (GEM231) targeting the type I regulatory subunit α of protein kinase A using a continuous infusion schedule in patients with refractory solid tumors. *Clin. Cancer Res.* **9,** 4069–4076.
98. Goel, S., Bulgaru, A., Desai, K., Martin, R. R., McKinlay, M., and Mani, S. (2003) Phase I trial using GEM 231, a second-generation antisense oligonucleotide targeting protein kinase A (PKA) RIα, in combination with docetaxel (D) in patients with advanced solid cancers. *Proc. Am. Soc. Clin. Oncol.* **22,** 210.
99. Bavlin, S. B. (1997) Tying it all together: epigenetics, genetics, cell cycle, and cancer. *Science* **277,** 1948–1949.
100. Fournel, M., Sapieha, P., Beaulieu, N., Besterman, J. M., and MacLeod, A. R. (1999) Down-regulation of human DNA-(cytosine-5) methyltransferase induces

cell cycle regulators p16^{ink4A} and p21$^{WAF/Cip1}$ by distinct mechanisms. *J. Biol. Chem.* **274**, 24,250–24,256.

101. Besterman, J., Younan, J., and Marquis, J. (2001) MG98 (DNMT-1-directed antisense oligodeoxynucleotide): toxicity studies in cynomolgus monkeys. *Clin. Cancer Res.* **7**, 269 (abstract).

102. Davis, A. J., Moore, M. J., Gelmon, K. A., et al. (2000) Phase I and pharmacodynamic study of human DNA methyltransferase (MeTase) antisense oligodeoxynucleotide (ODN), MG98, administered as 21-day infusion q4 weekly. *Clin. Cancer Res.* **6**, 257 (abstract).

103. Donehower, R., Stewart, D., Eisenhauer, E., et al. (2001) A phase I and pharmacokinetic (PK) study of MG98, a human DNA methyltransferase (Dnmt) mRNA inhibitor, given as a 2-hour twice weekly (BIW) infusion 3 out of every 4 wks. *Clin. Cancer Res.* **7**, 133 (abstract).

104. Stauder, G. M., Schlingensiepen, R., Goldbrunner, M., et al. (2002) Safety pharmacology and toxicity studies of the TGF-beta-2 antisense oligonucleotide AP 12009. *Proc. Am. Soc. Clin. Oncol.* **21**, 1897 (abstract).

105. Hau, P., Bogdahn, U., Schulmeyer, F., et al. (2002) TGF-beta-2 antisense oligonucleotide AP12009 administered intratumorally to patients with malignant glioma in a clinical phase I/II dose escalation study: safety and preliminary efficacy data. *Proc. Am. Soc. Clin. Oncol.* **21**, 109 (abstract).

106. Miyake, H., Nelson, C., Rennie, P. S., and Gleave, M. E. (2000) Acquisition of chemoresistant phenotype by overexpression of the antiapoptotic gene, testosterone-repressed prostate message-2 in prostate cancer xenograft models. *Cancer Res.* **60**, 2547–2554.

107. Zellweger, T., Miyake, H., Cooper, S., et al. (2001) Antitumor activity of antisense clusterin oligonucleotides is improved in vitro and in vivo by incorporation of 2'-O-(2-methoxy) ethyl chemistry. *J.Pharm. Exp. Ther.* **298**, 934–940.

7

Antisense Therapy in Clinical Oncology
Preclinical and Clinical Experiences

Ingo Tamm

1. Introduction

Nucleic acid molecules have emerged as versatile tools with promising utility as therapeutics for human diseases. The specificity of hybridization of an antisense oligonucleotide (AS-ODN) to the target mRNA makes the antisense strategy attractive for selectively modulating the expression of genes involved in the pathogenesis of malignant or nonmalignant diseases. One antisense drug has been approved for local therapy of cytomegalovirus (CMV) retinitis, and a number of AS-ODNs are currently being tested in clinical trials including oglionucelotides that target bcl-2, protein kinase C-α (PKC-α), and DNA methyltransferase. Clinical studies indicate that AS-ODNs are well tolerated and may have therapeutic activity. In this overview, we summarize therapeutic concepts, clinical studies, and new promising molecular targets to treat human cancer with AS-ODNs.

2. The Sense of Antisense in Cancer and Leukemia

There is currently new hope for developing selective anticancer drugs with fewer cytotoxic side effects than conventional cancer chemotherapy. This optimism is based on the identification of new cancer-associated molecular targets, which would allow the selective targeting of cancer cells while sparing normal cells *(1)*. Most of the proteins involved in the pathogenesis of cancer operate inside the cell and thus are not accessible to protein-based drugs. However, several approaches are available to specifically manipulate gene expression on the DNA or RNA level. One strategy is to use AS-ODNs to modify gene expression *(2)*.

From: *Methods in Molecular Medicine, Vol. 106: Antisense Therapeutics, Second Edition*
Edited by: I. Phillips © Humana Press Inc., Totowa, NJ

AS-ODNs are unmodified or chemically modified single-stranded DNA molecules of 13–25 nucleotides (nt) in length that are designed to specifically hybridize to the corresponding RNA by Watson–Crick binding. They inhibit mRNA function by several mechanisms, including modulation of splicing, inhibition of protein translation by disrupting ribosome assembly, and utilization of endogenous RNase H enzymes by AS-ODN. RNase H recognizes the mRNA–ODN duplex and cleaves the mRNA strand, leaving the AS-ODN intact. The released AS-ODN can bind to another target RNA *(2–6)* .

Because AS-ODNs inhibit gene expression in a sequence-specific manner, it is possible to selectively alter the expression of genes with closely related sequences. The antisense strategy allows the detailed analysis of signal transduction pathways, which often comprise families of highly homologous proteins. Furthermore, it may lead to the identification of new therapeutic targets and provide the corresponding drug at the same time. Today, a variety of potential targets have been analyzed in vitro and in animals with encouraging results **(Table 1)**. Because most tumor cells have a different pattern of gene expression compared to normal cells, AS-ODNs can theoretically be used to specifically target tumor-associated genes or mutated genes without altering gene expression of normal cells *(2)*.

In general, the critical steps in the rational drug design process are the identification of an appropriate target responsible for a certain disease and the development of a therapeutic agent with specific recognition and affinity to this target. For the majority of drugs in use to date, the mechanism of action is not well defined. By contrast, the specificity of Watson–Crick hybridization is the basis for "rational drug design" of AS-ODN leading to a new class of drugs with a wide range of potential clinical applications. One approved antisense drug and a number of clinical antisense trials demonstrate the feasibility of this approach, with some evidence for clinical efficiency *(7–9)*. At the same time, elucidation of the pathogenetic role of individual target proteins for certain diseases is rapidly progressing, most notably in the field of cancer research *(1,10)*.

3. Clinical Trials Testing AS-ODNs in Hematology and Oncology

The number of ongoing clinical antisense trials represents the growing interest in this technology **(Table 2)** *(2)*. Generally, systemic AS-ODN treatment is well tolerated and side effects are dose dependent. Dose-limiting toxicities are thrombocytopenia, hypotension, fever, and asthenia *(9,11)*. Furthermore, elevation of the liver enzymes aspartate aminotransferase and alanine aminotransferase, as well as complement activation and a prolonged activated partial thromboplastin time (aPTT) have been reported *(12)*.

Table 1
**Selection of Antisense Targets in Oncology Tested In Vitro
and in Animal Models**

Antisense target	Cell type analyzed	Biological end points
Bcl-2	B-cell lymphoma, melanoma, lung tumor	Apoptosis
Survivin	Cervical tumor, lung cancer	Apoptosis
MDM2	Multiple tumors	p53 activation
Bcl-XL	Endothelial cells, lung cancer cells	Apoptosis
RelA	Fibrosarcoma cell line	Cell adhesion, tumorigenicity
RAS	Endothelial cells, bladder cancer	CAM expression, proliferation
RAF	Endothelial cells, smooth muscle cells	CAM expression, proliferation
BCR-ABL	Primary progenitor bone marrow cells	Adhesion, proliferation
JNK1,2	Renal epithelial cells	Apoptosis
Telomerase	Prostate cell lines	Cell death
c-myc	Leukemia cell lines	Proliferation, apoptosis
c-myb	Leukemia cell lines	Proliferation

CAM = Cell Adhesion Molecule

In 1998, the first antisense drug (fomivirsen, Vitravene® Ciba Vision) was approved by the Food and Drug Administration (FDA) for the treatment of CMV-induced retinitis in patients with acquired immunodeficiency syndrome. The inhibitory constant (IC_{50}) of fomivirsen for CMV replication in vitro is 0.06 µM; for ganciclovir the IC_{50} is 30-fold higher (2 µM). Although fomivirsen is administered locally (intravitreal injection), the FDA's approval demonstrates the feasibility of AS-ODNs as therapeutic agents for the treatment of human diseases *(13)*.

3.1. bcl-2: Augmerosen, Oblimersen, G3139, Genasense™

bcl-2 family members represent promising targets for an antisense approach in oncology. The apoptosis inhibitor bcl-2 was discovered as a protooncogene found at the break points of t(14;18) chromosomal translocations in low-grade B-cell non-Hodgkin's lymphomas (NHL). Overexpression of bcl-2 was found in most follicular lymphomas, in some cases of diffuse large cell lymphomas, and in chronic lymphocytic leukemia (B-cell lymphatic leukemia [B-CLL]) *(14)*. The oncogenic impetus of increased bcl-2 expression was verified in bcl-2 transgenic mice. These mice accumulated excess noncycling mature B-lymphocytes *(14)*. High levels of bcl-2 are associated with relapse in acute myelogenous leukemia (AML) and acute lymphocytic leukemia *(11)*. The bcl-2

Table 2:
AS-ODNs in Clinical Trials or Approved in Hematology and Oncology

Compound	Company	Protein target	Indication	Development phase
Vitravene (fomivirsen)	Isis, Carlsbad, CA (www.isispharmaceuticals.com) Ciba Vision (www.cibavision.com)	CMV IE2	CMV retinitis	Approved
Genasense (augmerosen, oblimersen, G3139)	Genta, Berkeley Heights, NJ (www.genta.com)	BCL-2	Malignant melanoma	III
			B-cell NHL	II
			B-CLL	II/III
			Multiple myeloma	III
			AML	II
			NSCLC	II/III
			SCLC	I/II
			Prostate cancer	II
Affinitac (ISIS 3521, LY900003)	Isis, Carlsbad, CA (www.isispharmaceuticals.com) Eli Lilly, Indianapolis, IN (www.lilly.com)	PKC-α	Solid cancers: NSCLC, others	III
			NHL	II
ISIS 5132	Isis, Carlsbad, CA (www.isispharmaceuticals.com)	RAF kinase	Solid cancers: ovarian, others	II
ISIS 2503	Isis, Carlsbad, CA (www.isispharmaceuticals.com)	Ha-RAS	Colon cancer	II
			Metastastic breast cancer	II

Drug	Company	Target	Cancer	Phase
GEM 231	Hybridon, Worcester, MA (www.hybridon.com)	PKA	NSCLC Pancreatic cancer Solid cancers	II II I
MG 98	MethylGene, Montreal, Canada (www.methylgene.com)	DNA methyltransferase	Head and neck cance Metastatic renal cancer	II
GTI 2040	Lorus Therapeutics, Markham, Canada (www.lorusthera.com)	R2 component of ribonucleotide reductase	Renal cancer	II
GTI 2501	Lorus Therapeutics, Markham, Canada (www.lorusthera.com)	R1 component of ribonucleotide reductase	Lymphomas and solid cancers	I
AP 12009	Antisense Pharma GmbH, Regensburg, Germany (www.antisense-pharma.com)	TGF-β2	Malignant glioma	I/II

family has been implicated not only in the pathogenesis of cancer, but also in resistance to cancer therapy. Anticancer drugs and radiation ultimately kill cancer cells by inducing apoptosis, the suicide pathway of cells. bcl-2 blocks caspase activation in tumor cells at the mitochondrial level, which, in turn, prevents apoptosis induced by radiation and essentially all chemotherapeutic agents currently in clinical use *(15)*.

In a phase I study, the pharmacokinetics, toxicity, and therapeutic activity of an AS-ODN targeting bcl-2 were evaluated *(9)*. Twenty-one patients with bcl-2-positive-relapsed NHL received a 14-d sc infusion of an 18mer phosphoro-thioate oglionucelotide complementary to the first six codons of the bcl-2 open reading frame (G3139). Eight cohorts of patients received doses between 4.6 and 195.8 mg/m^2/d. No significant systemic toxicity was seen at doses up to 110.4 mg/m^2/d. All patients displayed skin inflammation at the sc infusion site. Dose-limiting toxicities in this study were thrombocytopenia, hypotension, fever, and asthenia. The maximum-tolerated dose was 147.2 mg/m^2/d. By standard criteria, there were one complete response, two minor responses, nine cases of stable disease, and nine cases of progressive disease. bcl-2 protein was reduced in 7 of 16 assessable patients, as determined by fluorescence-activated cell sorting. In two of these seven patients, reduced levels of bcl-2 were detected in tumor cells derived from lymph nodes, and in five patients in peripheral blood or bone marrow mononuclear cell populations. Expression of human leukocyte antigen (HLA), which was used as a control protein, was not affected by antisense therapy. From these results, it was concluded that bcl-2 antisense therapy is feasible, that it shows potential for antitumor activity in NHL, and that downregulation of bcl-2 protein but not HLA suggests a specific antisense mechanism *(9)*. However, it is important to note that bcl-2 was diminished in less than half of the treated patients. The mean inhibition of bcl-2 expression was moderate (24%), and the biological significance of this relatively small decline is uncertain *(2)*. Based on the results of this phase I study, a phase II trial is being conducted using G3139 in combination with standard chemotherapy for patients with relapsed, chemotherapy-resistant NHL.

In addition, a phase III randomized trial is testing whether the addition of G3139 to standard therapy consisting of fludarabin and cyclophosphamid is superior to standard therapy alone in B-CLL. A phase II trial is evaluating the efficacy of G3139 as a single agent in patients with B-CLL. Another phase III randomized trial is testing whether the addition of G3139 to high-dose dexamethasone is superior to standard therapy alone in multiple myeloma. A nonrandomized phase II trial is testing the single agent safety and activity of G3139 in patients with mantle cell lymphoma. In addition, it will assess the activity of G3139 in combination with rituximab plus CHOP (a combination of a monoclonal antibody [MAb] with chemotherapy) in patients who do not respond to G3139 alone.

Overexpression of bcl-2 is not only a hallmark of B-cell malignancies but also is found in other malignant tumors. A total of 35 patients with advanced cancer were treated in cohorts of three to six with 0.6–6.9 mg/kg/d of bcl-2 AS-ODN as a continuous infusion for 14–21 d. Again, bcl-2 AS-ODN was well tolerated. Current randomized trials are using the highest daily dose established in this study by shorter infusion periods (i.e., 7 mg/kg/d for 5–7 d) to enhance the antitumor activity of standard cytotoxic drugs *(16)*.

Human melanoma expresses bcl-2 in up to 90% of all cases *(17)*. Jansen et al. *(18)* showed that G3139 improves the chemosensitivity of human melanoma transplants in severe combined immunodeficient (SCID) mice *(18)*. In a phase I/II clinical study, this group tested the combination of G3139 and dacarbazine in patients with advanced malignant melanoma *(18)*. In a within-patient dose-escalation protocol, augmerosen was administered intravenously or subcutaneously to 14 patients with advanced malignant melanoma in daily doses of 0.6–6.5 mg/kg plus standard dacarbazine treatment (total doses up to 1000 mg/m^2 per cycle). In serial tumor biopsy samples, bcl-2 protein concentrations were measured by immunoblot, and apoptosis of tumor cells was assessed using the TUNEL assay. The combination regimen was well tolerated with no dose-limiting toxicity. Hematological abnormalities were mild to moderate. Lymphopenia was common, but no febrile neutropenia occurred. Higher doses of G3139 were associated with transient fever. Four patients had liver function abnormalities that resolved within 1 wk. Steady-state plasma concentrations of G3139 were obtained within 24 h and increased with the administered dose, as assessed by liquid chromatography. By d 5, daily doses of 1.7 mg/kg and higher led to a median decrease in bcl-2 protein expression of 40% in melanoma samples compared with baseline. Reduced bcl-2 expression was associated with increased apoptosis of tumor cells. Apoptosis was further enhanced after dacarbazine treatment. Six of 14 patients showed antitumor responses (complete in 1 patient, partial in 2 patients, and minor in 3 patients). The estimated median survival of all patients was more than 12 mo, which compares favorably with survival of stage IV malignant melanoma patients (usually 6–9 mo with and without therapy). This study is remarkable in that it is the first antisense trial in which downregulation of the target protein in the target tissue was shown. Based on the promising results of this study, the combination of dacarbazine and G3139 therapy in patients with malignant melanoma has received a fast-track approval by the FDA and is currently being evaluated in a phase III multicenter trial.

Phase I, II and III studies testing G3139 in combination with docetaxel in patients with advanced breast cancer, hormone-refractory prostate cancer, non–small cell lung cancer (NSCLC), and other solid tumors *(19–21)* are currently underway. No objective responses were seen in a phase I/II trial of G3139 and paclitaxel in chemorefractory small cell lung cancer (SCLC) *(22)*. Another phase I trial is studying the effectiveness of chemotherapy (carboplatin and etoposide) and G3139 in patients with extensive-stage SCLC.

The combination of G3139 and mitoxantrone has been tested in patients with hormone-refractory prostate cancer *(23)*. Patients were treated at seven dose levels of G3139 ranging from 0.6–5.0 mg/kg/d and mitoxantrone from 4–12 mg/m^2. In 26 patients treated so far, toxicities were transient and included neutropenia (grade 3), lymphopenia (grade 2), fatigue, arthralgias, and myalgias (all grade 1). No dose-limiting toxicities were reported for the dosages tested, two patients had a >50% prostate-specifc antigen response, one patient with symptomatic improvement in bone pain. Peripheral blood lymphcyte bcl-2 protein expression decreased in five of five patients administered G3139 at 5 mg/kd/d, as assessed by flow cytometry (mean change from baseline: –12.8%; SD: 16.4%).

Another phase I study has been started to test G3139 together with salvage chemotherapy comprising fludarabine, cytarabine, and granulocyte colony-stimulating factor (FLAG) in patients with refractory or relapsed acute myeloid leukemia or acute lymphoblastic leukemia. Four of 13 patients achieved complete remission and 2 were still in remission at d 93 and 180, respectively. Two patients achieved incomplete response with no evidence of disease but persistent neuropenia/thrombocytopenia. Of these six responders, three had prior high-dose AraC chemotherapy *(24)*. A nonrandomized phase II trial is testing whether the addition of G3139 to gemtuzumab ozogamin (Mylotarg®), a MAb conjugated with a toxin, is superior to Mylotarg alone in relapsed AML patients over the age of 60. The results of a phase I trial of combination treatment of G3139 and irinotecan has been reported in abstract form in patients with metastatic colorectal cancer *(25)*.

Thus, G3139 is being tested alone and in combination with chemotherapy in a variety of malignant diseases and is showing promising results.

3.2. c-raf-1: ISIS 5132

Other attractive targets for antisense therapy in oncology are raf kinases and ras. Raf kinases are serine/threonine kinases that regulate mitotic signaling pathways, most notably the mitogen-activated protein kinase pathway that transmits signals from ras. c-raf has been reported to bind to bcl-2 and to be involved in the regulation of apoptosis. The ras oncogene is dysregulated or mutated more frequently than any other oncogene studied in human cancer *(26,27)*. In several tumors, including breast and NSCLC, the expression of ras is a prognostic factor *(28)*. In pancreatic cancer for which standard therapy is strikingly ineffective, 95% of all cases show ras mutations *(27)*. This suggests that alterations in this pathway play a significant role in the pathogenesis of cancer *(7)*.

An AS-ODN directed to the 3' untranslated region of the c-raf mRNA (ISIS 5132) inhibited the growth of human tumor cell lines in vitro and in vivo in association with specific downregulation of target message expression. In a phase I trial, changes in c-raf-1 mRNA expression were analyzed in peripheral blood mononuclear cells (PBMCs) collected from patients with advanced cancers treated with ISIS 5132. Significant reductions of c-raf-1 expression from baseline were detected in 13 of 14 patients ($p = 0.002$). Two patients, both of whom had demonstrated tumor progression with previous cytotoxic chemotherapy, exhibited long-term stable disease in response to the AS-ODN therapy. The investigators suggested that PBMCs can be used to confirm antisense-mediated inhibition of the target protein in vivo *(29)*. However, one must consider that the decrease in c-raf-1 expression in total PBMCs could represent changes in the proportion of leukocyte populations owing to non-antisense-mediated immune stimulation and thus is not proof of an antisense-specific effect.

In a phase I trial, 31 patients with advanced malignancies received ISIS 5132 as a 2-h iv infusion three times weekly for three consecutive weeks with doses ranging from 0.5–6.0 mg/kg *(30)*. Clinical toxicities included fever and fatigue, neither of which were dose limiting. Two patients experienced prolonged disease stabilization for more than 7 mo. In both of these cases, this was associated with reduction in c-raf-1 expression in PBMCs.

Cunningham et al. *(31)* reported the results of a trial testing continuous iv infusion of ISIS 5132 for 21 d every 4 wk in 34 patients with a variety of solid tumors refractory to standard therapy. Toxicities up to 4.0 mg/kg were not dose limiting. Doses of 2.0–4.0 mg/kg are comparable with doses in mice at which activity was observed in human xenograft models. Grade 3 fever occurred in 2 of the 34 patients treated. One patient treated with 5.0 mg/kg had fever as a dose-limiting toxicity. Three grade 3 or 4 thrombocytopenia and one grade 3 leukopenia were observed. Two patients developed sepsis: one of them, while septic, manifested grade 4 thrombocytopenia, grade 4 hyperbilirubinemia, and a grade 3 elevation in aspartate aminotransferase; the other developed grade 4 thrombocytopenia. Leukopenia was mild, and no patient had neutropenia. One patient with therapy refractory ovarian cancer had a dramatic reduction in her CA-125 level (97%), and two other patients had prolonged disease stabilization for 9 and 10 mo, respectively.

Rudin et al. *(32)* reported phase I results of ISIS 5132 administered as a weekly 24-h iv infusion in 22 patients with advanced cancer. The trial defined a maximum tolerated dose of 24 mg/kg/wk on this schedule. No major responses were documented. In contrast to other trials of ISIS 5132, there was no consistent suppression of PBMC c-raf-1 mRNA level, possibly suggesting that the efficacy and toxicity profile of AS-ODNs are dependent on the schedule of administration *(32)*.

After the safety of ISIS 5132 was demonstrated in these phase I trials, several phase II trials were initiated. There was no evidence of single agent activity of ISIS 5132 in pretreated patients with recurrent ovarian cancer *(33)*. In this study, 22 patients were treated at a dose of 4 mg/kg/d by 21-d continuous intravenous infusion every 4 wk. ISIS 5132 was well tolerated with no grade 3 or 4 hematological or biochemical toxicity. There were six documented episodes of grade 3 nonhematological toxicity (two cases of lethargy, one of anorexia, two of pain, and one of shortness of breath). No objective clinical response was seen. Three patients had stable disease for a median of 3.8 mo, and the remaining evaluable patients had documented progressive disease. No patient had a decrease in CA-125 of 50% or more.

No objective responses were seen in a phase II trial for patients with lung cancer *(34)*. Twenty-two patients with progressive lung cancer (18 with NSCLC, 4 with SCLC) were treated with ISIS 5132 at 2 mg/kg/d, 21-d continuous infusion every 4 wk. Hematological toxicity did not exceed grade 2. Nonhematological toxicity was mild to moderate. Progressive disease was diagnosed in 10 patients, and 8 more patients were considered as treatment failures. No objective response was observed. Other phase II clinical studies including prostate and colon cancer are under way. Thus, there appears to be no clinical activity of ISIS 5132 as a single agent in these phase II trials.

3.3. H-ras: ISIS 2503

Abnormal expression of Ras proteins is frequently found with oncogenic transformation, making ras a promising therapeutic target. A 20-base phosphorothioate AS-ODN that binds to the translation initiation region of human H-ras mRNA (ISIS 2503) selectively reduced the expression of H-ras mRNA and protein in cell culture. Expression of other family members including N-ras, Ki-rasA, and Ki-rasB was not affected by ISIS 2503 in vitro. In a phase I trial, ISIS 2503 caused no dose-limiting toxicity at doses up to 10 mg/kg/d by 14-d continuous iv infusion every 3 wk. Four of 23 patients experienced stabilization of their disease for 6–10 cycles of therapy. No consistent decreases in H-ras mRNA levels were observed in peripheral blood lymphocytes *(35)*. A nontoxic dose of 6 mg/kg/d was selected for further study. In an interim report of ISIS 2503 as first line therapy for patients with previously untreated stage IV or recurrent colorectal carcinoma, 17 patients received 38 cycles. Toxicity was limited to grade 1 to 2 fever and grade 1 thrombocytopenia in three patients. Two cases of stable disease after 3+ and 6 cycles of therapy were observed *(36)*. ISIS 2503 in combination with chemotherapy is now in phase II clinical trials for the treatment of metastastic breast cancer, pancreatic cancer, and NSCLC. A multicenter phase II trial evaluated ISIS 2503 at 6 mg/kg/d by continuous iv infusion for 14 d (21-d cycle) in patients with advanced pancreatic cancer. The best responses were 2 cases of stable and 14 cases of progressive disease. The median time to progression was 2 mo. Toxicities were generally mild; the only grade 3 toxicities observed were asthenia and thrombocytopenia (one patient each) *(37)*.

Another multicenter phase II trial analyzed ISIS 2503 in stage IIIB/IV NSCLC. Seven of 20 evaluable patients achieved stable disease, and 13 progressed within the first three cycles. Again, there were no objective responses *(38)*.

3.4. PKC-α: LY900003, ISIS 3521, Affinitac™

Another target of AS-ODN is PKC-α. PKC-α belongs to a class of serine-threonine kinases whose involvement in oncogenesis is suggested by the fact that they are the major intracellular receptors for tumor-inducing phorbol esters. The results of a phase I study suggested that an AS-ODN directed against PKC-α (ISIS 3521) may be effective in the treatment of low-grade lymphoma *(11)*.

In a phase I trial, ISIS 3521 was delivered over a period of 21 d by continuous iv infusion followed by a 7-d rest period *(39)*. Doses were increased from 0.5 to 3.0 mg/kg/d. Twenty-one patients with incurable malignancies were treated in five patient cohorts. The maximum tolerated dose was 2.0 mg/kg/d, equivalent to pharmacologically active doses against human xenografts in mice. The dose-limiting toxicities were thrombocytopenia and fatigue at a dose of 3.0 mg/kg/d. Evidence of tumor response lasting up to 11 mo was observed in three of four patients with ovarian cancer. Another phase I study analyzed the combination of ISIS 3521 with 5-fluorouracil and leukovorin (5-FU/LV) in patients with advanced cancer. ISIS 3521 was tolerable when given with 5-FU/LV. Partial remission was seen in 2 of 14 patientsp; however, it is uncertain whether clinical activity is a result of enhanced drug interaction *(40)*.

A phase I/II trial of ISIS 3521 with carboplatin and paclitaxel in NSCLC has been reported. The phase I portion, which also included other malignant tumors, showed no dose-limiting toxicity at the highest dose level, consisting of 2.0 mg/kg/d, d 1–14 by continuous iv infusion of carboplatin (area under the curve 6, d 4) and paclitaxel (175 mg/m^2, d 4). There was no evidence of pharmacokinetic interactions between ISIS 3521 and either chemotherapy agent. In the expanded phase, patients received carboplatin, paclitaxel, and ISIS 3521 as above every 21 d. Toxicity consisted of grade 3 to 4 neutropenia (30 patients) and thrombocytopenia (13 patients), but no episodes of grade 3 to 4 neuropathy. Among 48 evaluable patients, 42% had a partial or complete response, and 17% progressed during treatment. The median overall survival was 19 mo with a 1-yr survival rate of 75% and a median time to progression of 6.6 mo *(41)*. Typical survival of patients with NSCLC with similar disease receiving carboplatin and paclitaxel alone is approx 8 mo.

Thus, in this trial the combination of ISIS 3521, carboplatin, and paclitaxel was well tolerated and showed promising activity in NSCLC. Based on these results, a 600-patient, randomized phase III clinical trial of ISIS 3521 in combination with chemotherapy for NSCLC was initiated and has already completed enrollment; no data are currently available.

Another phase II study examined the safety and efficacy of ISIS 3521/ LY900003 combined with docetaxel in patients with stage IIIB/IV, previously treated NSCLC. Response rates are available for 36 patients. The best responses were as follows: partial response in 5 patients (14%); 2 of whom failed prior paclitaxel; stable disease in 15 patients (42%); progressive disease in 16 patients (44%). Grade 3/4 hematological toxicities were neutropenia in 11 patients and thrombocytopenia in 9 patients. The most common other adverse events included grade 3/4 fatigue (three patients), infections (six patients), and neutropenic fever (five patients) *(42)*.

Another phase I/II trial initially evaluated the safety of 2 mg/kg/d ISIS 3521/ LY900003, administered by continous iv infusion on d 0–14 plus 80 mg/m^2 of cisplatin on d 0 and 1.000 mg/m^2 of gemcitabine on d 0 and 7. The combination was well tolerated, with grade 3 to 4 neutropenia and thrombocytopenia in 57 and 43% of cycles, respectively, in seven patients with advanced cancer. No pharmacokinetic interactions were observed. In the phase II portion, data are available for 43 advanced, previously untreated NSCLC patients. Grade 3 to 4 toxicities were as follows: thrombocytopenia (38 patients), neutropenia (19 patients), anemia, (6 patients), fatigue (11 patients), dehydration (7 patients), sepsis (3 patients). There was no grade 3 to 4 neuropathy or azotemia. Thirty-one patients were evaluable for response: complete response was seen in 1 patient (3%), partial response in 11 patients (35%), stable disease in 17 patients (55%), and progressive disease in 2 patients (7%) *(43)*.

ISIS 3521 failed to show significant activity in a phase II trial in patients with metastatic breast cancer (ECOG trial 3197). No objective responses were observed in the study population of 15 patients. Patients were treated with 2 mg/kg/d for 21 d every 28 d administered as continuous iv infusion. Median time to progression was 1.2 mo. Grade 3 to 4 toxicities included thrombocytopenia and infection *(44)*.

Alavi et al. *(45)* tested the efficacy, toxicity, and pharmacology of ISIS 3521 delivered as a 21-d continuous iv infusion in patients with recurrent high-grade astrocytomas. Toxicities were mild and reversible. There was no evidence of a clinical benefit. Median time to progression was 35 d after entering this protocol, and median survival was 93 d.

In a phase II multicenter trial, 29 patients with advanced relapsed or refractory low-grade or follicular NHL received 2 mg/kg/d of ISIS 3521/LY900003 by 21-d continous iv infusion repeated every 4 wk. Grade 3 to 4 thrombocytopenia occurred in seven patients (six with grade 3, one with grade 4), requiring a dose reduction in two. Two patients had grade 3 fatigue, requiring a dose reduction in one. One patient undergoing a dose reduction experienced signs of acute tumor lysis. Another patient developed self-limited toxic hepatitis. One patient had a mixed response. This patient had a 73% reduction in measurable disease, received radiation to three sc nodules, and continues with an excellent response after 8 mo of ongoing treatment. Other best responses included 16 patients with stable disease and 4 patients with progressive disease. Thus, ISIS 3521 treatment is well tolerated in patients with NHL. The observed modest activity is comparable with that observed with other AS-ODNs in NHL *(46)*.

3.5. c-myb:LR 3001 and Others

Autologous transplantation has become part of the routine management of many hematological malignancies. However, many patients relapse following the procedure. Gene-marking studies suggest that contaminating tumor cells that are inadvertently reinfused with the graft may contribute to relapse in AML and chronic myelogenous leukemia (CML). The myb product is a nuclear binding protein that controls the passage through G_1/S phase of the cell cycle, and it may play a critical role in hematopoietic cell development. Although expression is not restricted to leukemic cells, leukemic progenitors may be more susceptible to inhibition of c-myb than normal progenitors in vitro. To test this hypothesis, an AS-ODN targeted to the c-myb protooncogene was used to purge marrow autografts administered to allograft-ineligible CML patients *(47)*. CD34$^+$ marrow cells were purged with AS-ODNs for either 24 ($n = 19$) or 72 ($n = 5$) h. After purging, myb mRNA levels declined substantially in approx 50% of patients. Analysis of bcr/abl expression in long-term culture-initiating cells suggested that purging had been accomplished at a primitive cell level in more than 50% of patients and was oglionucleotide dependent. Nearly 50% of patients originally obtained a major cytogenetic response. Conclusions regarding clinical efficacy of oglionucleotide marrow purging cannot be drawn from this small pilot study. Nevertheless, these results lead to the speculation that enhanced delivery of oglionucleotide, targeted at critical proteins of short half-life, might lead to the development of more effective nucleic acid drugs and the enhanced clinical utility of these compounds in the future *(47)*.

3.6. DNA Methyltransferase:MG 98

Hypermethylation by the enzyme DNA methyltransferase has been postulated to inactivate tumor suppressor genes, resulting in neoplastic transformation and tumorigenesis. Agents that prevent or reverse DNA methylation might therefore restore normal growth control to cancer cells. MG 98 is a phosphorothioate AS-ODN that is a highly specific inhibitor of translation of the mRNA for human DNA methyltransferase with IC_{50} values of 50–70 nM in cell lines. Tumor growth delay and regression were observed with MG 98 in human lung and colon cancer xenografts. A phase I study examined MG 98 administered as a continuous 21-d iv infusion every 4 wk. At an interim report, nine patients with solid cancers received 10 courses at doses up to 240 mg/m^2/d. Dose-limiting grade 3 drug-related elevation of transaminases was encountered in two of two patients at the 240 mg/m^2/d dose level. Other toxicities were minimal. Biologically relevant concentrations for the inhibition of human DNA methyltransferase mRNA were achievable at the lowest dose level evaluated (40 mg/m^2/d) *(48)*.

Phase II trials are currently being conducted in patients with head and neck as well as metastatic renal cancer. A third phase II trial will analyze MG 98 in patients with myelodysplastic syndrome and AML. The studies aim to evaluate the clinical efficacy as well as the effects of MG 98 both on the methylation status of potential tumor suppressor genes and on suppression of DNA methyltransferase 1 (MeTase) mRNA levels in primary tumor biopsy samples.

3.7. PKA: GEM 231

A number of advanced chemistry modifications have been employed to improve the specificity, pharmacokinetics, and safety profiles of phosphorothioate oglionucleotide. These so-called mixed-backbone oglionucleotides permit oral and colorectal administration as a result of their increased in vivo metabolic stability. One of these compounds, GEM 231, is designed to interfere with the production of the RIα regulatory subunit of PKA, a cellular growth promoter that is increased within a wide variety of cancer cells. PKA is overexpressed in the majority of human cancers, correlating with worse clinicopathological features and prognosis in ovarian and breast cancer patients. After oral or ip administration, GEM 231 displayed dose-dependent in vivo antitumor activity in SCID and nude mice bearing xenografts of human cancers of the colon (LS174T), breast (MDA-MB-468), and lung (A549) *(49)*. In a phase I clinical study, preliminary data show that escalating doses of GEM 231 were well tolerated when given twice each wk by iv injections. Repeated doses of up to 360 mg/m^2 (equivalent to a range of 7–9 mg/kg) were administered over periods of up to 10 wk. There are ongoing phase I studies testing the safety of GEM 231 in combination with Taxotere® or Taxol® in patients with advanced cancers. Preliminary results suggest that GEM 231 produced only mild and reversible side effects and did not increase the side effects produced by the taxanes.

In a phase I study, GEM 231 was infused via a pump at 80 mg/m^2/d for 3 d (one patient), then for 5 d each wk at 80 (three patients), 120 (three patients), and 180 (two patients) mg/m²/d. At 180 mg/m²/d, both patients had dose-limiting toxicity: grade 3 to 4 transaminitis. Across all dose levels, thrombocytopenia, anemia, elevated alkaline phosphatase, and elevated SGOT and SGPT were observed. Treatment-related changes in aPTT were minor and seen in six patients. There was an apparent dose dependency for the occurrence of elevated transaminaases and aPTT. Serum extracellular PKA levels in treated patients demonstrated a marked decrease in a dose- and time-dependent manner *(50)*. Thus, this AS-ODN has an acceptable toxicity profile. No data on its clinical efficacy are available so far.

3.8. Transforming Growth Factor-β2: AP 12009

In a phase I/II dose escalation study, an AS-ODN is being administered intratumorally to patients with malignant glioma by high-flow microperfusion to evaluate safety and tolerability. AP 12009 is directed against transforming growth factor-β2 (TGF-β2), which has been recognized as an important tumor progression factor in malignant glioma and other tumors. In preclinical studies, AP 12009 reduced glioma cell proliferation and reversed TGF-β2-induced T-cell immunosuppression. No drug-related clinically relevant adverse events have been observed *(51)*.

4. New Targets for Antisense Therapy of Cancer

There are several new potential targets for specific antisense therapy of human cancer. Inhibitor of apoptosis (IAP) family proteins constitute a group of apoptosis suppressors that are conserved throughout animal evolution, with homologs in flies, worms, mice, and humans *(52,53)*. These proteins function in part as direct inhibitors of certain caspases *(54,55)*. Since caspases are central for most apoptotic pathways, it is not surprising that IAPs protect cells from several anticancer drugs as well as other IAPs.

cIAP2 at 11q21, and a novel gene, MLT at 18q21, are involved in t(11;18)(q21;q21) associated with mucosa-associated lymphoid tissue (MALT) lymphoma *(56)*. The translocation suggests a role for cIAP2 in the pathogenesis of MALT-lymphoma since this rearrangement is found in approx 50% of low-grade MALT-lymphomas *(57)*. AS-ODNs that target either the cIAP2/MLT break point or one of the two partners involved in the fusion protein in MALT-lymphoma cells could potentially alter cIAP2's antiapoptotic function and induce cell death in MALT-lymphoma cells.

Survivin is another IAP family member. Survivin is overexpressed in a large proportion of human cancers, providing evidence that altered expression of these proteins occurs during tumorigenesis *(58,59)*. In colorectal, gastric, breast, bladder, and lung cancers, as well as in diffuse large B-cell lymphoma, survivin expression is associated with shorter survival *(60–67)*. In neuroblastoma survivin expression correlates with higher stage of disease *(68)*. Interestingly, survivin is expressed in a cell cycle–dependent manner with highest levels in G2/M and rapid downregulation following cell-cycle arrest *(69)*. At the beginning of mitosis, survivin associates with the mitotic spindle, and disruption of this interaction results in a loss of its antiapoptotic function *(69)*. It has been suggested that survivin frees cyclin-dependent kinase 4 (CDK4) from the CDK inhibitor p16[INK4/CDKN2]. CDK4 then translocates into the nucleus, where it initiates the S-phase of the cell cycle *(70)*. The overexpression of survivin in cancer may thus overcome cell-cycle checkpoints and favor aberrant progression of transformed cells through mitosis. Therefore, survivin bridges apoptosis and cell cycle. Mutation of a conserved cysteine in the survivin BIR domain abolished survivin's cytoprotective abilities. However, the BIR mutant retained the ability to associate with microtubules similar to wild-type survivin and interfered with the function of endogenous survivin by competing for microtubule binding *(71)*. Thus, in contrast to p53, which links DNA replication in the S-phase of the cell cycle to apoptosis, survivin appears to couple the cell-suicide response to the checkpoint machinery involved in later cell-cycle steps (G2/M) *(71,72)*. An AS-ODN targeting nt 232–251 of human survivin mRNA was recently shown to induce apoptosis in lung cancer cell lines and sensitize cells to chemotherapy *(73)*. Moreover, blockade of survivin expression induces apoptosis in myeloma cell lines *(74)*. Isis is testing ISIS 23722, a second-generation AS-ODN against survivin, in preclinical studies.

Clusterin is a glycoprotein with a nearly ubiquitous tissue expression and an apparent involvement in various biological processes *(75)*. In vitro experiments have led to suggestions that clusterin functions in membrane lipid recycling, apoptotic cell death, and as a stress-induced secreted chaperone protein, among others. The question of whether clusterin is a multifunctional protein or deploys a single primary function influenced by cellular context remains to be explored. Clusterin acts as a cell-survival protein that is overexpressed in response to tumor-killing strategies, such as chemotherapy, hormone ablation, and radiation therapy. Overexpression of clusterin prolongs cell survival and leads to enhanced metastatic potential of cancer cells in vitro *(76)*. Antisense against clusterin significantly enhanced chemosensitivity in prostate and renal carcinoma cells in vitro *(77)*. OncoGenex and Isis will perform phase I/II clinical trials to assess the safety and efficacy of OGX-011 (ISIS 112989), a second-generation anticancer antisense drug, as a single agent and in combination with doxetaxel in men with localized and hormone refractory prostate cancer.

Other new targets for antisense therapy are involved in tumor cell proliferation, angiogenesis, and metastasis: growth factor receptor tyrosine kinases such as the epidermal growth factor receptor, transcription factors such as nuclear factor-κB, HER-2/neu, CDKs, and telomerase (proliferation); the vascular endothelial growth factor receptor and the basic fibroblast growth factor receptor (angiogenesis); and matrix metalloproteinases, angiogenin, and integrins (angiogenesis and metastasis) *(8)*.

5. Future Perspectives of AS-ODNs in Oncology

There are now various phase I trials showing that systemic treatment with AS-ODNs can be safely performed in cancer patients. However, the proof of clinical efficacy of AS-ODNs in the field of oncology is still missing. Large controlled trials are now needed to demonstrate that AS-ODNs are superior to other therapeutic approaches used today. Some phase II trials point to a modest therapeutic activity of AS-ODNs when combined with chemotherapy. Some studies show a downregulation of the target protein within the target tissue, suggesting that the principle of antisense may work in cancer patients. Again, further controlled studies will be needed to confirm antisense-mediated downregulation of the target protein in a larger number of patients and for other AS-ODNs. In the end, it may turn out that a sound proof of principle is virtually impossible in a clinical trial owing to limitations to the use of adequate controls. Once the mechanism for one AS-ODN is established in patients, the door is open for combination treatment with several AS-ODNs targeting a variety of oncogenes to overcome tumor escape and to improve therapeutic activity of this approach. Oglionucleotides with new backbone modifications other than phosphorothioate (i.e., 2'-*O*-methoxy-ethoxy, morpholino) or with methylated cytosines are potent inhibitors of target protein expression *(6,78)*. The future of antisense is likely to be based on such new-generation compounds.

Acknowledgments

Portions of this chapter are reprinted with permission from Elsevier Science (*Lancet*, 2001;358:489–497). This work was supported by a grant from the Deutsche Forschungsgemeinschaft.

References

1. Buolamwini, J. K. (1999) Novel anticancer drug discovery. *Curr. Opin. Chem. Biol.* **3**, 500–509.
2. Clark, R. E. (2000) Antisense therapeutics in chronic myeloid leukaemia: the promise, the progress and the problems. *Leukemia* **14**, 347–355.
3. Baker, B. F. and Monia, B. P. (1999) Novel mechanisms for antisense-mediated regulation of gene expression. *Biochim. Biophys. Acta* **1489**, 3–18.
4. Crooke, S. T. (1999) Molecular mechanisms of action of antisense drugs. *Biochim. Biophys. Acta* **1489**, 31–44.
5. Gewirtz, A. M. (2000). Oligonucleotide therapeutics: a step forward. *J. Clin. Oncol.* **18**, 1809–1811.

6. Koller, E., Gaarde, W. A., and Monia, B. P. (2000) Elucidating cell signaling mechanisms using antisense technology. *Trends Pharmacol. Sci.* **21,** 142–148.
7. Khuri, F. R. and Kurie, J. M. (2000) Antisense approaches enter the clinic. *Clin. Cancer Res.* **6,** 1607–1610.
8. Tamm, I., Dörken, B., and Hartmann, G. (2001) Antisense therapy in oncology: new hope for an old idea? *Lancet* **358,** 489–497.
9. Waters, J. S., Webb, A., Cunningham, D., et al. (2000) Phase I clinical and pharmacokinetic study of bcl-2 antisense oligonucleotide therapy in patients with non-hodgkin's lymphoma. *J. Clin. Oncol.* **18,** 1812–1823.
10. Brysch, W., Rifai, A., Tischmeyer, W., and Schlingensiepen, K.-H. (1996) Antisense-medited inhibition of protein synthesis, *Antisense Therapeutics* (Agrawal, S., ed.), Humana, Totowa, NJ, pp. 159–182
11. Cotter, F. E. (1999) Antisense therapy of hematologic malignancies. *Semin. Hematol.* **36,** 9–14.
12. Levin, A. A. (1999) A review of issues in the pharmacokinetics and toxicology of phophorothioate antisense oligonucleotides. *Biochim. Biophys. Acta* **1489,** 69–84.
13. de Smet, M. D., Meenken, C. J., and van den Horn, G. J. (1999) Fomivirsen - a phosphorothioate oligonucleotide for the treatment of CMV retinitis. *Ocul. Immunol. Inflamm.* **7,** 189–198.
14. Adams, J. M. and Cory, S. (1998) The bcl-2 protein family: arbiters of cell survival. *Science* **281,** 1322–1326.
15. Reed, J. C. (1995) Regulation of apoptosis by bcl-2 family proteins and its role in cancer and chemoresistance. *Curr. Opin. Oncol.* **7,** 541–546.
16. Morris, M. J., Tong, W. P., Cordon-Cardo, C., et al. (2002) Phase I trial of bcl-2 antisense oligonucleotide (G3139) administered by continous intravenous infusion in patients with advanced cancer. *Clin. Cancer Res.* **8,** 679–683.
17. Cerroni, L., Soyer, H. P., and Kerl, H. (1995) Bcl-2 expression in cutaneous malignant melanoma and benign melanocytic nevi. *Am. J. Dermatopathol.* **17,** 7–11.
18. Jansen, B., Wacheck, V., Heere-Reess, E., et al. (2000) Chemosensitization of malignant melanoma by BCL2 antisense therapy. *Lancet* **356,** 1728–1733.
19. Chen, H. X., Marshall, J. L., Trocky, N., et al. (2000) A phase I study of bcl-2 antisense G3139 (GENTA) and weekly docetaxel in patients with advanced breast cancer and other solid tumors. *Proc. Am. Soc. Clin. Oncol.* **19,** 178a.
20. de Bono, J. S., Rowinsky, E. K., Kuhn, J., et al. (2001) Phase I pharmacokinetic (pk) and pharmacodynamic (pd) trial of bcl-2 (Genasense) and docetaxel (D) in hormone refractory prostate cancer. *Proc. Am. Soc. Clin. Oncol.* **20,** 474a.
21. Tolcher, A. W., Kuhn, J., Basler, J., et al. (2000) A phase I, pharmacokinetic and biologic correlative study of G3139 (Bcl-2 antisense oligonucleotide) and Docetaxel in patients with hormone-refractory prostate cancer (HRPC). *Proc. Am. Soc. Clin. Oncol.* **19,** 527a.
22. Rudin, C., Otterson, G. A., George, C. M., Mauer, A. M., Szeto, L., and Vokes, E. E. (2001a) A phase I/II trial of genasense and paclitaxel in chemorefractory small cell lung cancer. *Proc. Am. Soc. Clin. Oncol.* **20,** 1283a.

23. Chi, K. N., Gleave, M. E., Klasa, R., et al. (2001) A phase I dose-finding study of combined treatment with an antisense bcl-2 oligonucleotide (Genasense) and mitoxantrone in patients with metastatic hormone-refractory prostate cancer. *Clin. Cancer Res.* **7,** 3920–3927.

24. Marcucci, G., Bloomfield, C. D., Balcerzak, et al. (2001) Biologic activity of G3139 (Genasense), a bcl-2 antisense (AS), in refractory (REF) or relapsed (REL) acute leukemia (AL). *Proc. Am. Soc. Clin. Oncol.* **20,** 1149a.

25. Ochoa, L., Kuhn, J., Salinas, R., et al. (2001) A phase I, pharmacokinetic, and biologic correlative study of G3139 and irinotecan (CPT-11) in patients with meta-static colorectal cancer. *Proc. Am. Soc. Clin. Oncol.* **21,** 297a.

26. Bollag, K. and McCormick, F. (1991) Regulators and effectors of ras protein. *Annu. Rev. Cell Biol.* **7,** 601–632.

27. Bos, J. L. (1989) Ras oncogenes in human cancer: a review. *Cancer Res.* **49,** 4682–4689.

28. Eckhardt, S. G., Rizzo, J., Sweeney, K. R., et al. (1999) Phase I and pharmaco-logic study of the tyrosine kinase inhibitor SU101 in patients with advanced solid tumors. *J. Clin. Oncol.* **17,** 1095–1104.

29. O'Dwyer, P. J., Stevenson, J. P., Gallagher, M., et al. (1999) c-raf-1 depletion and tumor responses in patients treated with the c-raf-1 antisense oligonucleotide ISIS 5132 (CGP 69846A). *Clin. Cancer Res.* **5,** 3977–3982.

30. Stevenson, J. P., Yao, K. S., Gallagher, M., et al. (1999) Phase I clinical/pharma-cokinetic and pharmacodynamic trial of the c-raf-1 antisense oligonucleotide ISIS 5132 (CGP 69846A). *J. Clin. Oncol.* **17,** 2227–2236.

31. Cunningham, C. C., Holmlund, J. T., Schiller, J. H., Geary, R. S., Kwoh, T. J., Dorr, A. & Nemunaitis, J. (2000). A phase I trial of c-Raf kinase antisense oligo-nucleotide ISIS 5132 administered as a continous intravenous infusion in patients with advanced cancer. *Clin Canc Res* **6**, 1626-1631.

32. Rudin, C. M., Holmlund, J., Fleming, G. F., et al. (2001b) Phase I trial of ISIS 5132, an antisense oligonucleotide inhibitor of c-raf-1, administered by 24-hour weekly infusion to patients with advanced cancer. *Clin. Cancer Res.* **7,** 1214–1220.

33. Oza, A. M., Eisenhauer, E., Swenerton, K., et al. (2000) Phase II study of c-raf kinase antisense oligonucleotide ISIS 5132 in patients with recurrent ovarian can-cer. *Proc. Am. Soc. Clin. Oncol.* **19,** 530a.

34. Coudert, B., Anthoney, A., Fiedler, W., et al. (2001) Phase II trial with ISIS 5132 in patients with small-cell (SCLC) and non–small cell (NSCLC) lung cancer: a European Organization for Research and Treatment of Cancer (EORTC). *Eur. J. Cancer* **37,** 2194–2198.

35. Cunningham, C. C., Holmlund, J. T., Geary, R. S., et al. (2001) A phase I trial of H-ras antisense oligonucleotide ISIS 2503 administered as a continuous intrave-nous infusion in patients with advanced carcinoma. *Cancer* **92,** 1265–1271.

36. Saleh, M., Posey, J., Pleasani, L., et al. (2000) A phase II trial of ISIS 2503, an antisense inhibitor of H-ras, as first line therapy for advanced colorectal carci-noma. *Proc. Am. Soc. Clin. Oncol.* **19,** 320a.

37. Perez, R. P., Smith, J. W., III, Alberts, S. R., et al. (2001) Phase II trial of ISIS 2503, an antisense inhibitor of H-ras, in patients (pts) with advanced pancreatic carcinoma (ca). *Proc. Am. Soc. Clin. Oncol.* **20,** 628a.

38. Dang, T., Johnson, D. H., Kelly, K., Rizvi, N., Holmlund, J., and Dorr, A. (2001) Multicenter phase II trial of an antisense inhibitor of H-ras (ISIS-2503) in advanced non–small cell lung cancer (NSCLC). *Proc. Am. Soc. Clin. Oncol.* **20,** 1325a.

39. Yuen, A. R., Halsey, J., Fisher, G. A., et al. (1999) Phase I study of an antisense oligonucleotide to protein kinase C-alpha (ISIS 3521/CGP 64128A) in patients with cancer. *Clin. Cancer Res.* **5,** 3357–3363.

40. Mani, S., Rudin, C. M., Kunkel, K., et al. (2002) Phase I clinical and pharmacokinetic study of protein kinase c-alpha antisense oligonucleotide ISIS 3521 administered in combination with 5-fluorouracil and leucovorin in patients with advanced cancer. *Clin. Cancer Res.* **8,** 1042–1048.

41. Yuen, A., Halsey, J., Fisher, G., et al. (2001) Phase I/II trial of ISIS 3521, an antisense inhibitor of PKC-alpha, with carboplatin and paclitaxel in non–small cell lung cancer. *Proc. Am. Soc. Clin. Oncol.* **20,** 1234a.

42. Moore, M. R., Saleh, M., Jones, C. M., et al. (2002) Phase II trial of ISIS 3521/LY900003, an antisense inhibitor of PKC-alpha, with docetaxel in non–small cell lung cancer (NSCLC). *Proc. Am. Soc. Clin. Oncol.* **21,** 297a.

43. Ritch, P. S., Belt, R., George, S., et al. (2002) Phase I/II trial of ISIS 3521/LY900003, an antinsense inhibitor of PKA-alpha with cisplatin and gemcitabine in advanced non–small cell lung cancer. *Proc. Am. Soc. Clin. Oncol.* **21,** 309a.

44. Gradishar, W. J., O'Neill, A., Cobleigh, M., Goldstein, L. J., and Davidson, N. E. (2001) A phase II trial with antisense oligonucleotide ISIS 3521/Cgp 64128a in patients (pts) with metastastic breast cancer (MBC): ECOG trial 3197. *Proc. Am. Soc. Clin. Oncol.* **20,** 171a.

45. Alavi, J. B., Grossman, S. A., Supko, J., et al. (2000). Efficacy, toxicity, and pharmacology of an antisense oligonucleotide directed against protein kinase C-alpha (ISIS 3521) delivered as a 21 day continous intravenous infusion in patients with recurrent high grade astrocytomas (HGA). *Proc. Am. Soc. Clin. Oncol.* **19,** 167a.

46. Emmanouilides, C. E., Saleh, A., Laufman, L., et al. (2002) Phase II trial of the efficacy and safety of ISIS 3521/LY900003, an antisense inhibitor of PKC-alpha, in patients with low-grade, non-Hodgkin's lymphoma. *Proc. Am. Soc. Clin. Oncol.* **21,** 282a.

47. Luger, S. M., O'Brien, S. G., Ratajczak, J., et al. (2002) Oligonucleotide-mediated inhibition of c-myb gene expression in autografted bone marrow: a pilot study. *Blood* **99,** 1150–1158.

48. Siu, L. L., Gelmon, K. A., Moore, M. J., et al. (2000) A phase I and pharmacokinetik (PK) study of the human DNA methyltransferase (Metase) antisense oligodeoxynucleotide MG98 given as a 21-day continous infusion every 4 weeks. *Proc. Am. Soc. Clin. Oncol.* **19,** 250a.

49. Wang, H., Cai, Q., Zeng, X., Yu, D., Agrawal, S., and Zhang, R. (1999) Antitumor activity and pharmacokinetics of a mixed-backbone antisense oligonucle-

otide targeted to the RIalpha subunit of protein kinase A after oral administration. *Proc. Natl. Acad. Sci. USA* **96**, 13,989–13,994.

50. Goel, S., Cho-Chung, Y. S., Nesterova, M. V., et al. (2002) Phase I study monitoring extracellular PKA (ECPKA) levels in a continous intravenous infusion (CIV) with GEM231, a second generation antisense oligonucleotide targeted against PKA RIalpha. *Proc. Am. Soc. Clin. Oncol.* **21**, 1b.

51. Hau, P., Bogdahn, U., Schulmeyer, F., et al. (2002) TGF-beta2 antisense oligonucleotide AP 12009 administered intratumorally to patients with malignant glioma in a clinical phase I/II dose escalation study: safety and preliminary efficacy data. *Proc. Am. Soc. Clin. Oncol.* **21**, 28a.

52. Deveraux, Q. and Reed, J. C. (1998) IAP family proteins—suppressors of apoptosis. *Genes Dev.* **13**, 239–252.

53. Tamm, I., Kornblau, S. M., Segall, H., et al. (2000b) Expression and prognostic significance of IAP-family genes in human cancers and myeloid leukemias. *Clin. Cancer Res.* **6**, 1796–1803.

54. Deveraux, Q., Takahashi, R., Salvesen, G. S., and Reed, J. C. (1997) X-linked IAP is a direct inhibitor of cell death proteases. *Nature* **388**, 300–303.

55. Takahashi, R., Deveraux, Q., Tamm, et al. (1998) A single BIR domain of XIAP sufficient for inhibiting caspases. *J. Biol. Chem.* **273(14)**, 7787–7790.

56. Dierlamm, J., Baens, M., Wlodarska, I., et al. (1999) The apoptosis inhibitor gene API2 and a novel 18q gene, MLT, are recurrently rearranged in the t(11;18)(q21;q21) associated with mucosa-associated lymphoid tissue lymphomas. *Blood* **93**, 3601–3609.

57. Baens, M., Maes, B., Steyls, A., Geboes, K., De Wolf-Peeters, C., and Marynen, P. (1999) Fusion between the apoptosis inhibitor gene API2 and a novel 18q gene MLT, rearranged in the t(11;18)(q21;q21), marks half of the gastro-intestinal MALT-type lymphomas without large cell proliferation. *Blood* **94**, 384a.

58. Ambrosini, G., Adida, C., and Altieri, D. C. (1997) A novel anti-apoptosis gene, survivin, expressed in cancer and lymphoma. *Nature Med.* **3**, 917–921.

59. Tamm, I., Wang, Y., Sausville, E., et al. (1998) IAP-family protein survivin inhibits caspase activity and apoptosis induced by Fas (CD95), Bax, caspases, and anticancer drugs. *Cancer Res.* **58(23)**, 5315–5320.

60. Lu, C. D., Altieri, D. C., and Tanigawa, N. (1998) Expression of a novel antiapoptosis gene, survivin, correlated with tumor cell apoptosis and p53 accumulation in gastric carcinomas. *Cancer Res.* **58**, 1808–1812.

61. Islam, A., Kageyama, H., Takada, N., et al. (2000) High expression of Survivin, mapped to 17q25, is significantly associated with poor prognostic factors and promotes cell survival in human neuroblastoma. *Oncogene* **19**, 617–623.

62. Kasof, G. M. and Gomes, B. C. (2000) Livin, a novel inhibitor-of-apoptosis (IAP) family member. *J. Biol. Chem.* **276**, 3238–3246.

63. Kawasaki, H., Altieri, D. C., Lu, C. D., Toyoda, M., Tenjo, T., and Tanigawa, N. (1998) Inhibition of apoptosis by survivin predicts shorter survival rates in colorectal cancer. *Cancer Res.* **58**, 5071–5074.

64. Monzo, M., Rosell, R., Felip, E., et al. (1999) A novel anti-apoptosis gene: re-expression of survivin messenger RNA as a prognosis marker in non–small-cell lung cancers. *J. Clin. Oncol.* **17(7),** 2100–2104.

65. Sarela, A. I., Macadam, R. C., Farmery, S. M., Markham, A. F., andGuillou, P. J. (2000) Expression of the antiapoptosis gene, survivin, predicts death from recurrent colorectal carcinoma. *Gut* **46,** 645–650.

66. Swana, H. S., Grossman, D., Anthony, J. N., Weiss, R. M., and Altieri, D. C. (1999) Tumor content of the antiapoptosis molecule survivin and recurrence of bladder cancer. *N. Engl. J. Med.* **341(6),** 452–453.

67. Tanaka, K., Iwamoto, S., Gon, G., Nohara, T., Iwamoto, M., and Tanigawa, N. (2000) Expression of survivin and its relationship to loss of apoptosis in breast carcinomas. *Clin. Cancer Res.* **6,** 127–134.

68. Adida, C., Berrebi, D., Peuchmaur, M., Reyes-Mugica, M., and Altieri, D. C. (1998) Anti-apoptosis gene, survivin, and prognosis of neuroblastoma. *Lancet* **351,** 882–883.

69. Li, F., Ackermann, E. J., Bennett, C. F., et al. (1999) Pleiotropic cell-division defects and apoptosis induced by interference with survivin function. *Nat. Cell Biol.* **1,** 461–466.

70. Suzuki, A., Hayashida, M., Ito, T., et al. (2000) Survivin initiates cell cycle entry by the competitive interaction with Cdk-4/p16INK4a and Cdk2/Cyclin E complex activation. *Oncogene* **19,** 3225–3234.

71. Li, F., Ambrosini, G., Chu, E. Y., Plescia, J., Tognin, S., Marchisio, P. C., and Altieri, D. C. (1998) Control of apoptosis and mitotic spindle checkpoint by survivin. *Nature* **396,** 580–584.

72. Reed, J. C. (1999) Survivin' cell-separation anxiety. *Nat. Cell Biol.* **1,** 199–200.

73. Olie, R. A., Simoes-Wüst, A. P., Baumann, B., et al. (2000) A novel antisense oligonucleotide targeting survivin expression induces apoptosis and sensitizes lung cancer cells to chemotherapy. *Cancer Res.* **6,** 2805–2809.

74. Tamm, I., Höhnemann, D., and Dörken, B. (2000a) Down-regulation of survivin by antisense oligonucleotides in plasmocytoma cells leads to increased apoptosis and sensitivity to chemotherapeutic drugs. *Blood* **96,** 462a.

75. Jones, S. E. and Jomary, C. (2002) Clusterin. *Int. J. Biochem. Cell Biol.* **34,** 427–431.

76. Miyake, H., Chi, K. N., and Gleave, M. E. (2000) Antisense TRPM-2 oligodeoxynucleotides chemosensitizes human androgen-independent PC-3 prostate cancer cells both in vitro and in vivo. *Clin. Cancer Res.* **6,** 1655–1663.

77. Zellweger, T., Miyake, H., July, L. V., Akbari, M., Kiyama, S., and Gleave, M. E. (2001) Chemosensitization of human renal cell cancer using antisense oligonucleotides targeting the antiapoptotic gene clusterin. *Neoplasia* **3,** 360–367.

78. Summerton, J. (1999) Morpholino antisense oligomers: the case for an RNase H–independent structural type. *Biochim. Biophys. Acta* **1489,** 141–158.

8

Radionuclide–Peptide Nucleic Acid in Diagnosis and Treatment of Pancreatic Cancer

Eric Wickstrom, Xiaobing Tian, Nariman V. Amirkhanov, Atis Chakrabarti, Mohan R. Aruva, Ponugoti S. Rao, Wenyi Qin, Weizhu Zhu, Edward R. Sauter, and Mathew L. Thakur

1.Introduction

1.1. Pancreatic Cancer

Cancer of the exocrine pancreas will attack an estimated 31,000 Americans in 2004, causing an estimated 30,000 deaths *(1)*. Cancer of the exocrine pancreas (hereafter simply referred to as pancreatic cancer) occurs more frequently in African-Americans than in European-Americans. The disease progresses rapidly. Pancreatic cancer is usually refractory to treatment, with a median survival time of 12 mo. Overall, only 1% of patients live more than 5 yr after diagnosis. However, if the disease is diagnosed when it is localized, the 5-yr survival is approx 20% *(2)*. It would be beneficial to detect pancreatic cancer at an early stage, when combination therapy with surgery and 5-fluorouracil (5-FU)-based chemotherapy might permit survival. In view of the current lack of a reliable method for early diagnosis, we propose noninvasive measurement of the oncogene expression profile of the cancer and subsequent ablation of those cells expressing the activating oncogenes.

There are hundreds of different types of cancer, and it was recently shown that at least five genes were mutated in each tumor *(3)*, confirming Knudson's *(4)* original statistical model for adult cancers. Based on these data, it was subsequently estimated that there are approx 200 different genes that are mutated in different cancers *(5)*. Furthermore, each gene can be mutated at different locations within the gene. We may presume, then, that every cancer will be

From: *Methods in Molecular Medicine, Vol. 106: Antisense Therapeutics, Second Edition*
Edited by: I. Phillips © Humana Press Inc., Totowa, NJ

unique. Using clinical symptoms to classify tumors therefore cannot provide all the needed information. With high-throughput microarray transcript profiling techniques, it is possible to screen a sample and determine the genes that are expressed. Therefore, expression profiling has attracted great interest in cancer biology because of its potential to revolutionize cancer diagnosis *(6,7)* by using cluster analysis to identify genes that characterize the transformed cell *(8,9)*. A notable example appeared in the most common subtype of non-Hodgkin's lymphoma, diffuse large B-cell lymphoma (DLBCL) *(7)*. Remarkably, DLBCL can be grouped into distinct subgroups based on the gene expression profile. The two groups that segregated out were germinal center B-like DLBCL and activated B-like DLBCL. The survival of patients in these two subgroups was significantly different, with the former having a significantly higher survival rate than the latter.

Expression profiling has also been used to classify cutaneous malignant melanoma *(8)* and breast cancer *(10)* as well as to identify genes that are important for metastasis *(9)*. Segregating distinct groups of patients with a different prognosis or those who will vs those who will not respond to treatment would optimize patient care. Unfortunately, molecular signatures have thus far not provided clear guidance to targeted patient care.

In at least some cases, targeting a pivotal oncogene has provided encouraging results, as in STI-571 treatment of chronic myelogenous leukemia associated with *BCR/ABL* crossing over, the Philadelphia chromosome *(11)*. Unfortunately, resistance to STI-571 arises quickly *(12)*. To put the developmental time line in perspective, an effective antisense oligonucleotide (AS-ODN) against that target was identified 10 yr earlier *(13)*. Similarly, reduction of elevated Bcl-2 protein by antisense DNA treatment against *BCL2* gene has displayed clinical responses in follicular lymphoma *(14)*, melanoma *(15)*, and prostate cancer *(16)*. Efforts to develop a small-molecule inhibitor of Bcl-2 protein have not yet reached the point of STI-571. Ninety percent of patients with pancreatic cancer carry twelfth codon activating mutations in their K-*RAS* oncogenes, and almost all diagnosed will die because they present with advanced disease. Efforts to develop a small-molecule inhibitor of mutant K-Ras protein over the past 22 yr have achieved little. Early detection of the molecular signatures of pancreatic cancer might permit life-saving intervention.

1.2. Oncogenes Associated with Pancreatic Cancer

Cancerous cells overexpress normal or mutated proteins derived from one or more of the 5000 genes involved in cell proliferation *(17)*. Such genes are called proto-oncogenes *(18)*. The implication is that the targets that must be attacked in pancreatic cancer cells are normal cellular genes that have sustained an activating lesion. K-*RAS*, *CCND1*, *HER2*, and *MYC* oncogenes, as well as the tumor suppressor p53, are frequently mutated or overexpressed in pancreatic cancer cells. Oncogene-targeted antisense DNA sequences specifically downregulate K-*RAS* *(19)*, *CCNDI* *(20)*, *HER2* *(19,21)*, *MYC* *(22–25)*, and p53 *(26,27)*, inhibiting cancer cell proliferation. Thus, we hypothesize that mutated or overexpressed K-*RAS*, *CCND1*, *HER2*, *MYC*, and p53 mRNAs are significant markers of oncogenic transformation that we may utilize to identify cancerous masses in the asymptomatic pancreas by external scintigraphic imaging.

1.2.1. K-RAS

Pancreatic cancer provides the most genetically clear-cut indication for antisense diagnosis. Ninety percent of the tumors carry a mutation in the twelfth codon of the K-*RAS* oncogene *(28)*. K-*RAS*, H-*RAS*, and N-*RAS* are among the most frequently mutated oncogenes detected in human tumors *(18)*. The mutational spectrums of the K-*RAS* oncogenes detected in colon, lung, and pancreatic tumors have been characterized. For example, the G-to-T and G-to-A mutations in the second base of codon 12 account for 65% of the K-*RAS* mutations observed in pancreatic tumors. Mutant K-*RAS* genes may be detected in circulating cells, bile, and pancreatic juice, as well as in biopsied tissue of patients with advanced disease *(29)*.

The human K-*RAS* protooncogene codes for an evolutionarily conserved G-protein, K-Ras p21, which binds guanine nucleotides with high affinity and is associated with the inner surface of the plasma membrane. A broad range of eukaryotes carries the *RAS* gene family, whose members code for immunologically related proteins of approx 21 kDa with 188 to 189 amino acid residues. Many varieties of Ras proteins have been found, mostly differing at their carboxy termini. The Ras:guanosine 5'-diphosphate (GDP) complex transduces signals from growth factors binding to cell-surface receptors *(30)*, whereupon the GDP is exchanged for guanosine 5'-triphosphate (GTP) to convert the inactive Ras:GDP complex to the active Ras:GTP complex *(31)*. The Ras:GTP complex transmits the proliferative signal downstream through a cascade of kinases to activate nuclear transcription factors for proliferative genes. The active GTP complex with Ras is restored to the inactive GDP complex by hydrolysis of GTP to GDP. The Ras protein itself possesses intrinsic guanosine 5'-triphosphatase (GTPase) activity. In vivo, however, this intrinsic activity is very slow unless enhanced by GAP (GTPase-activating protein). After the discovery of GAP, it was shown that the main biochemical difference between wild-type p21 and oncogenic Ras proteins with mutations in codon 12, 13, or 61 is the ability of GAP to induce GTP hydrolysis in the active Ras:GTP complex. GAP-induced hydrolysis can be as much as 1000 times faster with wild-type Ras than with these mutant forms of Ras *(32)*. These mutant forms thus remain in the active GTP form much longer than the wild type. The continuous transmission of a growth signal by the mutant forms is responsible for the oncogenic properties.

Inhibition of K-Ras protein appears to be at least a part of the antiproliferative mechanism of paclitaxel, a natural product that binds to microtubules, along which K-Ras may traverse on its way from the endoplasmic reticulum to the inner leaflet of the cell membrane *(33)*, which H-Ras does not. Additionally, prevention of K-Ras post-translational farnesylation by farnesyltransferase inhibitors has been shown to inhibit the growth of *RAS*-dependent tumors in immunocompromised mice *(34)*. Unfortunately, paclitaxel displays strong dose-limiting toxicity, which limits its efficacy *(35)*.

The high rate of K-*RAS* mutations makes it a reasonable target for antisense single photon emission computerized tomography (SPECT), positron emission tomography (PET), or magnetic resonance imaging (MRI) detection of early stage pancreatic cancer. Regarding therapy with antisense K-*RAS*, although the presence of the twelfth codon mutation does not necessarily imply causation, reducing the level of K-*RAS* gene expression might inhibit proliferation or reverse transformation in malignant cells transformed by mutated K-*RAS*. These results were observed in several cell lines and nude mouse models with antisense RNA *(36,37)* or ribozyme expression from vectors *(38)*, and by antisense DNA treatment *(19,39–42)*.

1.2.2. CCND1

Cyclin D1 (*BCL1, PRAD1, CCND1*) is a protooncogenic regulator *(43)* of the G1/S checkpoint in the cell cycle that has been implicated in the pathogenesis of several types of cancer, including pancreatic cancer. The cyclin D1 protein is overexpressed in up to 80% of tumors *(44,45)*. There is substantial evidence that critical regulatory steps occur during the cell cycle that determine whether or not the cell will synthesize new DNA and divide. These critical regulators of G1 are frequent targets for mutations *(46)*. Among the most frequently mutated genes are those that control the checkpoint (often called the restriction or R point) in late G1. The major regulator of this checkpoint appears to be pRb, the protein product of the retinoblastoma gene *(47)*. When hypophosphorylated, pRb inhibits cell growth by binding to and preventing the function of a number of transcription factors, including some in the E2F family *(46)*. Phosphorylation of pRb in mid to late G1 releases the transcription factor(s) bound by pRb that leads to DNA synthesis *(48)*. Two important regulators of G1 are p53 and cyclin D1. p53 appears to suppress cell division by stimulating the synthesis of a cyclin-dependent kinase (CDK) inhibitor, p21 *(49)*. Cyclin D1 appears to function upstream of pRb by binding to CDK4 or CDK6, leading to pRb phosphorylation by the CDK *(50)*. Overexpression of cyclin D1 in cultured cells leads to a more rapid transversion through the G1 phase of the cell cycle and entry into S phase *(51,52)*. Cyclin D1 cooperates with Ras protein *(43)* and complements a defective E1a adenoviral gene *(53)* to function as an oncogene.

Amplification is only one method by which the protein product can be overexpressed. Increased expression has also been observed due to gene rearrangement *(54,55)* in both parathyroid tumors (11q13 with 11p15) and B-cell tumors (11q13 with 14q32). Amplification detected by Southern blotting of *CCND1* has been observed in 25% of pancreatic cancers *(56)*, whereas in the same study using reverse transcription-polymerase chain reaction (RT-PCR), overexpression of *CCND1* mRNA was observed in 82% of the examined tissues. Using an immunostain, nuclear overexpression was observed in 68.4% of cancers, and protein accumulation correlated significantly with poor prognosis (median survival of 18.1 vs 10.5 mo; $p < 0.01$ by χ^2 test). A recent study *(57)* suggests that β-catenin may be one of the methods by which *CCND1* overexpression occurs in the absence of gene amplification. β-Catenin is known to play a role in intracellular signaling and can function as an oncogene when it binds to the promoter region of *CCND1* and transactivates genes after translocation to the nucleus. Reduced membranous expression of β-catenin and accumulation of β-catenin in the cytoplasm correlated significantly with *CCND1* overexpression (both $p < 0.0005$). There was a clear correlation between reduced membranous expression and ectopic cytoplasmic expression of β-catenin ($p < 0.0005$). In patients with adenocarcinoma of the pancreas, when cytoplasmic expression of β-catenin was absent, the 1-yr survival was 86.6%, whereas among patients with carcinomas showing cytoplasmic expression, only 35.7% survived 1 yr ($p < 0.01$).

These results suggest that β-catenin may be involved in the tumorigenesis of pancreatic cancer and exhibits its effects mainly by the transactivation of *CCND1*. Indeed, we observed that *CCND1* antisense DNA sequences specifically downregulate *CCND1*, and inhibit proliferation, in melanoma cells *(20)*. In view of our ability to suppress *CCND1* expression with antisense RNA vectors *(58)*, it is reasonable to hypothesize that SPECT, PET, or MRI detection of hybridized *CCND1* antisense peptide nucleic acid (PNA) oligonucleotides could identify malignant lesions before symptoms arise.

1.2.3. HER2

Amplification of the *HER2* gene was first identified as a marker of advanced-stage breast cancer and a prognostic marker *(59)*. Subsequent studies have demonstrated the overexpression and amplification of the *HER2* gene in pancreatic adenocarcinomas *(60)*. *HER2* gene encodes a 185-kDa protein, Her2, belonging to the receptor-tyrosine kinase family of cell-surface proteins *(61)*. The Her2 protein displays strong homology with epidermal growth factor receptor

(62), as do the closely related ErbB3 and ErbB4 receptors *(63,64)*. All three receptors are located on epithelial cell surfaces in sufficient proximity that they can be crosslinked. Activation of the intracellular tyrosine kinase activity is most likely the key step by which p185 confers transforming signals to second-messenger systems, such as the Ras pathways. The Her2 receptor does not appear to be expressed in normal adult cells, except for secretory epithelial cells, and may only be generally necessary at an early developmental stage *(65)*. While Her2 overexpression is associated with poor prognosis, this receptor may not be the key to malignant cell proliferation but, rather, to recurrence, a more formidable problem *(66)*.

There is significant evidence that Her2 status is helpful in predicting breast cancer response to therapy, whether hormonal, cytotoxic, or radiation *(67)*. Serum-based testing for circulating Her2 protein also shows promise; most, but not all, studies suggest that elevated levels of Her2 in serum correlate with decreased survival and absence of clinical response to hormone therapy *(67)*. A humanized monoclonal antibody (MAb) to Her2, Herceptin®, has a favorable toxicity profile and has some efficacy in Her2-positive metastatic breast cancer both alone *(68)* and in combination with chemotherapy *(69)*. However, Herceptin has not yet been tested in pancreatic cancer.

Although tissue-based detection of Her2 remains the clinical standard, there is disagreement as to the best method of measuring overexpression *(67)*. For comparative purposes, we have chosen to detect *HER2* amplification through a highly sensitive PCR technique, and Her2 overexpression by immunostaining, which allows us to evaluate the tissue *in situ*. One advantage of PCR is that it can detect low-level amplification in histologically benign tissue adjacent to malignant cells, whereas overexpression is not seen in this setting.

Antisense DNA has been applied against *HER2* and was observed to decrease p185 levels *(21,70,71)*, and it has prevented tumorigenesis by *HER2*-transformed SKOV3 ovarian carcinoma cells in immunocompromised mice *(19,72)*. To the extent that *HER2* is also overexpressed in pancreatic cancer *(60)*, its mRNA may also be a responsive target for antisense SPECT, PET, or MRI detection of oncogene expression in the pancreas.

1.2.4. MYC

The protooncogene *MYC* is amplified in approx 70% of human pancreatic tumors *(73)*. The target oncogene, *MYC*, expresses a nuclear protein, Myc, with an electrophoretic apparent molecular mass of 65 kDa (p65). Myc is a leucine zipper protein that binds with a small partner protein, Max. The resulting heterodimer binds specifically to the promoter element dGACCACGTGGTC, which occurs in the regulatory regions upstream of proliferative genes *(74)*. Expression of the *MYC* gene is normally controlled by a variety of transcriptional activating proteins *(75)*, and suppressor proteins such as p53 *(76)*. Significantly, Ras protein activation synergizes with Myc protein by increasing its stability *(77)*.

MYC was the first oncogene targeted by antisense DNA, specifically the MYC6 sequence *(22)*, and a number of *MYC* antisense sequences have displayed sequence-specific antisense activity in a variety of malignant cells in culture *(78,79)*, in animal hosts *(80)*, and in humans with solid tumors *(81)*. These results provide a reasonable basis for pursuing *MYC* mRNA as a logical target for probing with antisense SPECT, PET, or MRI in pancreatic cancer.

1.2.5. p53

Mutations in the p53 tumor suppressor gene are found in some fraction of all tumor types, in particular 40–60% of pancreatic ductal carcinomas, but not in pancreatitis *(82)*. There is a high association between p53 mutations and p53 protein accumulation *(83)*, such that many researchers now accept p53 overexpression as a valid surrogate for p53 mutations *(84)*. p53 mutations typically appear when *in situ* (intraductal) carcinoma develops, whereas K-*RAS* codon 12 mutations appear much earlier, even in apparently normal surrounding tissue *(85)*. Simultaneous mutations in both p53 and K-*RAS* have been associated with more rapid progression of pancreatic cancer *(86)*. This point is controversial, however, as a result of the rapid course of all forms of pancreatic cancer.

Targeted therapy of human cancer based on molecular alterations in p53 are being studied in both preclinical and clinical settings. Adenoviral mediated transfection of immortalized Li-Fraumeni cells increased the efficacy of photodynamic therapy in inducing apoptosis *(87)*. Adenovirus-mediated p53 gene therapy has been shown to enhance chemotherapy in preclinical models of human cancer *(88)* and is currently being conducted in clinical trials, with low toxicity and evidence of antitumor activity *(89)*. p53 overexpression leads to enhanced radiosensitivity of tumors, both after adenoviral transduction *(90)* and with a liposome-mediated vehicle for delivery of therapy *(91)*.

Oncogene-targeted antisense DNA phosphorothioate sequences have been observed to downregulate p53 specifically in malignant hematopoietic cells *(26,27)*, inhibiting leukemic cell proliferation *(26,27)*. Hence, p53 mRNA appears to be another logical target for antisense SPECT, PET, or MRI detection of oncogene expression in pancreatic cancer.

1.3. Imaging Cancer by Molecular Signature

Early detection of the molecular signatures of pancreatic cancer might permit life-saving intervention. Unfortunately, current imaging modalities each have their limitations. Anatomical imaging by computerized tomography (CT),

SPECT, or MRI provides precise anatomical detail, but nothing about function, let alone molecular signatures. Combinations of anatomical imaging with metabolic PET imaging have yielded variable results. Fluorescence and luminescence show promise for functional imaging but are severely limited in depth of penetration and are unlikely to be of use in visualizing human visceral tumors. Many of the constructs used for preclinical imaging, such as luciferase, have defined toxicity in humans, limiting their clinical usefulness.

Imaging gene expression, noninvasively, with high sensitivity and specificity, will provide a more powerful diagnostic tool than any currently available. None of the current modalities can image oncogene expression directly, and no other reliable method is currently available to measure levels of specific mRNAs in vivo. In contrast to indirect approaches, noninvasive administration of antisense SPECT, PET, or MRI hybridization probes specific for particular oncogene mRNAs will allow us to image transformed cells overexpressing each specific oncogene. We will use human AsPC1 pancreatic cancer cell xenografts on the hind legs of nude mice as our first model system. Because the pancreas is very close to the liver, gallbladder, and kidneys, distinguishing probe bound specifically in the pancreas of a patient from that bound nonspecifically in the liver and kidneys presents a vastly greater challenge than imaging the isolated xenograft. However, our observations in mice indicate that by 24 h after administration, technetium-99m (Tc-99m)-peptide-PNA retained in liver was less than that in the tumor xenograft *(92)*.

1.3.1. Tc-99m SPECT Imaging of Cancer

For imaging gene expression, we use Tc-99m, the metastable short-lived isotope of Tc-99, because of its universal availability from Mo-99 generators and its physical decay characteristics. It has a half-life of 6 h, which is long enough to permit imaging of gene expression, but not long enough to persist in the body and impart an unnecessary radiation dose to the patient. It also decays by emission of 140-keV γ rays (90%) that can be efficiently detected externally by a commonly available device, the γ camera. Currently, nearly 90% of all scintigraphic imaging procedures use Tc-99m.

Initially, we prepared Tc-99m-HYNIC-oligonucleotide phosphorothioate antisense to *HER2* mRNA and looked for accumulation of labeled oligonucleotide in SKBR3 cells that overexpress *HER2* mRNA, compared with a scrambled control, but found equivalent amounts of label in each cellular preparation (Basu, Wickstrom, and Thakur, unpublished). Similarly, a Tc-99m-HYNIC-oligonucleotide phosphorothioate antisense to *MYC* mRNA did not show differences in uptake among cell lines with high, normal, or low levels of *MYC* mRNA *(93)*. The Eisenhut group has also prepared and tested the cellular binding of *BCL2* antisense phosphorothioate *(94)* and PNA *(95)* sequences conjugated to [I-125]Tyr(3)-octreotate. On the other hand, a Tc-99m-MAG3-oli-gonucleotide phosphorothioate antisense to protein kinase A type I regulatory subunit α was reported to accumulate up to 100,000 copies per cell in serum-starved human LS174T colorectal cancer cells *(96)*, twice as much as a sense control, after 24 h of incubation with 4 nM oligonucleotide. The high reported accumulation, orders of magnitude greater than mRNA copy number, might be a result of serum starvation, and not directly relevant to physiological conditions.

We previously developed technology to label peptides with Tc-99m *(97,98)*, which is readily applicable to label PNAs with Tc-99m. This will save us the time and funds that would be required to develop new synthetic pathways to label PNAs with F-18 or other positron-emitting radionuclides. The tetrapeptide Gly-D-Ala-Gly-Gly (GDAGG) (**Fig. 3**, left) chelates Tc-99m firmly and efficiently *(97,98)*. Not only is the labeling efficiency high (>95%), but also the tracer is stable in vivo, even in acidic vesicles. The Gly$_4$ spacer on the N-terminus of the PNA (**Fig. 3**, right) is not likely to bind Tc-99m significantly. We use 4-aminobutyric acid (Aba) as a spacer between the primary peptide and GDAGG, which minimizes steric hindrance from the chelating moiety and the Tc-99m chelate. This technique is also suitable for labeling PNAs with Re-188 for therapeutic use. More important, the GDAGG-Aba chelating sequence can be included with a PNA-peptide during solid-phase synthesis, thereby eliminating the need for postsynthetic conjugation, purification, and characterization of the required compound (GDAGG-Aba-PNA-peptide) (**Fig. 3**).

Using our technique, several peptides, such as vasoactive intestinal peptide (VIP), have been labeled with Tc-99m in the Thakur laboratory and successfully evaluated in vitro, in experimental animals, and in humans *(97,98)*. [I-123]VIP has already been utilized for imaging pancreatic cancer sites *(99)*, as well as [In-111]PAM4 MAb against mucin-1 overexpressed in pancreatic cancer *(100)*.

1.3.2. Copper-64 PET Imaging of Cancer

The high sensitivity and high spatial resolution of noninvasive PET make it a good candidate for tumor imaging. The use of positron-emitting radionuclides such as F-18 is becoming increasingly popular for scintigraphic imaging by PET scanners or coincidence γ cameras. However, F-18 requires a cyclotron to produce, has a half-life of only 110 min, and is not yet commercially available in the fluoride form essential for synthesis of F-18-PNAs. Furthermore, imaging lesions requires a PET scanner or coincidence cameras, which are available at present in <5% of the estimated 5000 nuclear medicine centers in the United States. Although PET imaging with F-18-fluorodeoxyglucose or F-18-fluoroguanine derivatives *(101)* may in the future permit imaging of sites of cellular proliferation, it will not provide the identities of overexpressed genes. PET measurement of tumor uptake of an F-18-fluoroguanine derivative that is specifically phosphorylated by herpes simplex virus thymidine kinase (HSVTK) represents the closest approach so far to determining specific gene expression in a case in which the tumors were directly transformed by adenoviral vectors carrying HSVTK *(102)*. Similarly, PET measurement of tumor uptake of F-18-2'-fluoro-5'-iodo-uridine arabinoside in tumor cells retrovirally transformed with HSVTK expressed from a p53-controlled promoter indirectly implied elevated expression of p53 *(103)*.

For imaging gene expression, we propose to use copper-64 (Cu-64)-labeled hybridization probes. Cu-64 has a half-life of 12.7 h, emitting both positrons (β^+: 655 keV, 17.4%) and electrons (β^-: 573 keV, 30%). Cu(II) bound to N_2S_2 chelators are known to be stable ex vivo and in vivo and have been shown to be useful for PET imaging *(104–106)*, as well as therapeutic applications, when injected in large quantities *(107)*.

The tomographic imaging capability of PET scanners, combined with their high spatial resolution and the high sensitivity, and high tumor uptake of Cu-64-TP3982 might render that agent and Cu-64-peptide-PNA-peptides worthy for PET imaging of pancreatic tumors and therapeutically important.

1.3.3. Gadolinium-Induced Contrast MRI of Cancer

Anatomical imaging by CT or MRI can provide structural details of tumors but provides no information on the type or level of oncogene expression in cancer cells. For imaging gene expression, gadolinium (Gd)-induced shifts in the MRI signal can be used if Gd(III) is sufficiently concentrated in the tumor relative to the surrounding tissue *(108)*. This approach was first attempted with albumin conjugated to the chelator diethylenetriaminepentaacetic acid (DTPA), then equilibrated with Gd(III) to yield albumin-(DTPA-Gd)30-34 *(109)*. Greater contrast has been achieved with DTPA-polyamidoamine (PAMAM) generation-6 dendrimers *(110)*.

1.3.4. Rhenium-188 R5adiotherapy of Cancer

In recent years, rhenium-188 (Re-188) has emerged as one of the leading radionuclides for therapeutic applications. Re-188 has a $t_{1/2}$ of 16.9 h, and like Tc-99m, is available from longer-lived W-188 ($t_{1/2}$ = 69.4 d) from the Oak Ridge National Laboratory. Re-188 decays with emission of 2.12-MeV β particles with an abundance of 77% and 155-keV γ rays with an abundance of 15%. The penetration in soft tissue is 5 mm. The long half-life of the parent nuclide W-188 makes Re-188 more easily and conveniently available than other radionuclides of therapeutic importance. Finally, the chemical behavior of rhenium is similar to that of technetium, so that labeling methods developed for Tc-99m will probably translate to Re-188 with little or no modification.

The theoretical relationship of tumor size to curability with 22 β-emitting radionuclides of therapeutic potential was examined with a mathematical model *(111)*. The calculations indicated that the mean energy emitted per unit of accumulated activity of Re-188 was less than that emitted by Y-90, Pr-142, and Ir-194. Of these, only Y-90 is available commercially, and it is more difficult to manage. Unlike Re-188, Y-90 is not a generator-produced radionuclide. Furthermore, Re-188, like Tc-99m, has high specific activity and chelation chemistry very similar to that of Tc-99m. The γ-ray energy of 155 keV permits in vivo imaging for distribution studies and tumor uptake. These qualities render Re-188 an effective therapeutic radionuclide, and convenient for preparing Re-188-PNA-peptides for preclinical testing without modifications in our chemistry.

These advantages invited aggressive evaluation of Re-188 in targeted tumor therapy and metastatic bone disease in animals and patients *(111,112)*. Following the hypothesis that the negatively charged [Re-188]perrhenate might be transported preferentially into breast cancer cells through the mammary gland sodium-iodide symporter, [Re-188]perrhenate was evaluated in *HER2*-transformed breast tumor xenografts in nude mice *(113)*. Delivered radiation dose was greater than with equal quantities of I-131. Re-188 hydroxyethylidene diphosphonate relieves pain from bone metastases and is well tolerated *(114,115)*. It is not clear, however, if a similar approach could be applied to pancreatic cancer.

Finally, the use of Re-186- or Re-188-labeled receptor-specific peptides and antibodies for radionuclide tumor therapy is well accepted. The Thakur laboratory has shown that Re-186-MAbs, including F(ab')$_2$ fragments, could be prepared and successfully used to target embryonal carcinoma xenografts in mice *(116)*. Immunospecificity of the labeled biomolecules was not compromised by Re-186 labeling.

1.4. Assessment of Apoptosis Following Radiotherapy

Cell death can occur in a disorganized, nonphysiological fashion (necrosis) or in a carefully orchestrated sequence (apoptosis) that leaves little, if any, residue. Most cancer therapies initiate several physiological stimuli that trigger a programmed cellular set of events resulting in apoptotic cell death *(117)*. Apoptotic cell death eventually leads to the shrinkage of a lesion. These anatomical changes can be staged by many noninvasive imaging techniques, such as ultrasound, CT, MRI, or even scintigraphic imaging. However, these anatomical manifestations of lesions occur only slowly, and imaging of this process does not permit one to monitor the functional cellular changes that trigger cell apoptosis in the early state of a therapeutic intervention. Imaging apoptosis can lead researchers to determine noninvasively the effectiveness of a given therapeutic treatment at a very early stage before anatomical shrinkage of a treated lesion is apparent.

DNA fragmentation is one of the hallmarks of apoptosis *(118)*. Improved immunohistochemical techniques have permitted the identification of apoptosis *in situ* by specific labeling of nuclear DNA fragments (TUNEL) *(119)*. An earlier indicator of apoptosis, however, arises from membrane phospholipid rearrangements that elevate phosphatidylserine on the cell surface *(118)*. Phosphatidylserine is exposed well before cell necrosis and before the characteristic morphological changes of plasma membrane blebbing, vesicle formation, or cytoskeletal disruption. The phosphatidylserines in turn provide a high-affinity ($K_d = 7$ nM) binding site for annexin-V, a 36-kDa human protein *(120)*. Tc-99m-annexin-V has been prepared and utilized successfully to visualize cancer cells undergoing apoptosis *(121)*.

This method has specifically been applied to identify apoptosing cells following antisense treatment *(122)*. Furthermore, Tc-99m-annexin-V has been shown to image apoptosis in vivo *(123)*. Hence, we propose noninvasive monitoring of apoptosis induced in vivo by Re-188-PNA-peptides by Tc-99m-annexin-V scintigraphy. The onset of apoptosis induces membrane phospholipid rearrangements that elevate phosphatidylserine on the cell surface well before cell necrosis and before the characteristic morphological changes of plasma membrane blebbing, vesicle formation, or cytoskeletal disruption *(118)*.

1.5. AS-ODN Derivatives

The ability to turn off individual genes at will in growing cells provides a powerful tool for elucidating the role of a particular gene, for diagnosis and for therapeutic intervention. AS-ODNs (**Fig. 1**) were first conceived as alkylating complementary oligonucleotides directed against naturally occurring nucleic acids *(124)* and first successfully utilized against Rous sarcoma virus *(125)*. Since those proofs of principle, antisense DNA derivatives have been utilized to inhibit the expression of a wide variety of target genes, in viral, bacterial, plant, and animal systems; in cells *(78)*; in animals *(126)*; and in humans *(127)*.

1.5.1. Oligodeoxynucleotide Derivatives

Novel oligonucleotide analogs (**Fig. 2**) have been synthesized to act as antisense/antigene agents, to improve the biological stability, solubility, cellular uptake, and ease of synthesis *(128)*. The simplest oligodeoxynucleotide modification involves blocking the 3' terminus to prevent attack by 3' exonucleases, the predominant extracellular degradative mechanism for oligodeoxynucleotides *(129)*. Other modifications focus on protecting the internucleoside linkage by changing the phosphodiester linkages to phosphorothioates *(130)*,

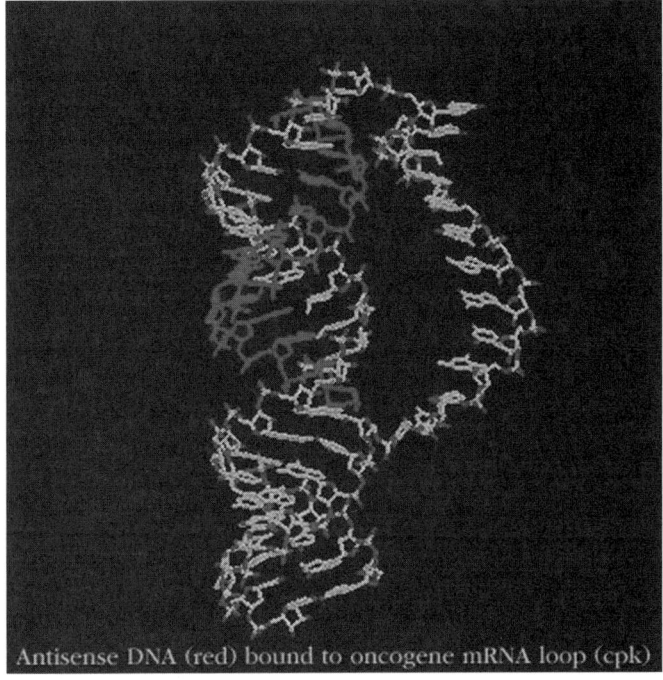

Fig. 1. Example of Tc-99-chelate-PNA-peptide.

methylphosphonates *(131)*, or boranophosphates *(132)*. Although these modifications increase the half-life of oligonucleotides in vivo, they also weaken hybridization to the RNA target sites due to the creation of chiral phosphorus diastereomers *(133)*. The deoxyribose may be modified to 2'-*O*-alkyl RNAs, such as 2'-*O*-methyl, strengthening hybridization and resisting nuclease attack *(134)*. Similar improvements result from preparing 3'-amino phosphoramidates *(135)* or morpholino phosphorodiamidates *(136)*.

Furthermore, the attachment of the base may be reversed from above the deoxyribose ring to below, changing the natural β-anomer to the α-anomer. The α-anomer achieves nuclease resistance without loss of base pairing *(137)*. Each of these structural changes affects not only nuclease susceptibility, but also cellular uptake, cellular trafficking, and RNase H activation *(128)*. Among the derivatives described, only phosphodiester, phosphorothioate, and boranophosphate DNAs direct RNase H degradation of hybridized RNA.

Phosphorothioate DNAs are the only derivative that has been administered so far to humans. Despite their efficacy, however, phosphorothioate DNAs exhibit less sequence specificity in their effects than do phosphodiesters or methylphosphonates *(78,138)*, due to significant binding to a spectrum of

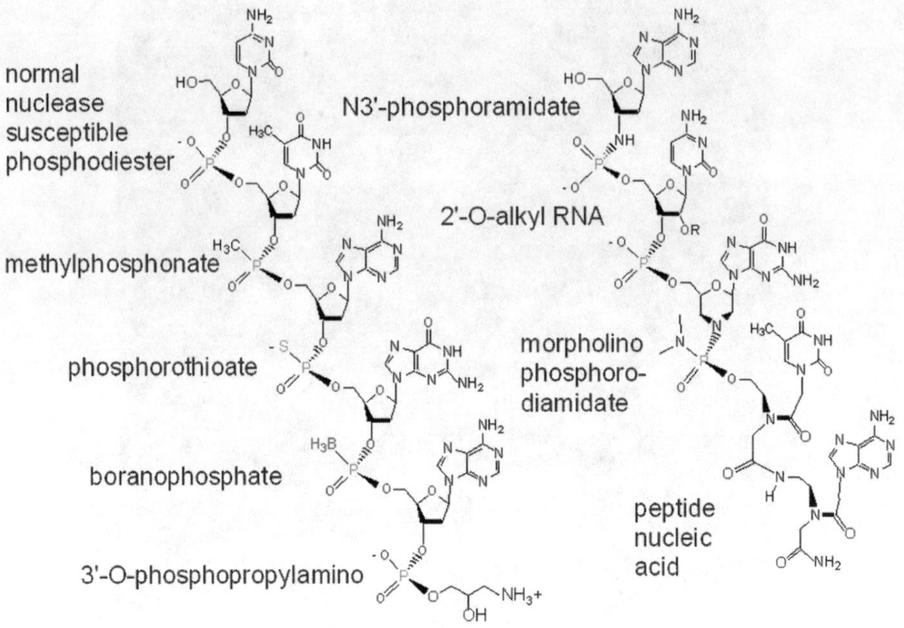

Fig. 2. Antisense DNA (red) base-paired to mRNA target.

plasma and cellular proteins *(139)*. The known modes of phosphorothioate toxicity invite further modifications to reduce sulfur content in therapeutic oligonucleotides.

Encouraging results have recently been obtained suggesting that greater potency and specificity might be possible with 2'-*O*-alkyl RNA/DNA/2'-*O*-alkyl RNA phosphorothioate chimeras *(140,141)*, PNA-peptide conjugates *(142)*, α/β/α-anomeric DAN chimeras *(143,144)*, DNA boranophosphates *(132,145)*, or peptide-DNA conjugates *(146,147)*.

Antisense DNA therapeutics have already overcome several hurdles *(148)*. These include uptake by cells, survival in cells, reaching of target mRNA in cells, blockage of gene function with predicted specificity, display of significant anticancer or antiviral effects as a result of gene blockage, nontoxicity in mice, and efficacy in mice. More than a dozen clinical trials are under way for a number of indications using phosphorothioates *(127)*. In 1998, the Food and Drug Administration issued its first approval of a therapeutic oligonucleotide, targeted against cytomegalovirus (CMV) *(149)*. A second sequence against *BCL2* oncogene is in Phase III trials. Antisense diagnostics have not advanced comparably, at least in part because charged antisense DNAs hybridized to RNA form a substrate for RNase H *(150)*, leading to the destruction of the message that one might wish to measure. That is not the case for PNAs (**Fig. 2**). We think, therefore, that PNAs can overcome the problems of charged DNA as antisense diagnostic probes.

1.5.2. Peptide Nucleic Acids (PNAs)

The most radical modifications are found in PNA, in which both the phosphodiester linkages and sugars are replaced with a peptide-like backbone of (*N*-2-aminoethyl) glycine units, with the bases directly attached by methylene-carbonyl linkers *(151)*. PNAs hybridize tightly to RNA, resist nuclease attack, and demonstrate antisense activity in vitro *(152)*. Compared with other oligonucleotide derivatives, PNAs display the highest melting temperature (T_m) for duplexes formed with single-stranded DNA or RNA *(153)*.

RNA hybridized to uncharged oligonucleotide derivatives, such as PNA, is not recognized by RNase H. Hence, PNA does not catalyze degradation of its bound RNA but inhibits mRNA translation solely by hybridization arrest, thus providing an opportunity for diagnostic application. Antisense activity in cells, however, requires microinjection of PNAs into nuclei. This stems from poor cellular uptake *(154)*, which was 10 times less efficient than uptake of phosphorothioates in a variety of mammalian cells *(155)*. To alleviate this situation, conjugation of receptor-specific ligands might improve cellular uptake *(142,156)*.

The insulin-like growth factor-1/insulin-like growth factor-1 receptor (IGF-1/IGF-1R) system plays a major regulatory role in development, cell-cycle progression, and the early phase of tumorigenicity *(157)*. The IGF-1R gene is amplified in approx 70% of human pancreatic tumors *(73)* and has been exploited as an antisense target in brain cancer *(158)*. Small cyclic peptides have been designed by molecular modeling as analogs of natural IGF-1. The most effective IGF-1 peptide analog, JB3, D-CSKAPKLPAAYC, inhibits the growth of certain cancer cell lines and competes with the natural ligand for binding to the IGF-1R *(159)*. The JB3 peptide has two Cys residues, one at each terminus, which are disulfide linked to form a loop with limited flexibility, favoring a conformation for binding to the receptor. The use of D-amino acids gave the peptide stability against cellular proteases. A reverse sequence was synthesized with respect to the normal L-amino acid sequence to account for the reversal of chirality.

Thus, we hypothesized that conjugation of the D-peptide analog with an antisense PNA against *IGF1R* mRNA *(159)* would provide cell-type specificity and increase cellular uptake by those cells overexpressing IGF-1R. To reduce the complexity of the synthesis, a smaller version of JB3, called JB9, cyclic D-CSKC, was selected for conjugation with the PNA (**Fig. 3**, right). Cellular uptake of the PNA-peptide conjugate, a control with two D-Ala resi-

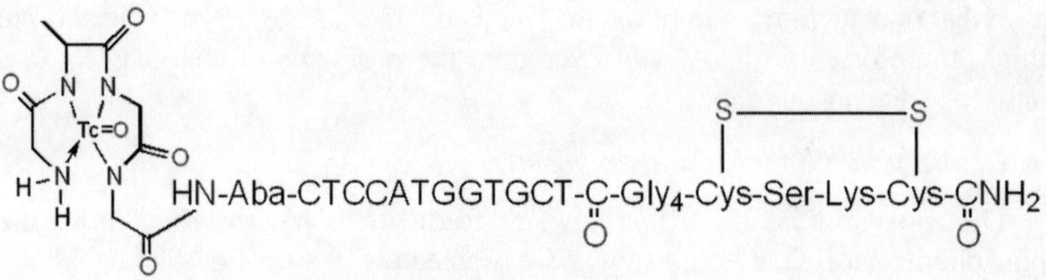

HN-Aba-CTCCATGGTGCT-C-Gly$_4$-Cys-Ser-Lys-Cys-CNH$_2$

Fig. 3. Oligonucleotide backbone derivatives.

dues in the peptide in place of D-Ser-Lys, and a control PNA without a peptide adduct were studied in three cell lines. Human Jurkat cells do not express IGF-1R *(160)* and thus serve as a negative control. Murine BALB/c3T3 cells express low levels of murine IGF-1R. Transfection of BALB/c3T3 cells with a human *IGF1R* gene yielded p6 cells, which overexpress IGF-1R *(161)*.

In p6 cells, 1 µ*M* of the specific PNA-peptide conjugate displayed dramatically higher uptake within 3 h than the control PNA or the control PNA-peptide *(142)*. This approach may allow cell-specific application of PNAs as gene expression diagnostics in vivo. In prostate cancer cells, for comparison, conjugation of dihydrotestosterone or a nuclear localization peptide to a *MYC* antisense PNA permitted some nuclear localization and Myc reduction in LNCaP cells expressing androgen receptor after 24 h exposure to 10 µ*M* PNA *(162)*. It would appear that use of a peptide analog specific for a cell-surface receptor is far more effective than a steroid capable of binding to a cytoplasmic protein after unassisted uptake.

2.Methods

2.1. Synthesis of Phosphorothioate Oligonucleotides

Phosphorothioate oligonucleotides were synthesized on a Millipore 8750 DNA synthesizer using standard phosphoramidite chemistry *(163)*. Phosphorothioates were prepared by sulfurization in lieu of oxidation *(164)*. Small-scale (1 µmol) preparations for cell culture studies were purified by C$_{18}$ reversed phase liquid chromatography *(22,165–168)*. Large-scale (15 µmol) preparations for animal studies were purified by *n*-butanol precipitation *(169)*, and purity was verified by gel electrophoresis and C$_{18}$ reversed-phase liquid chromatography *(22,165–168)*. Some sequences were analyzed by matrix-assisted laser desorption ionization–time-of-flight–mass spectroscopy (MALDI-TOF-MS) on a Ciphergen SELDI mass spectrometer with a 338-nm laser. Following purification, the DNAs were dissolved in deionized water, sterilized by filtration through 0.2-µm filters, and frozen at –80°C for storage.

2.2. Synthesis of Peptide-PNA-Peptide Chimeras

Chimeras were assembled using Fmoc PNA coupling (*170*) on an Applied Biosystems 8909 DNA synthesizer. Because PNA monomers couple more rapidly than amino acid monomers, three reaction conditions were tested for coupling between amino acid monomers to PNA and coupling between amino acid monomers: (1) standard single coupling, (2) double coupling, (3) single long coupling. We cleaved and deprotected the chimeras in $CF_3CO_2H:CH_2Cl_2$:*m*-cresol:Et_3SiH (85:5:9.5:0.5) for 2 h at room temperature. Chimeras were purified by preparative reverse-phase high-performance liquid chromatography (RP-HPLC) on a 10×250 mm Alltima C_{18} column eluted with a gradient over 25 min from 5 to 70% CH_3CN in aqueous 0.1% CF_3CO_2H at 1 mL/min and 50°C monitored at 260 nm. Purified chimeras were analyzed similarly on a 4.5×250 mm Alltima C_{18} column. MALDI-TOF-MS was carried out as in **Subheading 2.1.**

For N_4 chelation of Tc-99m, we extended GᴅAGG from the N-terminus of the PNA with an Aba spacer (*98*) to minimize steric hindrance between the chelator and the PNA. For Cu-64 chelation, we extended diaminopropanoic acid from the N-terminus of the PNA with an aminoethoxyethoxyacetic acid (AEEA) linker, then coupled two *S*-benzoyl thioglycolic acid residues to generate an N_2S_2 chelator. For Gd(III) chelation with DTPA or 1,4,7,10-tetraazacyclododecanetetraacetic acid (DOTA), we extended AEEA from the N-terminus of the PNA, followed by succinic anhydride, to which we coupled PAMAM-G3 with an average of 32 available amines. The isothiocyanates of DTPA or DOTA were then coupled to the remaining available PAMAM amines.

2.3. Labeling of Peptide-PNA-Peptide Chimeras

Purified peptide-PNA chimeras were labeled with Tc-99m (**Scheme 1**) essentially as described previously (*97,98*). Briefly, to 10 μg of chelator-PNA-peptide, 600 μL of 0.05 M Na_3PO_4 (pH 12), was added, followed by the addition of 75 μg of $SnCl_2 \cdot 2H_2O$ in 15 μL of 0.05 M HCl and 20–25 mCi of freshly eluted [Tc-99m] O^-_4 in 200 μL 0.15 M NaCl. The mixture was incubated for 15 min at 90°C, cooled to 22°C, and the pH was adjusted to approx 7 by the addition of 1 mL of 0.05 M NaH_2PO_4 (pH 4.6). Reaction mixtures were examined for free Tc-99m on a 4.5×250 mm reverse-phase C_{18} microbond HPLC column (Rainin, Emeryville, CA) coupled to a an ultraviolet detector, an NaI (Tl) radioactivity monitor, and a rate meter. The column was eluted with a 28-min gradient from 10 to 100% CH_3CN in aqueous 0.1% CF_3CO_2H at 1 mL/min and 25°C. Unchelated free Tc-99m, Rf 1.0, was determined by instant thin-layer chromatography on silica gel (ITLC-SG; Gelman, Ann Arbor, MI) developed with methylethyl ketone. Colloid formation, Rf 0.0, was determined using instant

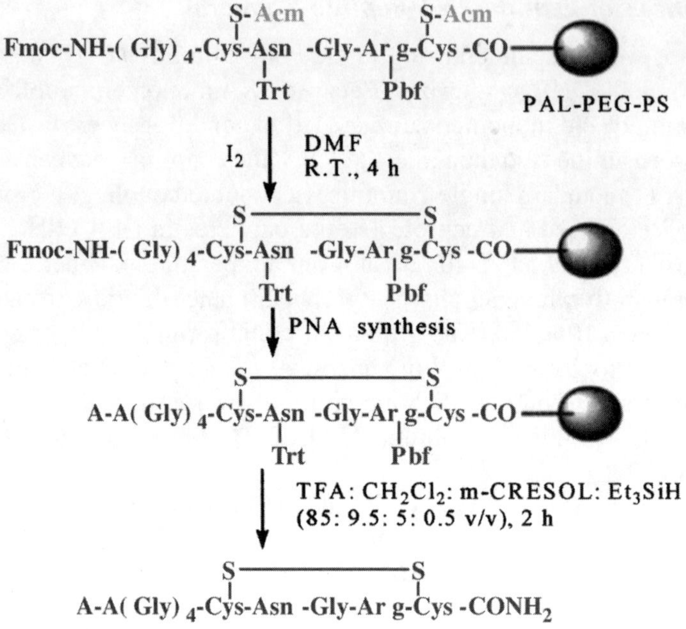

Scheme 1. Solid-phase cyclization cysteine residues. Two adenosine PNA residues were coupled following cyclization. DMF, dimethylformamide; TFA, trifluoroacetic acid. Acm, Trt, Phf are protecting groups.

TLC on silica gel (ITLC-SG; Gelman) developed with pyridine/HOAc/H$_2$O (3:5:1.5).

Chimeras were labeled with Cu-64 essentially as described previously *(97,98)*. [Cu-64]Cl$_2$ in 0.1 M HCl was added to purified chimera in 0.1 M ammonium citrate (pH 5.5), incubated for 60 min at 43°C, quenched with 1 mM EDTA, then purified by size-exclusion chromatography.

Chelator-PAMAM-G3-chimeras were labeled with Gd(III) essentially as described previously *(97,98)*. Aqueous 0.1 M GdCl$_3$ in 1 M HCl was added to purified chimera in 0.1 M Na acetate (pH 7) and incubated for 30 min at 40°C, then purified by reverse-phase liquid chromatography. Radiolabeling of these dendrimeric chimeras with low-specific-activity Gd-159 ($t_{1/2}$ = 18 h; β⁻, γ: 363 keV, 9%) was consistent with the prediction of 31 available DTPA residues per G3 dendrimer.

2.4. Cellular Assays with Oligonucleotide Derivatives and Chimeras

2.4.1. Cell Culture

The human AsPC1 pancreatic cell line was purchased from the American Type Culture Collection. AsPC1 cells were grown in a medium consisting of RPMI-1640 supplemented with 20% fetal bovine serum, 5 mM glutamine, 50 U/

mL of penicillin, and 5 µg/mL of streptomycin at 37°C in humidified air containing 5% CO_2. Human BT474 (ER–) or MCF7:IGF-1R (ER+) breast cancer cells, clone 17, transformed to express 1×10^6 IGF-1R/cell constitutively from a CMV promoter *(171,172)* were maintained in Dulbecco's modified Eagle's medium plus 5% calf serum, 50 U/mL of penicillin, 5 µg/mL of streptomycin, 2 mM glutamine, and 7.5 nM 17-β-estradiol (Sigma, St. Louis, MO) at 37°C under 5% CO_2.

2.4.2. Inhibition of Proliferation of Pancreatic Cell Lines by K-RAS Antisense Sequences

Aliquots of 1×10^5 cells were plated in six-well plates and were allowed to adhere to the plates for 24 h prior to oligonucleotide lipofection. Oligonucleotide complexes with lipofectamine were administered to cells as previously described *(21)* to a final concentration of 1.5 µM for 16 h, after which the medium was removed and replaced with fresh medium. The cells were then allowed to grow for 6 d, because Ras proteins exhibit a half-life of 20 h *(173)*, implying the need for long antisense treatment before seeing significant reduction of the protein. On the sixth day, cells were washed twice with phosphate-buffered saline (PBS), trypsinized, and counted. Viability was determined with trypan blue. Statistical analysis was carried out by applying the Kruskal–Wallis test in InStat 2.01 for Macintosh.

2.4.3. In Vitro Evaluation of Tc-99m-Chimeras

Each labeled chimera was evaluated for receptor specificity using a receptor displacement assay; cell-binding assays; or in the case of VIP, muscle relaxivity. Scatchard plot analyses were also performed, where appropriate, which led us to determine not only the K_d values but also the number of specific receptors expressed on the target. Because each of our peptides was an analog of the respective primary peptide, comparison of these data was made with an I-125-labeled primary peptide. The biological activity and receptor specificity of the chimeras were not altered by the Tc-99m ligand *(106,174)*.

2.5. Tissue Distribution and Imaging of Peptide-PNA-Peptide Chimeras

Human BT474 (ER–) or MCF7:IGF-1R (ER+) breast cancer cells (5 to 6×10^6) in 150 µL of tissue culture medium were implanted intramuscularly in the right thigh of nude mice using a sterile 27-gauge needle and 1-mL sterile syringe. Each injection included 10 µg of Matrigel (Becton-Dickinson). Tumors were allowed to grow to 1 cm or less in diameter. For 4-h distribution, 0.2–0.3 mCi of Tc-99m-peptide-PNA was administered to groups of five mice each through a lateral tail vein using a 27-gage needle. For 24-h distribution, 0.8–0.9 mCi of probe was administered. At 4 and 24 h postinjection, mice were lightly anesthetized and imaged using a Starcam (GE, Milwaukee, WI) γ camera equipped with a parallel hole collimator. For images 300,000 counts were recorded. Mice were then killed in a halothane gas chamber, blood was collected, and tissues were dissected. The tissues were washed free of any blood, blotted free of liquid, and weighed, and radioactivity associated with each tissue was counted in an automatic γ counter (Packard Series 5000; Meridien, CT) together with a standard radioactive solution of a known quantity prepared at the time of injection. The results were expressed as a percentage of injected dose per gram of tissue (% I.D./g). The data were evaluated statistically using Student's *t*-test. The protocol was approved by the institutional AAALAC-licensed Animal Care and Use Committee.

2.6. Tc-99m-Annexin V Detection of Apoptosis

2.6.1. Preparation of Chelator-Annexin-V

Human annexin-V produced by expression in *Escherichia coli* is obtained commercially from Sigma. Annexin-V is derivatized as described previously *(123)* with hydrozinonicotinamide (HYNIC) by gently mixing for 3 h at 22°C in the dark a mixture of 5.6 mg/mL of annexin-V in 20 m*M* HEPES (pH 7.4), 100 m*M* NaCl, and 22 μg of 6-HYNIC (Anor MED, Langley, British Columbia) dissolved in 18.5 μL of *N,N*-dimethyl formamide (42 m*M*). At the end of the incubation period, the reaction is quenched with 500 μL of 500 m*M* glycine in PBS (pH 7.4) and then dialyzed at 4°C against 20 m*M* Na citrate (pH 5.2), 100 m*M* NaCl overnight. The annexin-V-HYNIC solution is sedimented for 10 min at 15,000*g* to remove particulates and then aliquoted in 100-μg portions for storage at –70°C.

2.6.2. Labeling of Chelator-Annexin-V

Tc-99m labeling of HYNIC-annexin-V is accomplished as described previously *(123)* by incubating at 22°C for 1 h with 10 mCi of Tc-99m-glucoheptonate obtained commercially (Mallinckrodt Radiopharmacy, Philadelphia, PA). The reaction mixture is protected from light. Unbound Tc-99m is eliminated by gel filtration chromatography on Sephadex G-25 (Pharmacia, Piscataway, NJ) prewashed with 0.1% HAS in PBS. The labeled product is eluted with 0.1% HAS in PBS. The final product is examined for any free Tc-99m by instant ITLC on silica gel developed with 50% acetone in 0.9% NaCl.

2.6.3. Administration of Tc-99m-Annexin-V

An identical cohort of tumor-bearing mice treated with four dose levels of Re-188-PNA-peptide, as described above, are injected intravenously with 500–700 μCi of Tc-99m-HYNIC-annexin-V, imaged under a γ camera as described previously, and sacrificed at 4 and 24 h postinjection. Tissues are harvested and weighed, and the Tc-99m associated with them is determined using a γ counter as described above. Five animals per group are used. Tumor-bearing mice not treated with Re-188-PNA-peptide serve as controls. These studies are performed only after Re-188 ($t_{1/2}$ = 17 h) has completely decayed for 8 d (>11 half-lives) so that no cross-contribution of Re-188 radioactivity can occur. Results are evaluated using Student's *t*-test. Apoptotic cells are considered to be present in those tumors that have three times ($p < 0.05$) more Tc-99m/g than in tumors harvested from untreated mice.

3. Results

3.1. Antisense Probes of K-RAS Oncogene mRNA

Several laboratories have probed a variety of K-*RAS* mRNA sites for antisense efficacy *(19,39–42,175)*. We prepared and tested the phosphorothioate sequences given in **Table 1**.

3.1.1. Inhibition of Proliferation of Pancreatic Cell Lines by K-RAS Antisense Sequences

Figure 4 shows the effects of proliferation with the battery of oligonucleotides on AsPC1 cells. The KRASATGA antisense phosphorothioate was not effective in inhibiting proliferation in this cell line. An important observation in this set of experiments is that one particular sequence, KRASSA, was identified that inhibits pancreatic tumor cell proliferation by cells that harbor a mutated K-*RAS* oncogene in AsPC1 cells as well as in CFPAC-1 and MiaPaCa-2 cells (not shown) *(19)* with apparent sequence specificity.

3.1.2. Inhibition of K-Ras Protein Expression in AsPC1 Human Pancreatic Cancer Cells by K-RAS Antisense Sequence KRASSA

AsPC1 cells were treated for 72 h with KRASSS and KRASSA antisense oligonucleotides over a range of concentrations. Western blot analysis (**Fig. 5**) demonstrated reduction of both Ras p21 and the active farnesylated Ras p21 in response to the KRASSA antisense sequence, with dose response and sequence specificity. This validates an oligonucleotide approach to detecting K-*RAS* mRNA.

3.2.Synthesis of Peptide-PNA-Peptide Chimeras

To optimize amino acid coupling on an automated DNA synthesizer, we practiced synthesis of an amino acid–linker-PNA trimer, *N*-Gly-Aba-Ado. Double coupling, recommended by the manufacturer, was not as complete as a single long (1400 s) cycle (**Fig. 6**). The long cycle product was 98.72% of the A_{260}, and the purified peak displayed the predicted mass, 435.5 Daltons (**Fig. 7**). The long cycle was used for all of the amino acid couplings described below *(176)*.

Table 1
Antisense and Control Sequences Targeted Against K-RAS mRNA

K-*RAS* 5' upstream

KRASSS	sense	5'-dTCAGCGGGGCGGCGT-3'
KRASSA	antisense	5'-dAGTCGCCCCGCCGCA-3'
KRASSX	(four mismatches)	5'-dAGTCGAAAAGCCGCA-3'

K-*RAS* AUG site

KRASATGA	antisense	5'-dTTTATATTCAGTCAT-3'
KRASATGX	scrambled	5'-dTATTATGCTCTATTA-3'

K-*RAS* 12th codon site

KRASG12	antisense (G12)	5'-dTACGCCACCAGCTCC-3'
KRASV12	antisense (V12)	5'-dTACGCCAACAGCTCC-3'
KRASG12X	scrambled	5'-dCCACGACCACTCCGT-3'

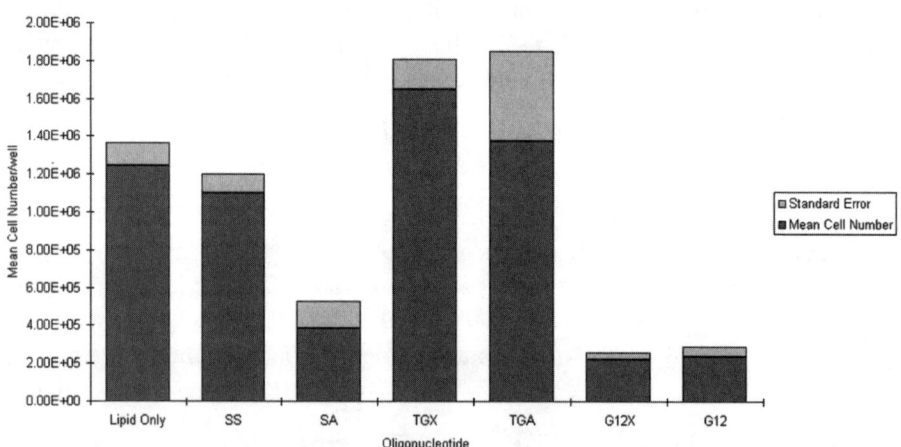

Fig. 4. Inhibition of AsPC1 cell proliferation by K-*RAS* antisense phosphorothioate sequences. Cells were lipofected with 1.5 μM DNA for 6 h, then incubated in fresh medium for 6 d. Sequence abbreviations are those shown in boldface in **Table 1**.

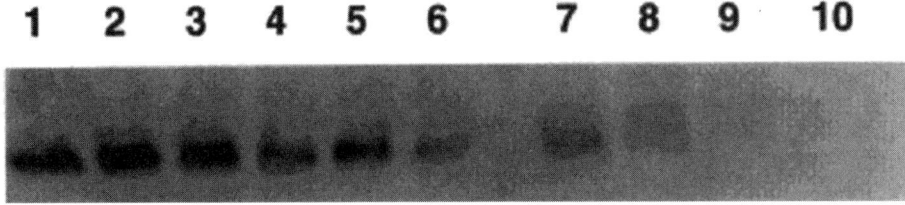

Fig. 5. Western blot of K-Ras antigens from AsPC1 cells treated with KRASSS or KRASSA for 72 h. Lane 1: untreated cells; lane 2: lipofection alone; lanes 3–6: 0.5 μ*M*, 1.0 μ*M*, 1.5 μ*M*, and 2.0 μ*M* KRASSS; lanes 7–10: 0.5 μ*M*, 1.0 μ*M*, 1.5 μ*M*, and 2.0 μ*M* KRASSA.

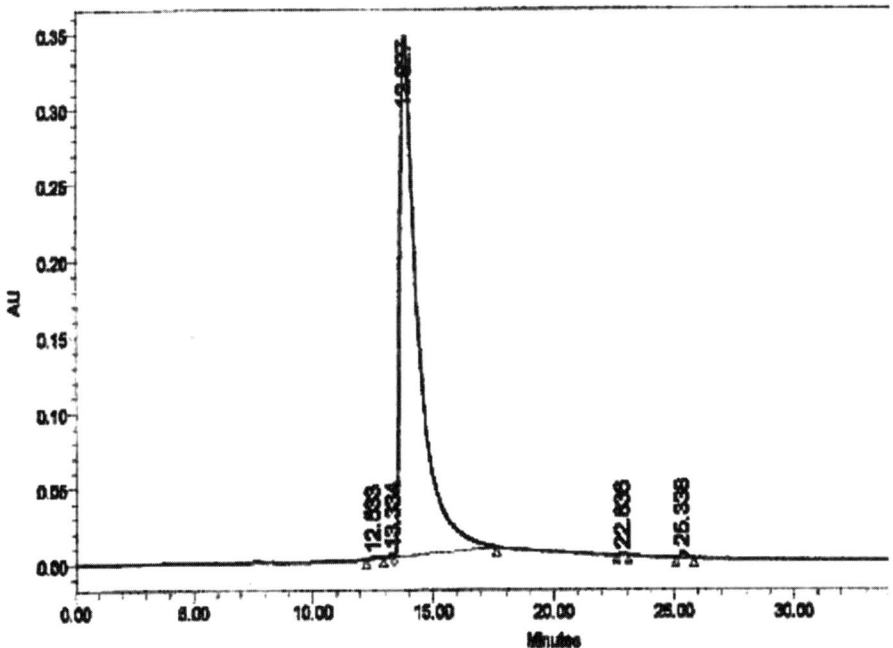

Fig. 6. Preparative C_{18} HPLC of *N*-Gly-Aba-Ado conjugate on 10 × 250 mm Alltima column eluted over 25 min from 5 to 69% CH_3CN in 0.1% aqueous CF_3CO_2H at 1 mL/ min, and 50°C monitored at 260 nm.

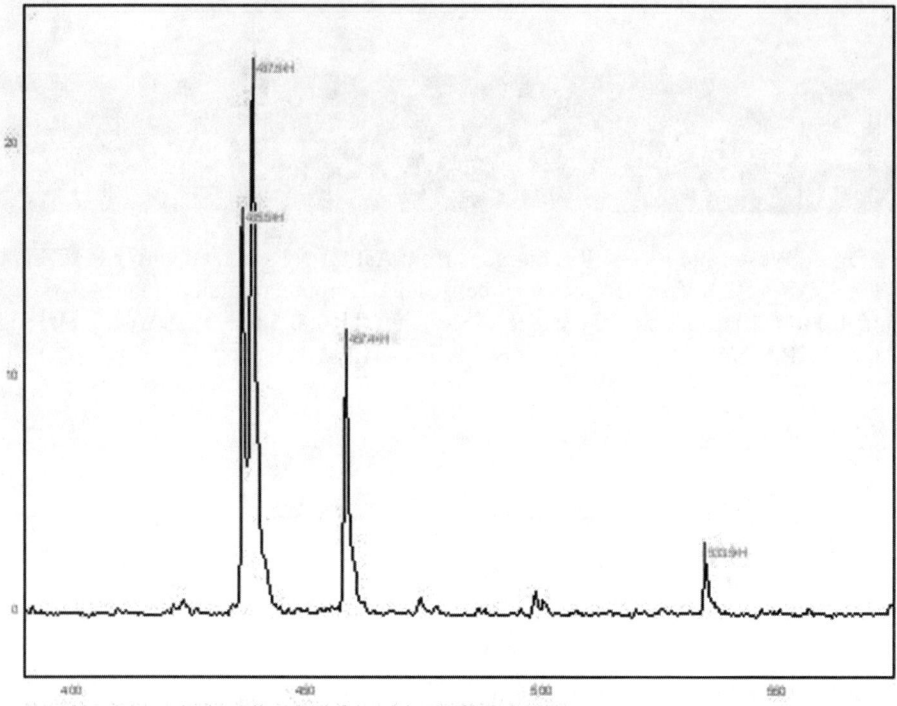

Fig. 7. MALDI-TOF mass spectrum of *N*-Gly-Aba-Ado chimera. Experimental mass: 435.5 Daltons; calculated mass: 434.4 Daltons.

For the challenge of preparing a labeled probe, we first synthesized *N*-Gly-D-Ala-Gly-Gly-Aba-GCATCGTCGCGG (WT3613), a chelator-PNA specific for *MYC* mRNA on a 2 μmol Fmoc-XAL-PEG-PS column in a single automated run. The overall yield following preparative RP-HPLC was 35.4%. The purified peak (**Fig. 8**) displayed a MALDI-TOF mass spectrum with a main peak at 3613.8 Daltons, in agreement with the predicted mass, 3614.5 Daltons (**Fig. 9**) *(176)*.

The IGF-1 analog D-(Cys-Ser-Lys-Cys), which enables PNA uptake by cells bearing receptors for IGF-1, includes a disulfide bridge between terminating cysteines. The cyclization and purification of such peptides typically limits overall yield of an oligonucleotide-peptide conjugate. Following conjugation and purification, a peptide with a pair of cysteines to be cyclized is usually assembled on solid phase, deprotected, cleaved, and chromatographically purified. The reduced cysteine thiols are then oxidized nonspecifically by air in a basic solution (pH 8.5) at high dilution to minimize dimerization and oligomerization *(142)*.

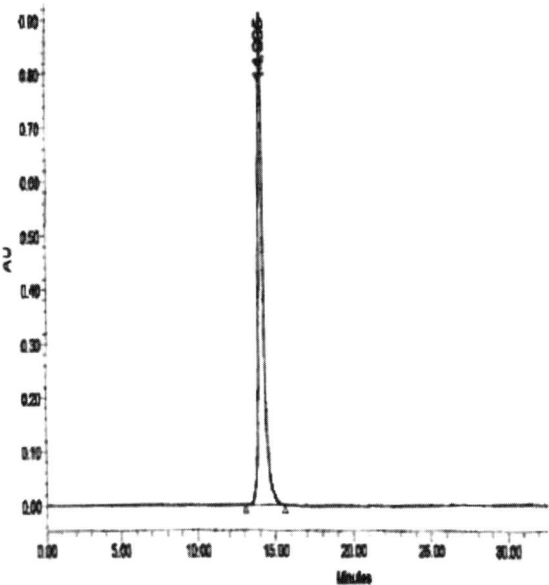

Fig. 8. Analytical C$_{18}$ HPLC of *N*-Gly-D-Ala-Gly-Gly-Aba-GCATCGTCGCGG PNA chimera as in **Fig. 6**.

Unfortunately, PNAs are poorly soluble when the pH is higher than 4, which also lowers yield. Deprotection/oxidation of the cysteines while the peptide chain is still on the support provides an alternative to the dilute aqueous route. Solid-phase cyclization is based on the assumption that for polymer-bound reactants intramolecular processes are preferred owing to pseudodilution *(177)*. Under the latter conditions, deprotection and cyclization of cysteine residues are achieved simultaneously. Thus, the solubility problem of PNAs can be avoided, and the preparative steps can be simplified, improving overall yield. We therefore tested different conditions of cyclization on solid phase in order to find a simple and efficient cyclization method for peptide-PNA-peptide chimeras that would obviate the need for postsynthetic cyclization. We protected the two cysteine side chains with the *S*-acetamidomethyl (Acm) instead of *S*-triphcnylmethyl (Trt) because Acm gives a higher yield of cyclization in (Me)$_2$NCHO *(178)*.

We optimized cyclization methods in a PNA dimer/peptide nonamer control sequence, *N*-Ado-Ado-Gly-Gly-Gly-Gly-D(Cys-Asn-Gly-Arg-Cys). Solution and solid-phase (**Scheme 1**) methods were compared. The maximum yield, 75%, was obtained on solid phase (**Fig. 10**). MS of the purified peak agreed with the predicted mass of the cyclized peptide, 1328.2 Daltons, which was

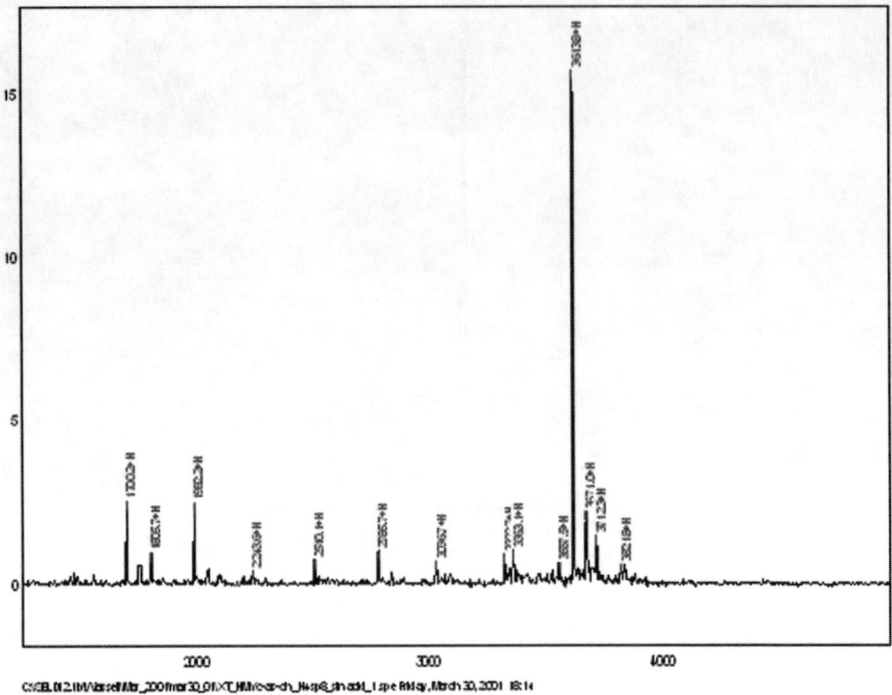

Fig. 9. MALDI-TOF mass spectrum of *N*-Gly-D-Ala-Gly-Gly-Aba-GCATCGT CGCGG PNA chimera. Experimental mass: 3613.8 Daltons; calculated mass: 3614.5 Daltons.

144 Daltons less than the linear, Acm-protected peptide (**Fig. 11**) *(176)*. Solid-phase cyclization *(178)* followed by PNA extension, cleavage, deprotection, and purification improves yield and saves significant effort, compared with postsynthetic aqueous cyclization and a second chromatographic purification *(142)*.

The final synthetic milestone was the synthesis and cyclization of *N*-Gly-D-Ala-Gly-Gly-Aba-GCATCGTCGCGG-(Gly)$_4$-D(Cys-Ser-Lys-Cys) (WT4261), a chelator-PNA-peptide specific for *MYC* mRNA and the IGF-1R. We used Fmoc-PAL-PEG-PS (0.2–0.3 mmol/g) for assembling the targeting peptides because it is compatible with PNA synthesis. Our experimental results suggested that if the loading of resin is higher than 0.3 mmol/g, cyclization on resin would lead to complicated products due to crossreaction between peptide chains. Three different reagents have been utilized for deprotection/oxidation of resin-bound peptide: (1) Tl(tfa)$_3$ *(179)*, (2) I$_2$ under acidic conditions *(180)*; (3) I$_2$ in (Me)$_2$NCHO *E (181)*. Tl(tfa)$_3$ is expensive, highly toxic, and requires

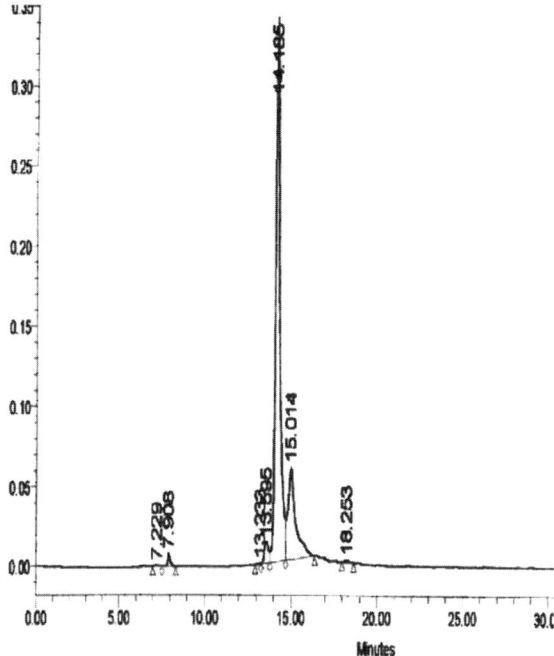

Fig. 10. Preparative C_{18} HPLC of cyclized *N*-Ado-Ado-(Gly)$_4$-D(Cys-Asn-Gly-Arg-Cys) conjugate as in **Fig. 6**.

anhydrous conditions. In addition, it is possible that Tl(tfa)$_3$ would react with amino or hydroxyl groups on nucleobases during the cyclization of cysteine residues. CF$_3$CO$_2$H or HOAc could not be used as the solvent owing to the acid lability of the PAL anchor. Hence, we tested two deprotection/oxidation conditions with I$_2$/(Me)$_2$NCHO: cyclization of cysteine residues (1) prior to addition of the Gly-D-Ala-Gly-Gly-Aba-PNA residues, and (2) after assembly of the complete peptide-PNA-peptide sequence.

For route 1, (Gly)$_4$-D(Cys-Ser-Lys-Cys)-resin was suspended in (Me)$_2$NCHO. Oxidation was carried out with I$_2$ (10 eq) for 4 h at room temperature (**Scheme 2**). The resin was washed with (Me)$_2$NCHO to remove excess iodine and dried under vacuum. Two micromole equivalents of dry resin were placed in an empty column for PNA extension and chelator coupling. Cleaved and deprotected sequence **4** was purified by preparative RP-HPLC at 50°C as above and gave an overall yield of 19.5%. Over 26 coupling cycles and one deprotection/oxidation step, the average yield per cycle was therefore 94%. MALDI-TOF-MS indicated a mass of 4261.8 Daltons, consistent with the predicted 4261.0 Daltons. The result indicated that the intramolecular disulfide

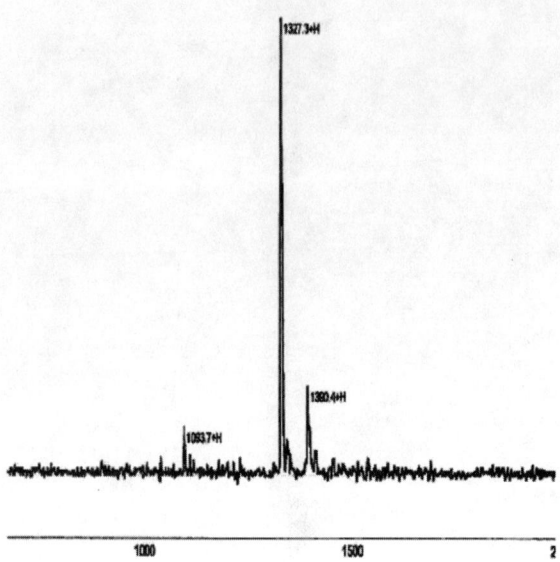

Fig. 11. MALDI-TOF mass spectrum of *N*-Ado-Ado-(Gly)₄-ᴅ(Cys-Asn-Gly-Arg-Cys) conjugate. Experimental mass: 1327.3 Daltons; calculated mass: 1328.2 Daltons.

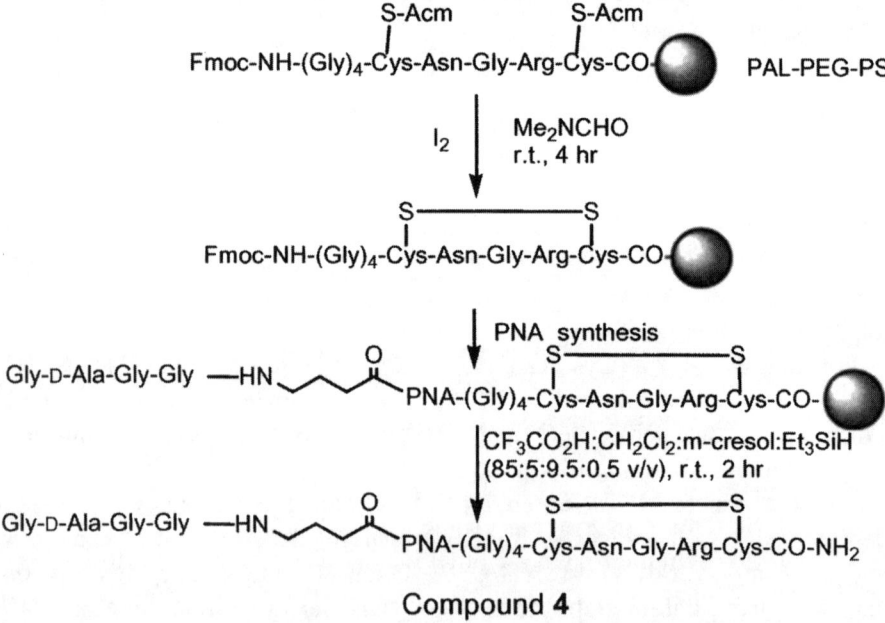

Scheme 2. Cysteine cyclization on solid phase before PNA expression.

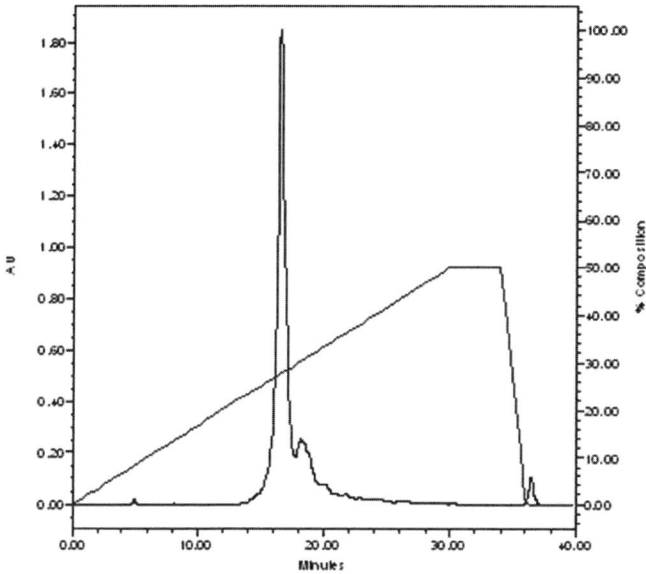

Fig. 12. Preparative C_{18} HPLC of cyclized **4**, *N*-Gly-D-Ala-Gly-Gly-Aba-GCATCG TCGCGG-(Gly)$_4$-D(CysAsnGlyArgCys) chimera **5** as in **Fig. 6**.

bond formed on the solid phase is stable under the conditions of PNA synthesis, and that the cyclization of peptides prior to extending PNA is feasible *(176)*.

For route 2, the question arises, will the additional Gly-D-Ala-Gly-Gly-Aba-PNA chain interfere with the formation of intramolecular disulfide bonds while the chimera remains bound to the support? To address this question, we synthesized five 12mer PNA sequences with 5' and 3' peptides by programmed continuous solid-phase synthesis, with cyclization after assembly but before cleavage (**Scheme 3**). Sequences **5, 6,** and **7** (**Table 2**) are complementary to *MYC* mRNA (antisense, antisense with control peptide, and mismatched sequences, respectively), sequence **8** targets *CCND1* mRNA, and sequence **9** targets *HER2* mRNA. After cleavage, the chimeras were purified by RP-HPLC (**Fig. 12**), and analyzed by MALDI-TOF mass spectroscopy (**Fig. 13**). The yields (**Table 2**) were higher than when cyclizing before PNA extension. Perhaps the coupling of the first PNA monomer to cyclic (Gly)$_4$-D(Cys-Ser-Lys-Cys) (**Scheme 2**) was not as efficient as coupling to linear (Gly)$_4$-D(Cys-Ser-Lys-Cys) (**Scheme 3**). On the other hand, it is possible that I$_2$ treatment of (Gly)$_4$-d(Cys-Ser-Lys-Cys)-bound resin created some bypro-ducts that reduced the efficiency of the subsequent coupling reaction. In any case, the

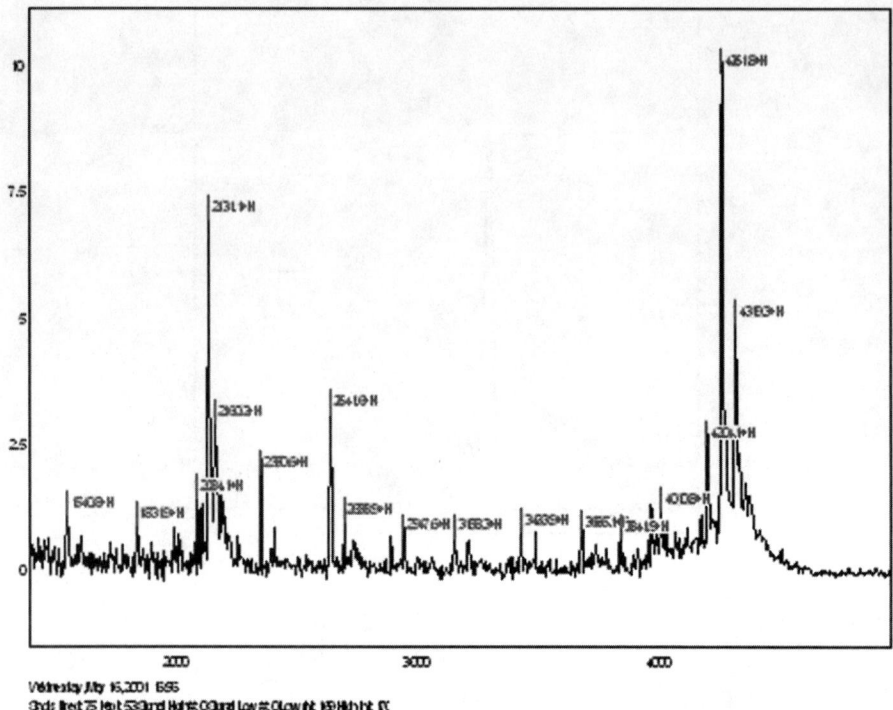

Wednesday July 16,2001 696
3rd: Iret 75 Kpt 530grd Hght #: 00grd Low #: 0 Low ht: 169 High ht: 111

Fig. 13. MALDI-TOF mass spectrum of *N*-Gly-D-Ala-Gly-Gly-Aba-GCATCGTC GCGG-(Gly)$_4$-D(Cys-Ser-Lys-Cys) chimera **5**. Experimental mass: 4261.8 Daltons; calculated mass: 4261.0 Daltons.

results imply that the full sequence of peptide-PNA-peptide anchored to its support does not impair the formation of intramolecular disulfide bonds *(176)*.

To eliminate potential ambiguity in the site of Tc-99m labeling, we then replaced the Gly$_4$ spacer, whose structure is similar to the chelator peptide Gly-D-Ala-Gly-Gly, with an (EtO)$_2$Ac spacer. Simultaneously, we solved the problem of Fmoc-Aba monomer decomposing to Fmoc and insoluble Aba by using the entire chelator peptide as a single synthon, instead of coupling the five amino acids sequentially. We expected that the two measures would improve the overall yields of peptide-PNA-peptide chimeras because the number of coupling cycles would be reduced from 25 to 17. Sequences **10–12** were prepared with cyclization on solid phase after PNA extension (**Table 3**), but MALDI-TOF mass spectra displayed peaks of [M+14]+ and [M+106]+ besides [M+1]+. The intensity ratio [M+1]+/[M+14]+ was about 3:2, whereas the intensity of [M+106]+ peak was much lower. There were also many unidentified peaks.

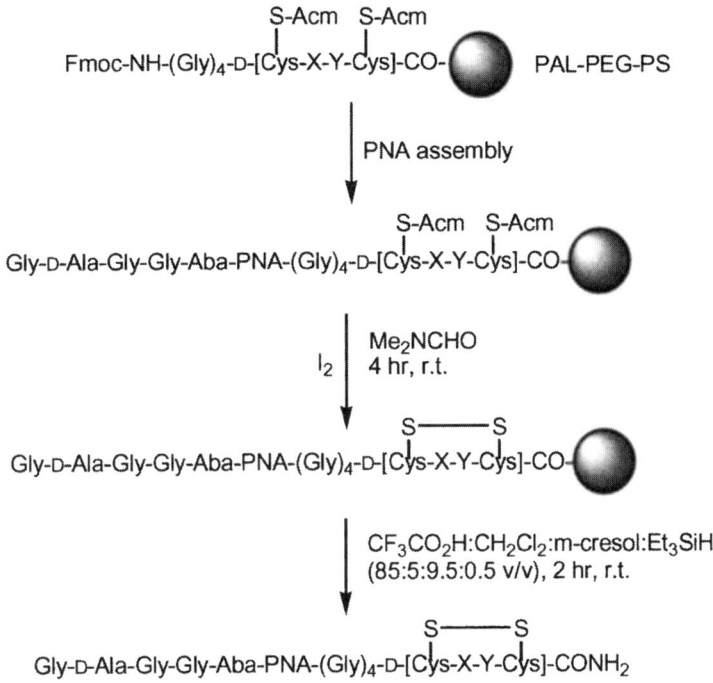

Scheme 3. Cysteine cyclization on solid phase after PNA expression.

Table 2
Yields From Cyclization on Resin After Assembly

Sequence	X	Y	PNA (5'→3')	Yield (%)
5	Ser	Lys	GCATCGTCGCGG	32.9
6	Ala	Ala	GCATCGTCGCGG	37.6
7	Ser	Lys	GCATGTCTGCGG	21.6
8	Ser	Lys	CTGGTGTTCCAT	39.1
9	Ser	Lys	CATGGTGCTCAC	57.6

Table 3
Yields with (EtO)$_2$Ac Spacer

Sequence	X	Y	R	PNA (5'→ 3')	Yield (%)
10	Ser	Lys	H	GCATCGTCGCGG	—
11	Ala	Ala	H	GCATCGTCGCGG	—
12	Ser	Lys	H	GCATGTCTGCGG	—
13	Ser	Lys	Ac	GCATCGTCGCGG	27.2
14	Ala	Ala	Ac	GCATCGTCGCGG	26.8
15	Ser	Lys	Ac	CTGGTGTTCCAT	30.6
16	Ala	Ala	Ac	CTGGTGTTCCAT	34.0

R—N(H)—Gly-D-Ala-Gly-Gly-Aba —**PNA**—HN∼∼O∼∼O∼C(O)—D-[Cys-X-Y-Cys]-CONH$_2$ (with S—S bridge)

When we tried to prepare fluorescein-labeled **10**, Fmoc-Lys(ε-fluorescein)-OH could not be coupled to the peptide-PNA-peptide. Instead, acetylated product **13** was obtained, which gave a sharp HPLC peak with a MALDI-TOF [M+1]$^+$ mass of 4220 Da. Thus, it would appear that acetylation or a similar capping reaction is necessary prior to Cys cyclization with these particular chimeras. Based on related studies on the mechanism of deprotection/oxidation of cysteine residues *(180,181)*, a possible mechanism of formation of the two peaks [M+14]$^+$, 4233 Da, and [M+106]$^+$, 4325 Da, is proposed in **Scheme 4**. Perhaps the flexibility and hydrophilicity of the (EtO)$_2$Ac spacer, relative to the Gly$_4$ spacer, favored the postulated intramolecular S$_N$2 reaction through route A and contributed to the formation of the byproducts (**Scheme 4**). Compounds **13–16** were assembled, cyclized, acetylated by capping, cleaved/deprotected, and purified by RP-HPLC as above. In all cases, MALDI-TOF mass spectra indicated that acetylated full-length chimeras were the major products, and no methylated or *m*-cresolyl products were detected *(176)*.

Melting temperature analysis was performed to determine the influence of the peptide moiety on the hybridization efficiency of the antisense PNA for its complementary RNA strand. Measurements were carried out in triplicate with equimolar 2.5 µM PNAs and complementary 12mer RNAs in 10 mM Na$_2$HPO$_4$, 1.0 M NaCl, 0.5 mM EDTA, pH 7.0. The T_m of the parent unmodified PNA GCATCGTCGCGG (**17**) duplex with RNA was found to be 79.0 ± 1.2°C; mismatched chimera **7** showed no hybridization with its complementary RNA

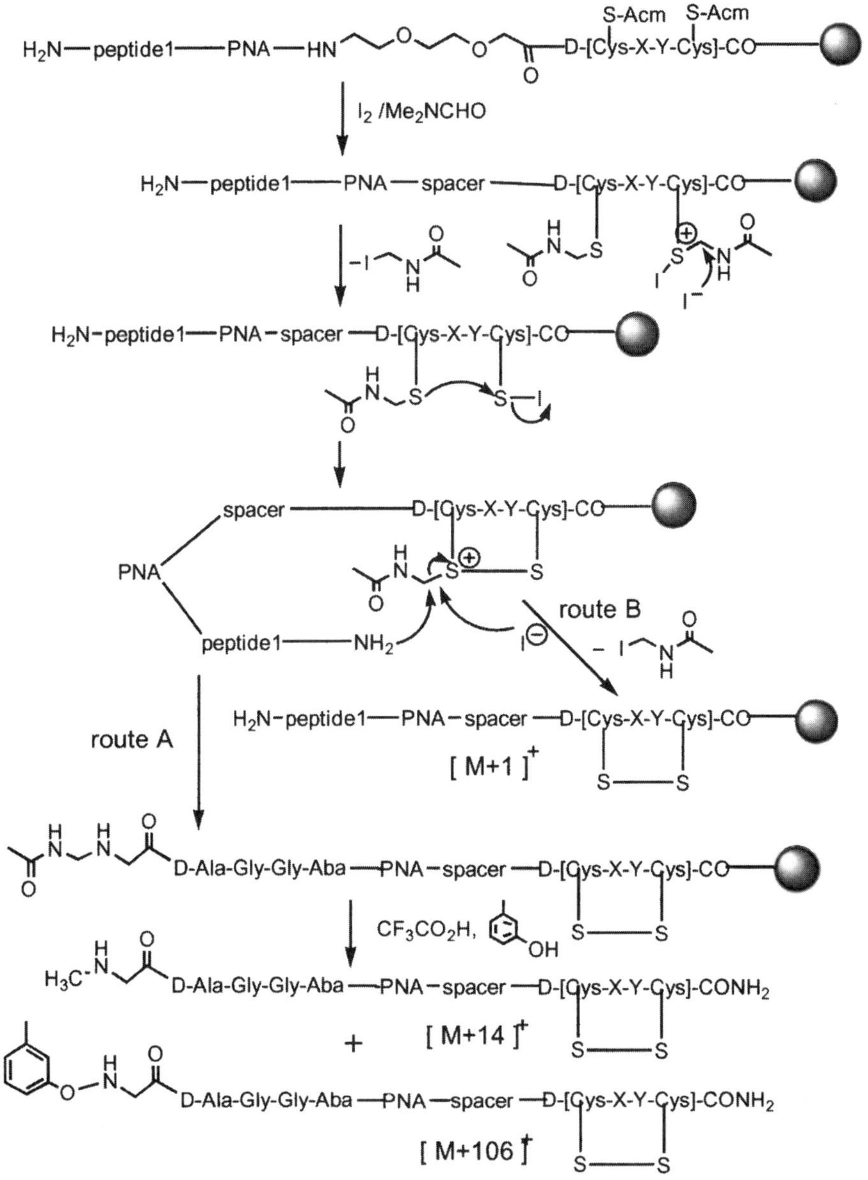

Scheme 4. Possible mechanism of byproduct formation.

Table 4
K-RAS, CCND1, HER2, MYC, and p53 Antisense
and Mismatch PNA Sequences

	Sequence	
K-*RAS* antisense	5'-GCCAACAGCTCC	Codons 10–13
K-*RAS* mismatch	5'-GCCTTGTGCTCC	Four central mismatches
CCND1 antisense	5'-CTGGTGTTCCAT	Codons 1–4
CCND1 mismatch	5'-CTGGACAACCAT	Four central mismatches
HER2 antisense	5'-CATGGTGCTCAC	Codons –3 to 1
HER2 mismatch	5'-CATGCACTTCAC	Four central mismatches
MYC antisense	5'-GCATCGTCGCGG	Codons –3 to 1
MYC mismatch	5'-GCATGTCTGCGG	Four central mismatches
p53 antisense	5'-CCCCCTGGCTCC	Exon 10
p53 mismatch	5'-CCCCTACCCTCC	Four central mismatches

strand; and the T_m values of chimeras **2** (80.5 ± 0.5°C), **5** (82.4 ± 0.3°C), and **6** (81.3 ± 0.2°C) were undiminished. These results indicate that the pair of peptide moieties at the N and C termini of the 12mer PNA do not interfere with hybridization efficiency *(176)*.

A full set of antisense and mismatch chimeras are being synthesized (**Table 4**) to probe the other oncogene targets described above.

3.3. Labeling of Peptide-PNA-Peptide Chimeras

3.3.1. Tc-99m Labeling of Peptide-PNA-Peptide Chimeras

Purified peptide-PNA chimeras were labeled with Tc-99m as described above. Unchelated free Tc-99m, Rf 1.0, was 1.5%, determined by instant TLC on silica gel (ITLC-SG; Gelman) developed with methylethyl ketone. Colloid formation, Rf 0.0, was 1.0%, determined using instant TLC also on silica gel (ITLC-SG; Gelman) developed with pyridine/HOAc/H$_2$O (3:5:1.5). These preparations were stable at 22°C for more than 4 h, as determined by HPLC (**Fig. 14**), and were stable to challenges with 100 molar excesses of DTPA, HAS, or cysteine. Similar results were obtained for the *MYC* mismatch PNA (WT3629). Specific activities of the preparations were 7–10 Ci/µmol *E(92)*.

3.3.2. Cu-64 Labeling of Peptide-PNA-Peptide Chimeras

Recently, in Thakur Laboratory *(106)*, we have labeled vasoactive intestinal peptide (VIP) analog TP3982 with Cu-64 (**Fig. 15**). Free Cu-64, 3.4%, eluted at 4.3 min, and the Cu-64-TP3982, 96.6%, at 8.0 min. VIP, a 28-amino-acid endogenous hormone, is a muscle relaxant that has a high affinity for VPAC1

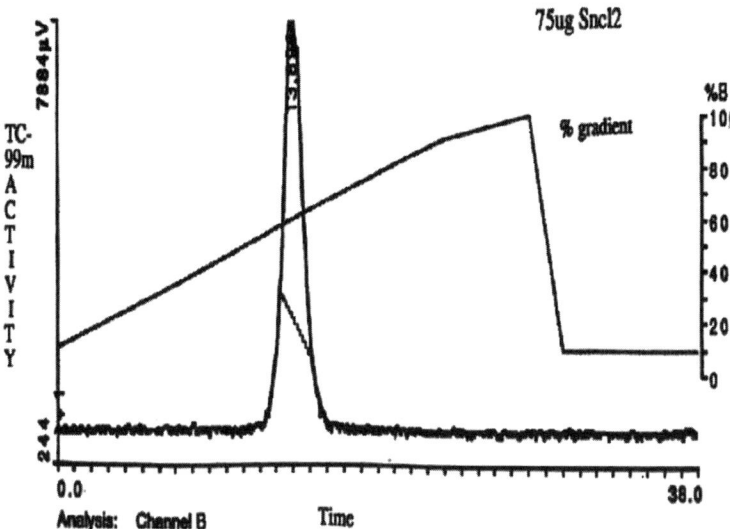

Fig. 14. Analytical HPLC of Tc-99m-Gly-D-Ala-Gly-Gly-Aba-GCATCGTCGCGG PNA (WT3613) on a 4.5 × 250 mm C_{18} column eluted with a 28-min gradient from 10 to 100% CH_3CN in aqueous 0.1% CF_3CO_2H at 1 mL/min and 25°C. The right y-axis shows % CH_3CN and the left y-axis shows radiometric γ emission.

and VPAC2, oncogene receptors that are expressed in high density on many types of human cancer cells. The biological activity of TP3982 was not compromised by the addition of the chelation sequence and Cu-64. The Cu-64 labeling yields were >95%, and the Cu-64-TP3982 presented as a single radioHPLC peak.

Both peptides and peptide-PNA-peptide chimeras are labeled according to **Scheme 5 (Fig. 15)**.

3.3.3. Gd(III) Labeling of Peptide-PNA-Peptide Chimeras

Gd(III), unlabeled and Gd-159, are chelated to peptides and peptide-PNA-peptide chimeras as in **Scheme 6**. An example of Gd-159 labeling of a PAMAM model is shown in **Fig. 16**. Quantitation implied complete labeling of the predicted 31 free DOTA chelators.

3.4. Tissue Distribution and Imaging of Peptide-PNA-Peptide Chimeras

3.4.1. Tc-99m-Chimera Imaging and Distribution

We administered the PNA-free control Tc-99m-Gly-D-Ala-Gly-Gly-Aba-Gly_4-D-(Cys-Ser-Lys-Cys) (WT990), the chelator plus IGF-1 analog, to human BT474 breast cancer xenografts to determine whether the probe bound to the

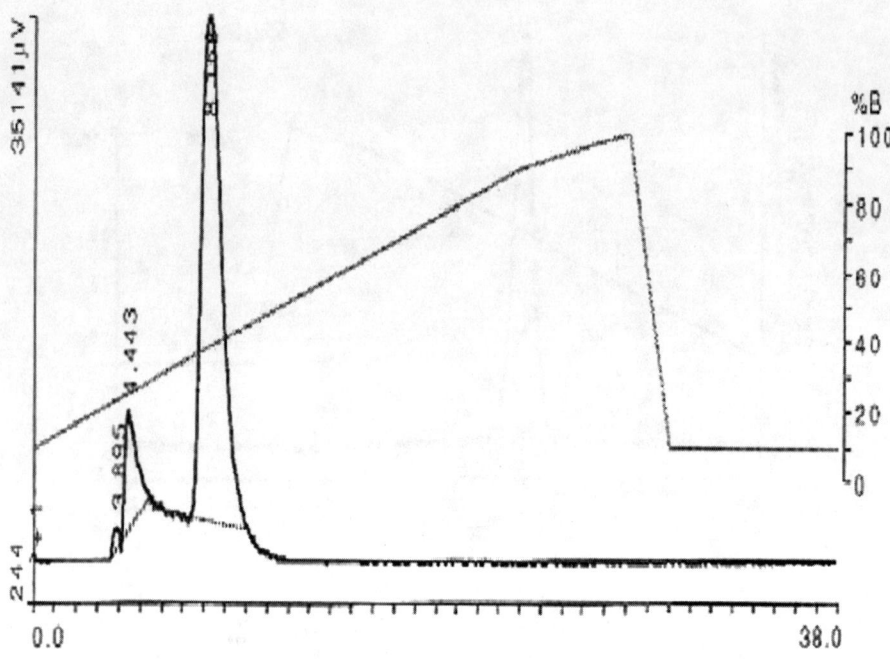

Fig. 15. Analytical HPLC of Cu-64-TP3982 as in **Fig. 14**.

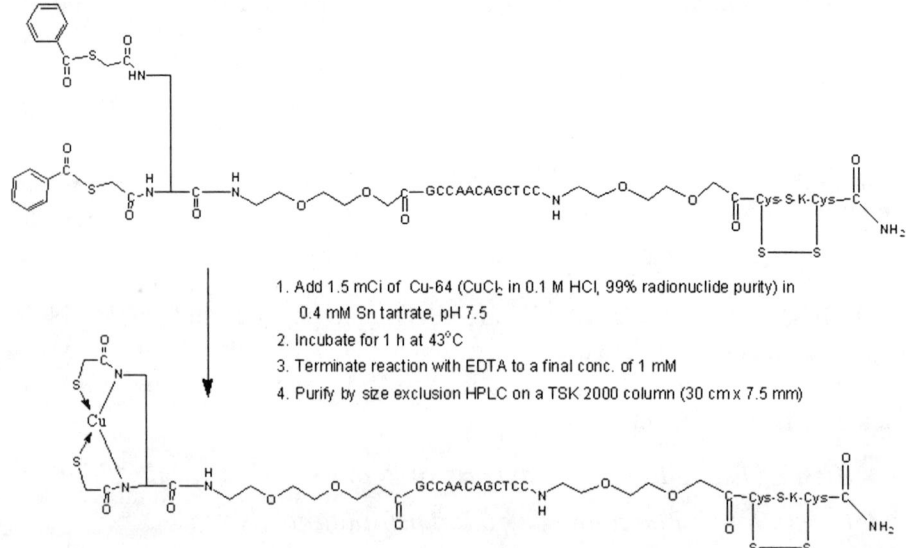

Scheme 5. Labeling of chimeras with Cu-64.

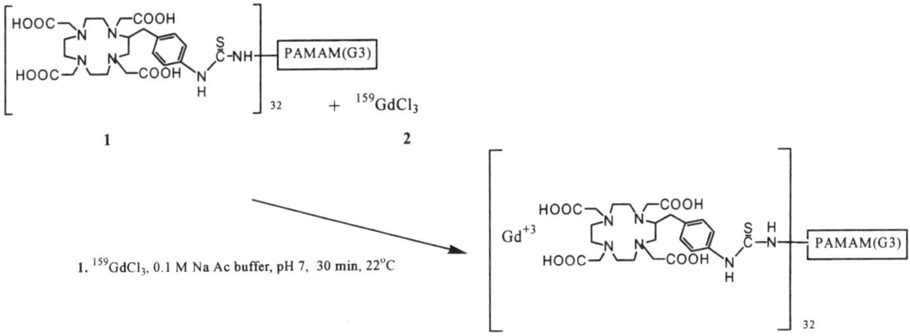

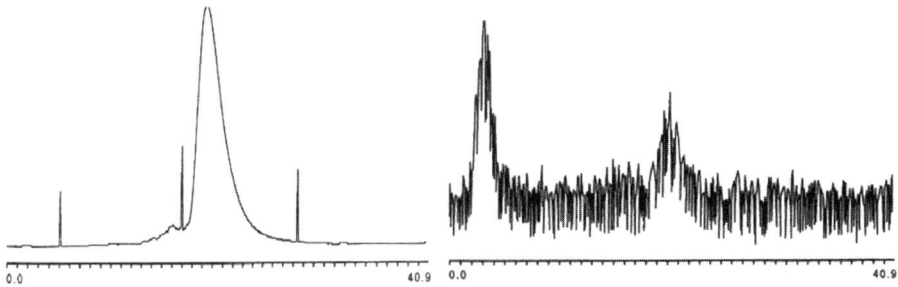

Scheme 6. Radiolabeling of $(DOTA)_{32}$-PAMAM(3G) complex with Gd-159.

Fig. 16. Gd-159 of 1.4 nmol of $(DOTA)_{32}$-PAMAM(3G) was labeled as in **Scheme 6**, then analyzed by analytical HPLC as in **Fig. 14**. Left, absorption at 250 nm; right, radioactivity.

tumors despite low expression of IGF-1R. At 4 and 24 h postinjection, mice were lightly anesthetized, imaged using a Starcam (GE) γ camera equipped with a parallel hole collimator. For images 300,000 counts were recorded. Scintigraphic imaging of the mice did not display significant concentration of label in the tumors at 4 h (**Fig. 17**) nor at 24 h postinjection. These results correlate with the low expression of IGF-1R on BT474 cells *(92)*.

We then administered the *MYC* PNA probe Tc-99m Gly-D-Ala-Gly-Gly-Aba-GCATCGTCGCGG (WT3613), without the IGF-1 analog, to MCF7:IGF-1R xenografts to determine whether the probe without IGF-1 analog would localize more to tumors with high expression of *MYC* mRNA than would the four-mismatch *MYC* PNA probe Tc-99m-Gly-D-Ala-Gly-Gly-Aba-GCTTCCT CCCGC (WT3629). Scintigraphic imaging of the mice did not display a significant concentration of label in the tumors at 4 h (**Fig. 18**) nor at 24 h. These

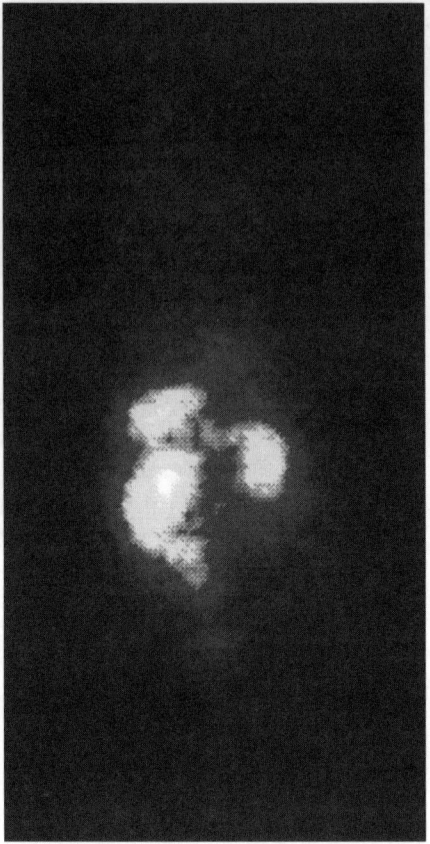

Fig. 17. Scintigraphic imaging of nude mice bearing BT474 tumors (IGF-1R–) 4 h after administration of PNA-free Tc-99m-Gly-D-Ala-Gly-Gly-Aba-Gly$_4$-D-(Cys-Ser-Lys-Cys) (WT990).

results correlate with the lack of the D-(Cys-Ser-Lys-Cys) IGF-1 analog on the Tc-99m-sequences *(92)*.

Tissue distribution data obtained at 4 h postinjection of the *MYC* antisense PNA (WT3613) and the corresponding mismatch sequence (WT3629) in groups of five mice each bearing human MCF7:IGF-1R breast tumor xenografts are shown in **Table 5** and **Fig. 19**. Both agents cleared through urinary excretion, and renal uptake was highest among all tissues. The uptake of radioactivity in the remaining normal tissues was relatively small, but the tumor uptake with antisense PNA at 4 h postinjection was twofold greater ($p < 0.01$) than the corresponding mismatch PNA. This early control was not predicted to show signifi-

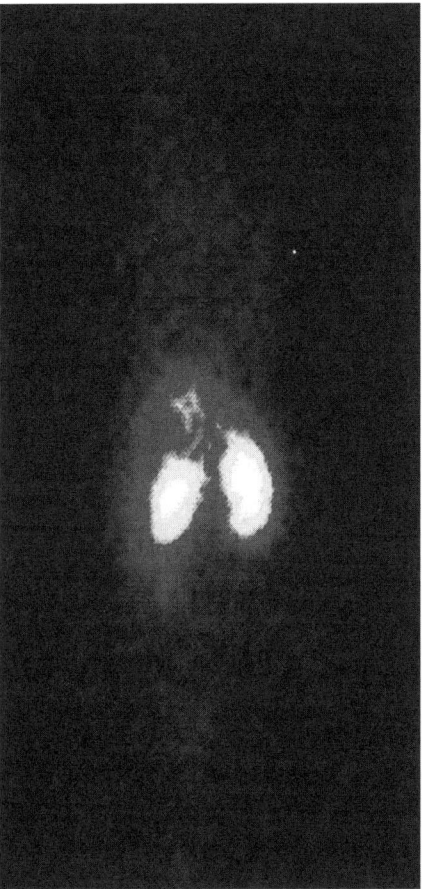

Fig. 18. Scintigraphic imaging of nude mice bearing MCF7:IGF-1R tumors (IGF-1R+, ER+) 4 h after administration of Tc-99m-GdAGGAba-GCATCGTCGCGG (WT3613).

cant differences due to the absence of the IGF-1 analog. Nevertheless, the apparent two-fold difference may reflect the low level of PNA uptake that occurs without an assisting ligand *(92)*.

Urine was collected up to 4 h postinjection from five mice injected with *MYC* antisense Tc-99m-Gly-D-Ala-Gly-Gly-Aba-GCATCGTCGCGG (WT3613) for assessment of probe stability. Analytical HPLC of radioactivity in the combined deproteinized, lyophilized urine revealed a void volume peak of free Tc-99m with 17% of the radioactivity, and a probe peak with 83% of the radioactivity (**Fig. 20**). Breakdown fragments were not detected, consistent with the model that the PNA and the chelator peptide are resistant to proteases and nucleases.

Table 5
Tissue Distribution (% I.D./g + SD) of MYC Antisense Tc-99m-
PNA (WT3613) and Mismatch (WT3629) at 4 h Postinjection
in Nude Mice (n = 5) Bearing Human MCF7:IGF-1R Breast
Tumor Xenografts

Tissue	Antisense	Mismatch	p Value
Muscle	0.28 ± 0.02	0.12 ± 0.04	0.01
Intestine	0.29 ± 0.09	0.20 ± 0.063	0.08
Heart	0.29 ± 0.05	0.11 ± 0.01	0.01
Lungs	0.84 ± 0.14	0.39 ± 0.08	0.01
Blood	0.73 ± 0.52	0.36 ± 0.15	0.12
Spleen	1.36 ± 0.46	1.35 ± 0.75	0.99
Kidneys	8.25 ± 0.83	5.27 ± 1.98	0.12
Liver	3.57 ± 0.08	1.28 ± 0.23	0.01
Tumor	0.49 ± 0.10	0.22 ± 0.04	0.01
Tumor/muscle ratio	1.81 ± 0.51	2.13 ± 0.54	0.35
Tumor/blood ratio	1.74 ± 2.87	0.84 ± 0.91	0.45

At 24 h postinjection, radioactivity for both agents declined in all tissues, including the tumors (**Table 6**). The 24-h Tc-99m distributions were indistinguishable, with nonspecifically bound probe much lower in liver yet still significant in the kidneys *(92)*. This is consistent with most Tc-99m agents, whether antibodies, peptides, or antisense.

3.4.2. Cu-64-TP3982 Distribution

Cu-64-TP3982 was administered to immunocompromised nude mice bearing exprimental human T47D breast tumors *(106)*. PET imaging at 24 h postinjection revealed a strong tumor signal (**Fig. 21**). The Cu-64-TP3982 was stable in vitro and in vivo and displayed significantly greater uptake than Tc-99m-TP3982 in exprimental human breast tumors grown in immunocompromised nude mice. We anticipate comparable results from Cu-64-PNA-peptide probes in pancreatic tumors.

3.4.3. Re-186 Chimera Distribution

With respect to Re-188 therapeutic conjugates, we previously prepared a nuclear histone-specific Re-186-TNT-F(ab')$_2$ MAb fragment *(116)* and observed localization to embryonal carcinoma xenografts in mice (**Fig. 22**). Comparable results are expected for Re-188 chimeras directed against pancreatic cancer oncogene mRNAs.

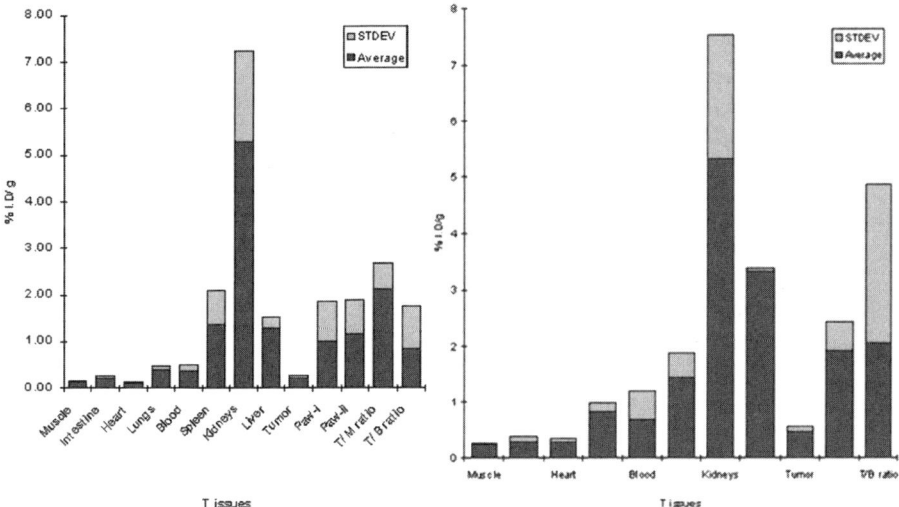

Fig. 19. Tissue distributions of Tc-99m-c-*MYC* PNA antisense (WT3613) or mismatch (WT3629) in nude mice bearing MCF7:IGF-1R (ER+, IGF-1R+) xenografts 4 h after administration. Left, antisense; right, mismatch.

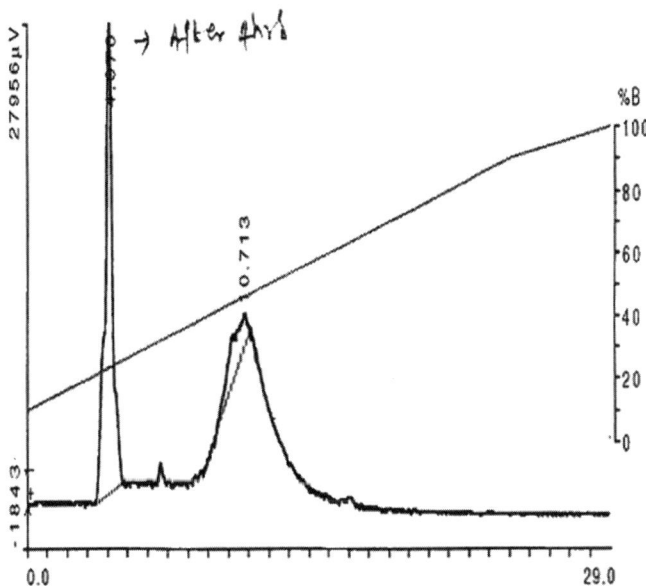

Fig. 20. Analytical HPLC of Tc-99m-Gly-D-Ala-Gly-Gly-Aba-GCATCGTCGCGG PNA (WT3613) that was recovered from mouse urine 4 h after administration as in **Fig. 14**.

Table 6
Tissue Distribution (% I.D./g ± SD) of MYC Antisense
Tc-99m-PNA (WT3613) and Mismatch (WT3629) at 24 h
Postinjection in Five Nude Mice (n = 5) Bearing Human
MCF7:IGF-1R Breast Tumor Xenografts

Tissue	Antisense	Mismatch	p Value
Muscle	0.09 ± 0.01	0.05 ± 0.01	0.01
Intestine	0.12 ± 0.01	0.09 ± 0.03	0.03
Heart	0.09 ± 0.02	0.07 ± 0.01	0.01
Lungs	0.13 ± 0.06	0.17 ± 0.03	0.46
Blood	0.10 ± 0.01	0.12 ± 0.05	0.76
Spleen	0.63 ± 0.35	0.86 ± 0.43	0.34
Kidneys	8.60 ± 3.23	30.1 ± 0.96	0.04
Liver	1.13 ± 0.51	0.73 ± 0.11	0.08
Tumor	0.21 ± 0.02	0.17 ± 0.08	0.34
Tumor/muscle ratio	2.88 ± 1.66	3.56 ± 0.03	0.51
Tumor/blood ratio	2.08 ± 0.43	1.65 ± 0.03	0.55

4. Preclinical Translation

4.1. Tc-99m Chimeras

The distribution data at 4 h postinjection in mice bearing human breast tumor xenografts are encouraging, because they suggest that the uptake of *MYC* antisense Tc-99m-PNA in tumor xenografts in mice was significantly ($p < 0.01$) greater than that of the corresponding *MYC* mismatch probe, even without the benefit of a receptor-specific ligand. The relatively low uptake in other normal tissues at both 4 and 24 h postinjection, except that in the kidneys and liver, is also helpful in tumor delineation. Consistent with most Tc-99m agents, whether antibodies, peptides, or antisense, radioactivity for both agents, in nearly all tissues, declined significantly by 24 h. Tumor/blood and tumor/muscle ratios, however, increased, perhaps indicating that 24 h postinjection could be an optimal time for imaging. The reasons for the decline in radioactivity are not well understood and could be either the result of the in vivo instability of the agent, metabolic degradation, or simply the oxidation of Tc-99m following the subsequent washout from cells or tissues.

These results suggest that Tc-99m-peptide-PNA probes might be applied for imaging gene expression in tumors, if cellular uptake could be enhanced. One of the primary requirements for an oligonucleotide analog to be successful as an antigene/antisense agent is for it to be taken up by the cells in reasonable quantity so that it can reach its target in sufficient concentration. Because the PNAs suffer from poor cellular uptake *(154,155)*, they have not yet been

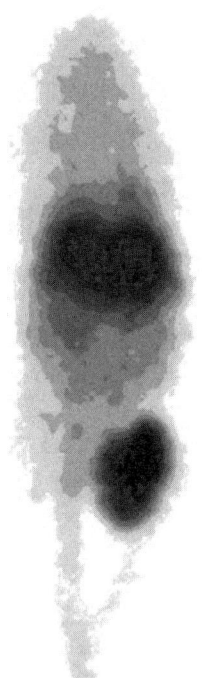

Fig. 21. PET image of nude mouse bearing human T47D breast tumor xenograft 24 h after injection of 100 µCi of Cu-64-TP3982.

well developed as an antigene/antisense therapeutic agent. To improve cellular uptake, PNAs have been conjugated with cell-penetrating positively charged peptides, such as the SV40 nuclear localization sequence *(182,183)*, or homologous basic sequences, such as transportan, penetratin, and TAR-binding peptides *(184)*. Cell penetration in these cases is not receptor dependent and thus not cell specific. Typical conjugation routes begin with purified moieties and suffer from low yields *(185)*. To elevate cellular uptake of an antisense PNA targeted to the mRNA for IGF-1R but only in cells that overexpress that particular mRNA *(142)*, the IGF-1R PNA was extended from a peptide moiety (D-CSKC) modeled to imitate an IGF-1R-binding domain of IGF-1 *(159)*. The specific PNA-peptide conjugate displayed 5–10 times greater uptake than control PNA, but only in cells overexpressing IGF-1R *(142)*. Combining a ligand specific for a receptor overexpressed on malignant cells with Tc-99m-peptide-PNAs specific for characteristic oncogene mRNAs might provide a sensitive, oncogene-specific probe for early detection of tumors by scintigraphic imaging.

We will test the specificity of uptake and mRNA hybridization of our Tc-99m-PNA-peptide probes in normal human cells vs transformed AsPC1 human

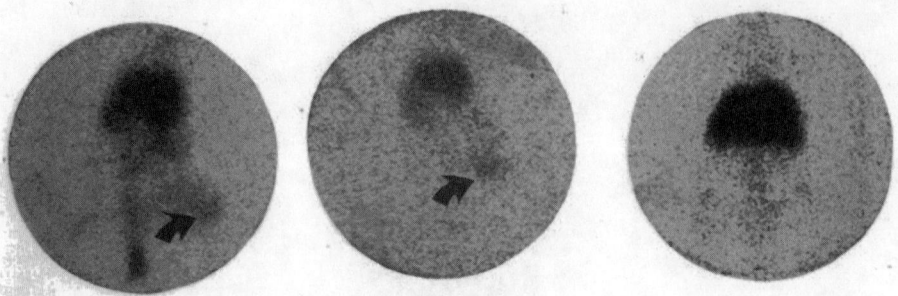

Fig. 22. Scintifigraphic imaging of mice bearing embryonal carcinoma xenografts (arrow) on right hind leg after administration of Re-186-TNT-F(ab')$_2$. Left, 4 h postinjection; center, 24 h postinjection; right, no tumor visualized 4 h after administration of Re-186-HSA as a negative control.

pancreatic cancer cells. If our initial set of probes and control sequences does not consistently display at least a threefold excess of gene-specific probes in cultured cells, compared with control sequences, then we will test adjacent mRNA sequence targets, or alternative oligonucleotide backbone derivatives, until the desired level of specificity is attained.

We will administer oncogene-specific Tc-99m-PNA-peptide probes intravenously to cohorts of nude mice bearing AsPC1 human pancreatic cancer xenografts and determine the sensitivity and specificity of scintigraphic imaging of the targeted oncogene mRNAs in the tumors relative to the nonspecific signals expected in the liver, gallbladder, and kidneys 4 and 24 h after administration. If our initial set of probes and control sequences does not consistently display at least a threefold excess of γ emission from tumors exposed to gene-specific probes, compared with control sequences, then we will test adjacent mRNA sequence targets, or alternative oligonucleotide backbone derivatives, until the desired level of specificity is attained.

4.2. Cu-64 Chimeras

Using the methods described above, we will prepare Cu-64-PNA-peptides with complementary sequences known to hybridize specifically to mRNAs for activated K-*RAS* mutated in the twelfth codon, *CCND1*, *HER2*, *MYC*, and mutant tumor suppressor p53, and capable of binding to the receptor for IGF-1. We will test the specificity of uptake and mRNA hybridization of the Cu-64-PNA-peptide probes in normal human epithelial cells vs K-*RAS* mutant AsPC1 human pancreatic cancer cells and K-*RAS* wild-type BT474 human breast cancer cells. We will administer Cu-64-PNA-peptide probes to cohorts of nude mice bearing AsPC1 human pancreatic cancer xenografts and determine the sensitivity and specificity of PET imaging of K-*RAS*, *CCND1*, *HER2*, *MYC*, and p53 mRNAs in the tumors relative to the nonspecific signals expected in the liver, gallbladder, and kidneys. We will compare the imaging results to real-time QRT-PCR measurements of those mRNAs in tumor cells removed from the subjects and to the same mRNAs in liver, gallbladder, and kidneys.

4.3. Gd(III) Chimeras

Similarly, we will prepare Gd_{256}-PNA-peptides capable of binding to the cell-surface receptor for IGF-1, with complementary sequences hybridizing specifically to mRNAs for the following oncogenes: activated K-*RAS* mutated in the twelfth codon, *CCND1*, *HER2*, *MYC*, and mutant tumor suppressor p53. We will test the specificity of uptake and mRNA hybridization of the Gd_{256}-PNA-peptide probes in normal human cells vs transformed AsPC1 human pancreatic cancer cells. We will measure T1 of water in cells treated with Gd_{256}-PNA-peptides to determine whether T1 increases in the case of cell-specific peptides and oncogene-specific PNAs. We will administer oncogene-specific Gd_{256}-PNA-peptide probes intravenously to cohorts of nude mice bearing AsPC1 human pancreatic cancer xenografts and determine the sensitivity and specificity of MRI of the targeted oncogene mRNAs in the tumors relative to the nonspecific signals expected in the liver, gallbladder, and kidneys. We will compare the imaging results with radioactive [Gd-159]Gd_{256}-PNA-peptide tissue distribution measurements, and with real-time QRT-PCR measurements of those mRNAs in tumor cells removed from the subjects and to the same mRNAs in the liver, gallbladder, and kidneys 4 and 24 h after administration.

4.4. Re-188 Chimeras

Even if Re-188-chimera treatment successfully ablates the pancreatic tumor xenografts, radiation killing might leave behind a scar zone that, in a patient, will always be identified as atypical in CT or MRI scans. While that is a more favorable outcome than death, suspicious masses might oblige physicians to order successive rounds of Tc-99m-PNA imaging on follow-up visits to assess the oncogene mRNA status of the nodules. Another potential paradox concerns the possibility of successful tumor ablation but excess toxicity. We will administer an oncogene-specific Re-188-PNA-peptide intravenously to mice with clearly imaged tumors in order to ablate precisely the tissues that took up the specific oncogene-imaging probe. Tumor-bearing mice will be treated with 5-FU as a positive therapeutic control. If the Re-188 derivatives of probes found effective and specific in detection experiments exhibit greater toxicity (weight loss >10%) than 5-FU at the initial Re-188 dose levels listed above, then the Re-188 dose will be reduced in stages until acceptable toxicity is attained, or the mice will be treated solely by antisense inhibition with unlabeled PNA-peptides. By analogy with clinically effective oligonucleotide phosphorothioates, tumor-bearing mice would be treated intraperitoneally every other day with 1, 2, and 4 mg/kg of unlabeled PNA-peptides for 14 d, then observed and measured for 21 additional d as above.

4.5. Tc-99m-Annexin-V Detection of Apoptosis

Maximum signal-to-noise ratios occur when most of the cell-surface phosphatidylserine bound to labeled annexin-V is still intact, but most unbound labeled annexin-V is excreted. That time point is predicted to occur within 12 h of administration but must be determined experimentally through serial observations of subjects over 24 h after administration. Our goal for signal: background ratio is 3; that would provide a comfortable margin for diagnostic applications.

While synthesis of the Tc-99m-annexin-V probes should be a straightforward repetition of published methods, the specificity of Tc-99m-annexin-V binding to apoptosing tumors in this system remains to be seen. We will administer Tc-99m-annexin-V intravenously to mice that have been treated with oncogene-specific Re-188-PNA-peptides, or mismatches, or 5-FU, in order to detect apoptosis by scintigraphic imaging in targeted tumors. If our preparations of Tc-99m-annexin-V do not consistently display at least a threefold excess of γ emission from tumors exposed to the 5-FU drug control, let alone PNA-peptides, then we will test alternate reagents for detecting apoptosis noninvasively. Remaining tumors or necrotic remnants will be analyzed by QRT-PCR to determine the remaining levels of oncogene mRNAs, as well as caspase-8 and caspase-10 mRNAs and proteins as early apoptotic indicators *(186)*. Thus, we will obtain a direct comparison of tumor-imaging results with tumor mRNA levels following therapy.

5. Conclusion

Consistent with our hypothesis, we were able to label PNA constructs with Tc-99m rapidly and efficiently. The preparations were stable thermodynamically and were eluted as a single, sharp chromatographic peak, indicating that no byproducts were formed during labeling with Tc-99m. Thus, we have observed that PNA-peptides designed to bind to IGF-1R on malignant cells are taken up specifically and concentrated in nuclei, and that Tc-99m can be chelated to PNA-peptides designed to bind to IGF-1R. We will therefore apply Tc-99m-PNA-peptides for SPECT imaging, Cu-64-PNA-peptides for PET imaging, and Gd_n-dendrimer-PNA-peptides for MRI, to detect the onset of activated K-*RAS, MYC, CCND1, HER2, IGF1R*, and p53 expression during the earliest stages of pancreatic cancer, attack the transformed cells with Re-188-PNA-peptides, and assess the resulting degree of apoptosis in transformed tissues with Tc-99m-annexin-V.

References

1. ACS, (2003) in *Cancer Facts and Figures,* American Cancer Society, pp. 18–22.
2. Evans, D. B., Abbruzzese, J. L., and Willett, C. G. (2001) Cancer of the pancreas, in *Cancer: Principles and Practice of Oncology,* DeVita, (V. T., Hellman, S., and Rosenberg, S. A., eds.), Lippincott, Williams, & Wilkins, Philadelphia, p. 1126–1161.
3. Hilgers, W. and Kern, S. E. (1999) Molecular genetic basis of pancreatic adenocarcinoma. *Genes Chromosomes Cancer* **26(1),** 1–12.
4. Knudson, A.G. Jr., (1970) Genetics and cancer. *Postgrad. Med.* **48(5),** 70–74.
5. Wooster, R. (2001) Cancer classification with DNA microarrays is less more? *Trends Genet.* **16(8),** 327–329.
6. Perou, C. M., et al. (1999) Distinctive gene expression patterns in human mammary epithelial cells and breast cancers. *Proc. Natl. Acad. Sci. USA,* **96(16),** 9212–9217.
7. Alizadeh, A., et al. (1998) Probing lymphocyte biology by genomic-scale gene expression analysis. *J. Clin. Immunol.* **18(6),** 373–379.
8. Barlund, M., et al. (2000) Detecting activation of ribosomal protein S6 kinase by complementary DNA and tissue microarray analysis. *J. Natl. Cancer Inst.* **92(15),** 1252–1259.
9. Clark, E. A., et al. (2000) Genomic analysis of metastasis reveals an essential role for RhoC. *Nature* **406(6795),** 532–535.
10. Perou, C. M., et al. (2000) Molecular portraits of human breast tumours. Nature **406(6797),** 747–752.
11. Druker, B. J., et al. (2001) Activity of a specific inhibitor of the *BCR-ABL* tyrosine kinase in the blast crisis of chronic myeloid leukemia and acute lymphoblastic leukemia with the Philadelphia chromosome. *N. Engl. J. Med.* **344(14),** 1038–1042.
12. Hochhaus, A., et al. (2001) Roots of clinical resistance to STI-571 cancer therapy. *Science* **293(5538),** 2163.
13. Szczylik, C., et al. (1991) Selective inhibition of leukemia cell proliferation by *BCR-ABL* antisense oligodeoxynucleotides. *Science* **253(5019),** 562–565.
14. Webb, A., et al. (1997) BCL-2 antisense therapy in patients with non-Hodgkin lymphoma. *Lancet* **349(9059),** 1137–1141.
15. Jansen, B., et al. (2000) Chemosensitisation of malignant melanoma by *BCL2* antisense therapy. *Lancet* **356(9243),** 1728–1733.
16. Chi, K. N., et al. (2001) A phase I dose-finding study of combined treatment with an antisense *BCL-2* oligonucleotide (Genasense) and mitoxantrone in patients with metastatic hormone-refractory prostate cancer. *Clin. Cancer Res.* **7(12),** 3920–3927.
17. Fong, L. Y., et al. (2000) Muir-Torre-like syndrome in Fhit-deficient mice. *Proc. Natl. Acad. Sci. USA* **97(9),** 4742–4747.
18. Bishop, J. M. (1991) Molecular themes in oncogenesis. *Cell* **64(2),** 235–248.
19. Wickstrom, E. and Tyson, F. L. (1997) Differential oligonucleotide activity in cell culture versus mouse models. in *Oligonucleotides as Therapeutic Agents.* Wiley, London.
20. Sauter, E. R., et al. (2000) Prolonged response to antisense cyclin D1 in a human squamous cancer xenograft model. *Clin. Cancer Res.* **6(2),** 654–660.

21. Vaughn, J. P., et al. (1995) Antisense DNA downregulation of the *ERBB2* oncogene measured by a flow cytometric assay. *Proc. Natl. Acad. Sci. USA* **92(18),** 8338–8342.

22. Wickstrom, E. L., et al. (1988) Human promyelocytic leukemia HL-60 cell proliferation and *c-myc* protein expression are inhibited by an antisense pentadecadeoxynucleotide targeted against c-*myc* mRNA. *Proc. Natl. Acad. Sci. USA* **85(4),** 1028–1032.

23. Wickstrom, E., Bacon, T. A., and Wickstrom, E. L. (1992) Down-regulation of c-*MYC* antigen expression in lymphocytes of Eμ-c-*myc* transgenic mice treated with anti-c-*myc* DNA methylphosphonates. *Cancer Res.* **52(24),** 6741–6745.

24. Huang, Y., et al. (1995) Prevention of tumor formation in a mouse model of Burkitt's lymphoma by 6 weeks of treatment with anti-c-*myc* DNA phosphorothioate. *Mol. Med.* **1(6),** 647–658.

25. Smith, J. B. and Wickstrom, E. (1998) Antisense c-*myc* and immunostimulatory oligonucleotide inhibition of tumorigenesis in a murine B-cell lymphoma transplant model. *J. Natl. Cancer Inst.* **90(15),** 1146–1154.

26. Bishop, M. R., et al. (1997) *Ex vivo* treatment of bone marrow with phosphorothioate oligonucleotide OL(1)p53 for autologous transplantation in acute myelogenous leukemia and myelodysplastic syndrome. *J. Hematother.* **6(5),** 441–446.

27. Bayever, E., et al. (1994) Selective cytotoxicity to human leukemic myeloblasts produced by oligodeoxyribonucleotide phosphorothioates complementary to p53 nucleotide sequences. *Leuk. Lymphoma* **12(3–4),** 223–231.

28. Almoguera, C., et al. (1998) Most human carcinomas of the exocrine pancreas contain mutant c-K-*ras* genes. *Cell* **53(4),** 549–554.

29. Tada, M., et al. (1993) Detection of *ras* gene mutations in pancreatic juice and peripheral blood of patients with pancreatic adenocarcinoma. *Cancer Res.* **53(11),** 2472–2474.

30. Lowy, D. R. and Willumsen, B. M. (1993) Function and regulation of *ras. Ann. Rev. Biochem.* **62,** 851–891.

31. Downward, J., et al. (1990) Identification of a nucleotide exchange-promoting activity for p21ras. *Proc. Natl. Acad. Sci. USA* **87(15),** 5998–6002.

32. Gibbs, J. B., et al. (1988) Purification of *ras* GTPase activating protein from bovine brain. *Proc. Natl. Acad. Sci. USA* **85(14),** 5026–5030.

33. Thissen, J. A., et al. (1997) Prenylation-dependent association of Ki-*Ras* with microtubules. Evidence for a role in subcellular trafficking. *J. Biol. Chem.* **272(48),** 30362–30370.

34. Oliff, A. (1999) Farnesyltransferase inhibitors: targeting the molecular basis of cancer. *Biochim. Biophys. Acta* **1423(3),** C19–30.

35. Seidman, A. D. (1999) Single-agent paclitaxel in the treatment of breast cancer: phase I and II development. *Semin. Oncol.* **26(3 Suppl 8),** 14–20.

36. Mukhopadhyay, T., et al. (1991) Specific inhibition of K-*RAS* expression and tumorigenicity of lung cancer cells by antisense RNA. *Cancer Res.* **51(6),** 1744–1748.

37. Georges, R. N., et al. (1993) Prevention of orthotopic human lung cancer growth by intratracheal instillation of a retroviral antisense K-*RAS* construct. *Cancer Res.* **53(8),** 1743–1746.

38. Kashani-Sabet, M., et al. (1994) Suppression of the neoplastic phenotype *in vivo* by an anti-*RAS* ribozyme. *Cancer Res.* **54(4),** 900–902.

39. Aoki, K., et al. (1995) Liposome-mediated *in vivo* gene transfer of antisense K-*RAS* construct inhibits pancreatic tumor dissemination in the murine peritoneal cavity. *Cancer Res.* **55(17),** 3810–3816.

40. Kawada, M., et al. (1997) Inhibition of anchorage-independent growth of *RAS*-transformed cells on polyHEMA surface by antisense oligodeoxynucleotides directed against K- ras. *Biochem. Biophys. Res. Commun.* **231(3),** 735–737.

41. Okada, F., et al. (1998) Impact of oncogenes in tumor angiogenesis: mutant K-*RAS* up-regulation of vascular endothelial growth factor/vascular permeability factor is necessary, but not sufficient for tumorigenicity of human colorectal carcinoma cells. *Proc. Natl. Acad. Sci. USA* **95(7),** 3609–3614.

42. Kita, K., et al. (1999) Growth inhibition of human pancreatic cancer cell lines by anti-sense oligonucleotides specific to mutated K-*RAS* genes. *Int. J. Cancer* **80(4),** 553–558.

43. Hinds, P. W., et al. (1994) Function of a human cyclin gene as an oncogene. *Proc. Natl. Acad. Sci. USA* **91,** 709–713.

44. Bartkova, J., et al. (1995) Abnormal patterns of D-type cyclin expression and G1 regulation in human head and neck cancer. *Cancer Res.* **55(4),** 949–956.

45. Adelaide, J., et al. (1995) Oesophageal cancer and amplification of the human cyclin D gene CCND1/PRAD1. *Br. J. Cancer* **71(1),** 64–68.

46. Helin, K. and Harlow, E. (1992) The retinoblastoma protein as a transcriptional repressor. *Trends Cell Biol.* **3,** 43–46.

47. Goodrich, D. W. and Lee, W. -H. (1993) Molecular characterization of the retinoblastoma susceptibility gene. *Biochim. Biophys. Acta* **1155,** 43–61.

48. Nevins, J. R. (1992) E2F; a link between the Rb tumor suppressor protein and viral oncogenesis. *Science* **258,** 424–429.

49. Deng, C., et al. (1995) Mice lacking p21CIP1/WAF1 undergo normal development, but are defective in G1 checkpoint control. *Cell* **82,** 675–684.

50. Matsushime, H. D., et al. (1994) D-type cyclin-dependent kinase activity in mammalian cells. *Mol. Cell Biol.* **14,** 2066–2076.

51. Jiang, W., et al. (1993) Overexpression of cyclin D1 in rat fibroblasts causes abnormalities in growth control, cell cycle progression and gene expression. *Oncogene* **8(12),** 3447–3457.

52. Quelle, D. E., et al. (1993) Overexpression of mouse D-type cyclins accelerates G1 phase in rodent fibroblasts. *Genes Dev.* **7(8),** 1559–1571.

53. Lovec, H., et al. (1994) Oncogenic activity of cyclin D1 revealed through cooperation with Ha- ras: link between cell cycle control and malignant transformation. *Oncogene* **9(1),** 323–326.

54. Arnold, A., et al. (1989) Molecular cloning and chromosomal mapping of DNA rearranged with the parathyroid hormone gene in a parathyroid adenoma. *J. Clin. Invest.* **83(6),** 2034–2040.

55. Motokura, T. and Arnold, A. (1993) *PRAD*1/cyclin D1 proto-oncogene: genomic organization, 5' DNA sequence, and sequence of a tumor-specific rearrangement breakpoint. *Genes Chromosomes Cancer* **7(2),** 89–95.
56. Gansauge, S., et al. (1997) Overexpression of cyclin D1 in human pancreatic carcinoma is associated with poor prognosis. Cancer Res. **57(9),** 1634–1637.
57. Qiao, Q., et al. (2001) Reduced membranous and ectopic cytoplasmic expression of beta-catenin correlate with cyclin D1 overexpression and poor prognosis in pancreatic cancer. *Int. J. Cancer* **95(3),** 194–197.
58. Sauter, E. R., et al. (1999) Antisense cyclin D1 induces apoptosis and tumor shrinkage in human squamous carcinomas. *Cancer Res.* **59(19),** 4876–4881.
59. Slamon, D. J., et al. (1987) Human breast cancer: correlation of relapse and survival with amplification of the *HER-2/neu* oncogene. *Science* **235(4785),** 177–182.
60. Dugan, M. C., et al. (1997) *HER-2/neu* expression in pancreatic adenocarcinoma: relation to tumor differentiation and survival. *Pancreas* **14(3),** 229–236.
61. Goldman, R., et al. (1990) Heterodimerization of the erbB-1 and erbB-2 receptors in human breast carcinoma cells: a mechanism for receptor transregulation. *Biochemistry* **29(50),** 11,024–11,028.
62. Schechter, A. L., et al. The *neu* gene: an *erbB*-homologous gene distinct from and unlinked to the gene encoding the EGF receptor. *Science* **229(4717),** 976–978.
63. Kraus, M. H., et al. (1989) Isolation and characterization of *ERBB3*, a third member of the *ERBB*/epidermal growth factor receptor family: evidence for overexpression in a subset of human mammary tumors. *Proc. Natl. Acad. Sci. USA* **86(23),** 9193–9197.
64. Alimandi, M., et al. (1995) Cooperative signaling of ErbB3 and ErbB2 in neoplastic transformation and human mammary carcinomas. *Oncogene* **10(9),** 1813–1821.
65. Kokai, Y., et al. (1987) Stage- and tissue-specific expression of the *neu* oncogene in rat development. *Proc. Natl. Acad. Sci. USA* **84(23),** 8498–8501.
66. Press, M. F., et al. (1993) Her-2/neu expression in node-negative breast cancer: direct tissue quantitation by computerized image analysis and association of overexpression with increased risk of recurrent disease. *Cancer Res.* **53(20),** 4960–4970.
67. Ross, J. S. and Fletcher, J. A. (1998) The *HER-2/neu* Oncogene in Breast Cancer: Prognostic Factor, Predictive Factor, and Target for Therapy. *Oncologist* **3(4),** 237–252.
68. Cobleigh, M. A., et al. (1998) Efficacy and safety of Herceptin (humanized anti-human Her-2 antibody) as a single agent in 222 women with *HER2* overexpression who relapsed following chemotherapy for metastatic breast cancer. *Proc. Am. Soc. Clin. Oncol.* **17,** 97.
69. Slamon, D., et al. (1998) Addition of Herceptin (humanized anti-human Her2 antibody) to first line chemotherapy for *HER2* overexpressing metastatic brest cancer markedly increases anticancer activity: a randomized, multinational controlled phase III trial. *Proc. Am. Soc. Clin. Oncol.* **17,** 98.
70. Bertram, J., et al. (1994) Reduction of *erbB2* gene product in mamma carcinoma cell lines by *erbB2* mRNA-specific and tyrosine kinase consensus phos-

phorothioate antisense oligonucleotides. *Biochem. Biophys. Res. Commun.* **200(1),** 661–667.

71. Vaughn, J. P., et al. Inhibition of the erbB-2 tyrosine kinase receptor in breast cancer cells by phosphoromonothioate and phosphorodithioate antisense oligonucleotides. *Nucleic Acids Res.* **24(22),** 4558–4564.

72. Pirollo, K. F., et al. (1997) Evidence supporting a signal transduction pathway leading to the radiation-resistant phenotype in human tumor cells. *Biochem. Biophys. Res. Commun.* **230(1),** 196–201.

73. Armengol, G., et al. (2000) DNA copy number changes and evaluation of *MYC*, *IGF1R*, and *FES* amplification in xenografts of pancreatic adenocarcinoma. *Cancer Genet. Cytogenet.* **116(2),** 133–141.

74. Blackwood, E. M. and Eisenman, R. N. (1991) Max: a helix-loop-helix zipper protein that forms a sequence-specific DNA-binding complex with Myc. *Science* **251(4998),** 1211–1217.

75. Moberg, K. H., et al. (1992) Three distinct elements within the murine c-*myc* promoter are required for transcription. *Oncogene* **7(3),** 411–421.

76. Moberg, K. H., Tyndall, W. A., and Hall, D. J. (1992) Wild-type murine p53 represses transcription from the murine c-myc promoter in a human glial cell line. *J. Cell Biochem.* **49(2),** 208–15.

77. Sears, R., et al. (2000) Multiple Ras-dependent phosphorylation pathways regulate Myc protein stability. *Genes Dev.* **14(19),** 2501–2514.

78. Wickstrom, E. (1991) *Prospects for Antisense Nucleic Acid Therapy of Cancer and AIDS.* Wiley-Liss, New York.

79. Watson, P. H., Pon, R. T., and Shiu, R. P. (1991) Inhibition of c-*myc* expression by phosphorothioate antisense oligonucleotide identifies a critical role for c-*myc* in the growth of human breast cancer. *Cancer Res.* **51(15),** 3996–4000.

80. Smith, J. B. and Wickstrom, E. (2000) Preclinical antisense DNA therapy of cancer in mice. *Methods Enzymol.* **314,** 537–580.

81. Gelmon, K. A., et al. (2001) A dose escalation phase I study of c-*MYC* antisense in combination with cisplatin in the treatment of solid tumours and lymphomas, in *Proceedings of the AACR-NCI-EORTC International Conference on Molecular Targets and Cancer Therapeutics.* Miami Beach, FL.

82. Yamaguchi, K., et al. (1999) Ki-*ras* codon 12 point mutation and p53 mutation in pancreatic diseases. *Hepatogastroenterology* **46(28),** 2575–2581.

83. Thor, A. D., et al. (1992) Accumulation of p53 tumor suppressor gene protein: an independent marker of prognosis in breast cancers. *J. Natl. Cancer Inst.* **84(11),** 845–855.

84. Thor, A. D., et al. (1998) erbB-2, p53, and efficacy of adjuvant therapy in lymph node–positive breast cancer. *J. Natl. Cancer Inst.* **90(18),** 1346–1360.

85. Kasuya, K., et al. (1997) p53 protein overexpression and K-*RAS* codon 12 mutation in pancreatic ductal carcinoma: correlation with histologic factors. *Pathol. Int.* **47(8),** 531–539.

86. Dong, M., et al. (2000) Ki-*RAS* point mutation and p53 expression in human pancreatic cancer: a comparative study among Chinese, Japanese, and Western patients. *Cancer Epidemiol. Biomarkers Prev.* **9(3),** 279–284.

87. Tong, Z., Singh, G., and Rainbow, A. J. The role of the p53 tumor suppressor in the response of human cells to photofrin-mediated photodynamic therapy [in process citation]. *Photochem. Photobiol.* **71(2),** 201–210.

88. Gurnani, M., et al. (1999) Adenovirus-mediated p53 gene therapy has greater efficacy when combined with chemotherapy against human head and neck, ovarian, prostate, and breast cancer. *Cancer Chemother. Pharmacol.* **44(2),** 143–151.

89. Nemunaitis, J., et al. (2000) Adenovirus-mediated p53 gene transfer in sequence with cisplatin to tumors of patients with non-small-cell lung cancer [in process citation]. *J. Clin. Oncol.* **18(3),** 609–622.

90. Broaddus, W. C., et al. (1999) Enhanced radiosensitivity of malignant glioma cells after adenoviral p53 transduction. *J. Neurosurg.* **91(6),** 997–1004.

91. Xu, L., et al. (1999) Transferrin-liposome-mediated systemic p53 gene therapy in combination with radiation results in regression of human head and neck cancer xenografts. *Hum. Gene Ther.* **10(18),** 2941–2952.

92. Rao, P. S., et al. (2003) ^{99m}Tc-peptide-peptide nucleic acid probes for imaging oncogene mRNAs in tumours. *Nucl. Med. Commun.* **24(8),** 857–863.

93. Stalteri, M. A. and Mather, S. J. (2000) In vitro studies on 99m-Tc-labeled HYNIC-conjugated oligonucleotides. *Nucl. Med. Commun.* **21,** 374.

94. Mier, W., et al. (2000) Preparation and evaluation of tumor-targeting peptide-oligonucleotide conjugates. *Bioconjug. Chem.* **11(6),** 855–860.

95. Mier, W., et al. (2001) Preparation and preclinical development of tumor-targeting peptide-PNA conjugates. *J. Labelled Compounds Radiopharm.* **42(5),** 115P.

96. Zhang, Y. M., et al. (2001) In vitro investigations of tumor targeting with (99m)Tc-labeled antisense DNA. *J. Nucl. Med.* **42(11),** 1660–1669.

97. Thakur, M. L., et al. (2000) Imaging vascular thrombosis with ^{99m}Tc-labeled fibrin alpha-chain peptide. *J. Nucl. Med.* **41(1),** 161–168.

98. Thakur, M. L., et al. (2000) ^{99m}Tc-labeled vasoactive intestinal peptide analog for rapid localization of tumors in humans. *J. Nucl. Med.* **41(1),** 107–110.

99. Eisenhut, M. and Haberkorn, U. (2000) [^{123}I]VIP receptor scintigraphy in patients with pancreatic adenocarcinomas. *Eur. J. Nucl. Med.* **27(11),** 1589–1590.

100. Gold, D. V., et al. (2001) Localization of pancreatic cancer with radiolabeled monoclonal antibody PAM4. *Crit. Rev. Oncol. Hematol.* **39(1–2),** 147–154.

101. Namavari, M., et al. (2000) Synthesis of 8-[(18)F]fluoroguanine derivatives: in vivo probes for imaging gene expression with positron emission tomography. *Nucl. Med. Biol.* **27(2),** 157–162.

102. Hustinx, R., et al. (2000) Imaging in vivo herpes simplex virus thymidine kinase gene transfer and expression in tumors using positron emission tomography. *J. Nucl. Med.* **41(5),** 264P.

103. Ponomarev, V., et al. (2000) PET imaging of p53 gene expression in tumors. *J. Nucl. Med.* **41(5),** 263P–264P.

104. Fujibayashi, Y., et al. (1999) Comparative studies of Cu-64-ATSM and C-11-acetate in an acute myocardial infarction model: ex vivo imaging of hypoxia in rats. *Nucl. Med. Biol.* **26(1),** 117–121.

105. Green, M. A., et al. Copper-62-labeled pyruvaldehyde bis(N4-methylthio-semicarbazonato)copper(II): synthesis and evaluation as a positron emission tomography tracer for cerebral and myocardial perfusion. *J. Nucl. Med.* **31(12),** 1989–1996.

106. Thakur, M. L., et al. PET imaging of oncogene overexpression using Cu-64-labeled peptide. *J. Nucl. Med.* **45,** in press.

107. Lewis, M. R., et al. (2003) In vivo evaluation of pretargeted 64Cu for tumor imaging and therapy. *J. Nucl. Med.* **44(8),** 1284–1292.

108. Stevenson, J. P., et al. (2003) Phase I trial of the antivascular agent combretastatin A4 phosphate on a 5-day schedule to patients with cancer: magnetic resonance imaging evidence for altered tumor blood flow. *J. Clin. Oncol.* **21(23),** 4428–4438.

109. Schmiedl, U., et al. (1987) Albumin labeled with Gd-DTPA as an intravascular, blood pool-enhancing agent for MR imaging: biodistribution and imaging studies. *Radiology* **162(1 Pt. 1),** 205–210.

110. Kobayashi, H., et al. (2001) Comparison of the macromolecular MR contrast agents with ethylenediamine-core versus ammonia-core generation-6 polyamidoamine dendrimer. *Bioconjug. Chem.* **12(1),** 100–107.

111. O'Donoghue, J. A., Bardies, M., and Wheldon, T. E. (1995) Relationships between tumor size and curability for uniformly targeted therapy with beta-emitting radionuclides. *J. Nucl. Med.* **36(10),** 1902–1909.

112. Blower, P. J., et al. (1998) Pentavalent rhenium-188 dimercaptosuccinic acid for targeted radiotherapy: synthesis and preliminary animal and human studies. *Eur. J. Nucl. Med.* **25(6),** 613–621.

113. Dadachova, E., et al. (2002) Rhenium-188 as an alternative to Iodine-131 for treatment of breast tumors expressing the sodium/iodide symporter (NIS). Nucl. Med. Biol. **29(1),** 13–18.

114. Roka, R., et al. (2000) Clinical experience with rhenium-188 HEDP therapy for metastatic bone pain. *Orv. Hetil.* **141(19),** 1019–1023.

115. Li, S., et al. (2001) Rhenium-188 hedp to treat painful bone metastases. *Clin. Nucl. Med.* **26(11),** 919–922.

116. John, E., et al. (1993) Rhenium-186-labeled monoclonal antibodies for radioimmunotherapy: preparation and evaluation. *J. Nucl. Med.* **34(2),** 260–267.

117. Alnemri, E. S., et al. (1992) Involvement of *BCL-2* in glucocorticoid-induced apoptosis of human pre-B-leukemias. *Cancer Res.* **52(2),** 491–495.

118. Martin, S. J., et al. (1995) Early redistribution of plasma membrane phosphatidylserine is a general feature of apoptosis regardless of the initiating stimulus: inhibition by overexpression of Bcl-2 and Abl. J. Exp. Med. **182(5),** 1545–1556.

119. Gavrieli, Y., Sherman, Y., and Ben-Sasson, S. A. (1992) Identification of programmed cell death in situ via specific labeling of nuclear DNA fragmentation. *J. Cell Biol.* **119(3),** 493–501.

120. van Heerde, W. L., de Groot, P. G., and Reutelingsperger, C. P. (1995) The complexity of the phospholipid binding protein Annexin V. *Thromb. Haemost.* **73(2),** 172–179.

121. Blankenberg, F. G., et al. (1998) In vivo detection and imaging of phosphatidylserine expression during programmed cell death. *Proc. Natl. Acad. Sci. USA* **95(11),** 6349–6354.

122. Lau, Q. C., Brusselbach, S., and Muller, R. (1998) Abrogation of c-Raf expression induces apoptosis in tumor cells. *Oncogene,* **16(14),** 1899–1902.

123. Blankenberg, F. G., et al. (1999) Imaging of apoptosis (programmed cell death) with ^{99m}Tc annexin V. *J. Nucl. Med.* **40(1),** 184–191.

124. Belikova, A. M., Zarytova, V. F., and Grineva, N. I. (1967) Synthesis of ribonucleosides and diribonucleoside phosphates containing 2-chloroethylamine and nitrogen mustard residues. *Tetrahedron Lett.* **37,** 3557–3562.

125. Zamecnik, P. C. and Stephenson, M. L. (1978) Inhibition of Rous sarcoma virus replication and cell transformation by a specific oligodeoxynucleotide. *Proc. Natl. Acad. Sci. USA* **75(1),** 280–284.

126. Agrawal, S. (1996) Antisense therapeutics in *Methods in Molecular Medicine,* (Walker, J. M., ed.), Humana, Totowa, NJ p. 276.

127. Wickstrom, E. (1998) *Clinical Trials of Genetic Therapy with Antisense DNA and DNA Vector,* Marcel Dekker, New York.

128. Wickstrom, E. (1992) Strategies for administering targeted therapeutic oligodeoxynucleotides. *Trends Biotechnol.* **10(8),** 281–287.

129. Zendegui, J. G., et al. (1992) In vivo stability and kinetics of absorption and disposition of 3' phosphopropyl amine oligonucleotides. *Nucleic Acids Res.* **20(2),** 307–314.

130. Stec, W. J., et al. (1991) Novel route to oligo(deoxyribonucleoside phosphorothioates). Stereocontrolled synthesis of P-chiral oligo(deoxyribo-nucleoside phosphorothioates). *Nucleic Acids Res.* **19(21),** 5883–5888.

131. Miller, P. S. (1991) Oligonucleoside methylphosphonates as antisense reagents. *Biotechnology (NY)* **9(4),** 358–362.

132. Shaw, B. R., et al. (2000) Boranophosphate backbone: a mimic of phosphodiesters, phosphorothioates, and methyl phosphonates. *Methods Enzymol.* **313,** 226–257.

133. Lebedev, A. V. and Wickstrom, E. (1996) The chirality problem in P-substituted oligonucleotides, in *Perspectives in Drug Discovery and Design,* (Trainor, G., ed.), ESCOM Science Publishers, Leiden, pp. 17–40.

134. Iribarren, A. M., et al. (1990) 2'-*O*-alkyl oligoribonucleotides as antisense probes. *Proc. Natl. Acad. Sci. USA* **87(19),** 7747–7751.

135. Gryaznov, S., et al. (1996) Oligonucleotide N3'—>P5' phosphoramidates as antisense agents. *Nucleic Acids Res.* **24(8),** 1508–1514.

136. Summerton, J. and Weller, D. (1997) Morpholino antisense oligomers: design, preparation, and properties. *Antisense Nucleic Acid Drug Dev.* **7(3),** 187–195.

137. Bacon, T. A., et al. (1988) alpha-Oligodeoxynucleotide stability in serum, subcellular extracts and culture media. *J. Biochem. Biophys. Methods* **16(4),** 311–318.

138. Ho, P. T., et al. (1991) Non-sequence-specific inhibition of transferrin receptor expression in HL-60 leukemia cells by phosphorothioate oligodeoxynucleotides. *Antisense Res. Dev.* **1(4)**, 329–342.

139. Agrawal, S. (1996) Antisense oligonucleotides: towards clinical trials. *Trends Biotechnol.* **14(10)**, 376–387.

140. Monia, B. P., et al. (1993) Evaluation of 2'-modified oligonucleotides containing 2'-deoxy gaps as antisense inhibitors of gene expression. *J. Biol. Chem.* **268(19)**, 14,514–14,522.

141. Agrawal, S., et al. (1997) Mixed-backbone oligonucleotides as second generation antisense oligonucleotides: in vitro and in vivo studies. *Proc. Natl. Acad. Sci. USA* **94(6)**, 2620–2625.

142. Basu, S. and Wickstrom, E. Synthesis and characterization of a peptide nucleic acid conjugated to a D-peptide analog of insulin-like growth factor 1 for increased cellular uptake. *Bioconjug. Chem.* **8(4)**, 481–488.

143. Tan, T. M., et al. (1998) Biologic activity of oligonucleotides with polarity and anomeric center reversal. *Antisense Nucleic Acid Drug Dev.* **8(2)**, 95–101.

144. Aramini, J.M. and M.W. Germann, *NMR studies of DNA duplexes containing alpha-anomeric nucleotides and polarity reversals.* Biochem Cell Biol, 1998. **76**(2-3): p. 403-410.

145. Rait, V. K. and Shaw, B. R. (1999) Boranophosphates support the RNase H cleavage of polyribonucleotides [in process citation]. *Antisense Nucleic Acid Drug Dev.* **9(1)**, 53–60.

146. Basu, S. and Wickstrom, E. (1995) Solid phase synthesis of a D-peptide-phosphorothioate oligodeoxynucleotide conjugate from two arms of a polyethylene glycol-polystyrene support. *Tetrahedron Lett.* **36**, 4943–4946.

147. Hughes, J., et al. (2000) In vitro transport and delivery of antisense oligonucleotides [in process citation]. *Methods Enzymol.* **313**, 342–358.

148. Scherer, L. J. and Rossi, J. J. (2003) Approaches for the sequence-specific knockdown of mRNA. *Nat. Biotechnol.* **21(12)**, 1457–1465.

149. Marwick, C. (1998) First "antisense" drug will treat CMV retinitis. *JAMA* **280(10)**, 871.

150. Walder, R. Y. and Walder, J. A. (1988) Role of RNase H in hybrid-arrested translation by antisense oligonucleotides. *Proc. Natl. Acad. Sci. USA* **85(14)**, 5011–5015.

151. Nielsen, P. E., et al. (1993) Peptide nucleic acids (PNAs): potential antisense and anti-gene agents. *Anticancer Drug Des.* **8(1)**, 53–63.

152. Hanvey, J. C., et al. (1992) Antisense and antigene properties of peptide nucleic acids. *Science* **258(5087)**, 1481–1485.

153. Egholm, M., et al. (1993) PNA hybridizes to complementary oligonucleotides obeying the Watson- Crick hydrogen-bonding rules [see comments]. *Nature* **365(6446)**, 566–568.

154. Bonham, M. A., et al. (1995) An assessment of the antisense properties of RNase H–competent and steric-blocking oligomers. *Nucleic Acids Res.* **23(7)**, 1197–1203.

155. Gray, G. D., Basu, S., and Wickstrom, E. Transformed and immortalized cellular uptake of oligodeoxynucleoside phosphorothioates, 3'-alkylamino oligo-deoxynucleotides, 2'-O-methyl oligoribonucleotides, oligodeoxynucleoside methylphosphonates, and peptide nucleic acids. *Biochem. Pharmacol.* **53(10)**, 1465–1476.

156. Good, L. and Nielsen, P. E. (1998) Inhibition of translation and bacterial growth by peptide nucleic acid targeted to ribosomal RNA. *Proc. Natl. Acad. Sci. USA* **95(5)**, 2073–2076.

157. Baserga, R. (1995) The insulin-like growth factor I receptor: a key to tumor growth? *Cancer Res.* **55(2)**, 249–252.

158. Andrews, D. W., et al. (2001) Results of a pilot study involving the use of an antisense oligodeoxynucleotide directed against the insulin-like growth factor type I receptor in malignant astrocytomas. *J. Clin. Oncol.* **19(8)**, 2189–2200.

159. Pietrzkowski, Z., et al. (1992) Inhibition of cellular proliferation by peptide analogues of insulin- like growth factor 1. *Cancer Res.* **52(23)**, 6447–6451.

160. Lal, R. B., et al. (1993) Over expression of insulin-like growth factor receptor type-I in T-cell lines infected with human T-lymphotropic virus types-I and -II. *Leukemia Res.* **17(1)**, 31–35.

161. Pietrzkowski, Z., et al. (1992) Roles of insulinlike growth factor 1 (IGF-1) and the IGF-1 receptor in epidermal growth factor–stimulated growth of 3T3 cells. *Mol. Cell Biol.* **12(9)**, 3883–3889.

162. Boffa, L. C., et al. (2000) Dihydrotestosterone as a selective cellular/nuclear localization vector for anti-gene peptide nucleic acid in prostatic carcinoma cells [in process citation]. *Cancer Res.* **60(8)**, 2258–2262.

163. Sinha, N. D., et al. (1984) Polymer support oligonucleotide synthesis XVIII: use of beta-cyanoethyl- N,N-dialkylamino-/N-morpholino phosphoramidite of deoxynucleosides for the synthesis of DNA fragments simplifying deprotection and isolation of the final product. *Nucleic Acids Res.* **12(11)**, 4539–4557.

164. Iyer, R. P., et al. (1990) The automated synthesis of sulfur-containing oligodeoxyribonucleotides using 3H-1,2-benzodithiol-3-one 1,1-dioxide as a sulfur-transfer reagent. *J. Org. Chem.* **55(11)**, 4693–4699.

165. Wickstrom, E. L., et al. (1986) HL60 cell proliferation inhibited by an anti-c-*myc* pentadecadeoxynucleotide. *Fed. Proc.* **45**, 1708.

166. Wickstrom, E. L., et al. (1989) Anti-c-*myc* DNA increases differentiation and decreases colony formation by HL-60 cells. *In Vitro Cell. Dev. Biol.* **25(3 Pt. 1)**, 297–302.

167. Bacon, T. A. and Wickstrom, E. (1991) Daily addition of an anti-c-*myc* DNA oligomer induces granulocytic differentiation of human promyelocytic leukemia HL-60 cells in both serum-containing and serum-free media. *Oncogene Res.* **6(1)**, 21–32.

168. Bacon, T. A. and Wickstrom, E. (1991) Walking along human c-*myc* mRNA with antisense oligodeoxynucleotides: maximum efficacy at the 5' cap region. *Oncogene Res.* **6(1)**, 13–19.

169. Sawadogo, M. and Van Dyke, M. W. (1991) A rapid method for the purification of deprotected oligodeoxynucleotides. *Nucleic Acids Res.* **19(3)**, 674.

170. Nielsen, P. E. and Egholm, M. (1999) Peptide nucleic acids: protocols and applications. Horizon Scientific Press, Norfolk, England.

171. Guvakova, M. A. and Surmacz, E. Tamoxifen interferes with the insulin-like growth factor I receptor (IGF-IR) signaling pathway in breast cancer cells. *Cancer Res.* **57(13)**, 2606–2610.

172. Mauro, L., et al. (2001) IGF-I receptor-induced cell-cell adhesion of MCF-7 breast cancer cells requires the expression of junction protein ZO-1. *J. Biol. Chem.* **276(43)**, 39,892–39,897.

173. Ulsh, L. S. and Shih, T. Y. (1984) Metabolic turnover of human c-rasH p21 protein of EJ bladder carcinoma and its normal cellular and viral homologs. *Mol. Cell Biol.* **4(8)**, 1647–1652.

174. Pallela, V. R., et al. (1999) ^{99m}Tc-labeled vasoactive intestinal peptide receptor agonist: functional studies. *J. Nucl. Med.* **40(2)**, 352–360.

175. Bennett, C. F., et al. (1996) Pharmacology of antisense therapeutic agents, in *Antisense therapeutics*, (Agrawal, S, ed.) Humana, Totowa, NJ, pp. 13–46.

176. Tian, X. and Wickstrom, E. (2002) Continuous solid-phase synthesis and disulfide cyclization of peptide-PNA-peptide chimeras. *Org. Lett.* **4(23)**, 4013–4016.

177. Masur, S. and Jayalekshmy, P. (1978) Chemistry of polymer-bound o-benzyne. Frequency of encounter between substituents on crosslinked polystyrenes. *J. Am. Chem. Soc.* **101**, 677–683.

178. Albericio, F., et al. (1991) Cyclization of disulfide-containing peptides in solid-phase synthesis. *Int. J. Pept. Protein Res.* **37(5)**, 402–413.

179. Yajima, H., et al. (1988) New strategy for the chemical synthesis of proteins. *Tetrahedron* **44**, 805–819.

180. Garcia-Echevarria, C., et al. (1993) Design, synthesis, and complexing properties of (1Cys-1'Cys, 4Cys-4'Cys)-dithiobis (Ac-L-1Cys-L-Pro-D-Val-L-4Cys-NH2). The first example of a new family of ion-binding peptides. *J. Am. Chem. Soc.* **115**, 11,663–11,670.

181. Kamber, B., et al. (1980) The synthesis of cystine peptides by iodine oxidation of S-tritylcysteine and S-acetamidomethylcysteine peptides. *Helv. Chim. Acta* **63**, 899–915.

182. Branden, L. J., Mohamed, A. J., and Smith, C. I. (1999) A peptide nucleic acid-nuclear localization signal fusion that mediates nuclear transport of DNA. *Nat. Biotechnol.* **17(8)**, 784–787.

183. Cutrona, G., et al. (2000) Effects in live cells of a c-*myc* anti-gene PNA linked to a nuclear localization signal. *Nat. Biotechnol.* **18(3)**, 300–303.

184. Soomets, U., Hallbrink, M. and Langel, U. Antisense properties of peptide nucleic acids. *Front. Biosci.* **4**, D78–76.

185. Pooga, M., et al. (1998) Cell penetrating PNA constructs regulate galanin receptor levels and modify pain transmission in vivo. *Nat. Biotechnol.* **16(9)**, 857–861.

186. Kischkel, F. C., et al. (2001) Death receptor recruitment of endogenous caspase-10 and apoptosis initiation in the absence of caspase-8. *J. Biol. Chem.* **276(49)**, 46,639–46,646.

9

Suppression of Pancreatic and Colon Cancer Cells by Antisense K-*ras* RNA Expression Vectors

Kazunori Aoki, Shumpei Ohnami, and Teruhiko Yoshida

1. Introduction

Various genetic abnormalities accumulate during multistage carcinogenesis. The changes include structural aberrations such as point mutation, amplification, and deletion, as well as functional alterations such as abnormal level and timing of gene expression *(1,2)*. Although the type and incidence of genetic alteration differ according to the type of human cancer, the most striking example may be the mutations of the K-*ras* gene, whose activation by point mutation is found at characteristically high frequencies of 70–90% in pancreatic cancer *(3–7)*. That incidence of gene mutation is followed by 40–50% in colon and thyroid tumors and 20–30% in lung cancers *(4)*. More than 90% of K-*ras* mutations are located at codon 12, with the remainder at codons 13 and 61 *(3–7)*. Considering the prominent transforming activity of the mutated K-*ras* oncogene on NIH3T3 cells *(8,9)*, K-*ras* may play a critical role in the expression of the malignant phenotype and serve as a good target in gene therapy for these cancers. It is expected that the suppression of K-*ras* activation provides some benefits for patients with advanced pancreatic and colorectal cancers.

Specific suppression of *ras* family oncogene expression has been examined in several cancers. Suppression of H-*ras* expression by antisense oligonucleotide, antisense RNA, or ribozyme led to inhibition of the neoplastic phenotype of bladder carcinoma cells and NIH3T3 cells transformed by H-*ras* oncogene *(10–14)*. The antisense K-*ras* retroviral vector infection was useful in suppressing the tumorigenicity of lung cancer in a nude mouse orthotopic

From: *Methods in Molecular Medicine, Vol. 106: Antisense Therapeutics, Second Edition*
Edited by: I. Phillips © Humana Press Inc., Totowa, NJ

transplantation model *(15,16)*. Adenovirus (Ad)-mediated gene transfer of antisense K-*ras* produced an inhibition of growth and colony formation of lung cancer cells *(17)*. We showed that the liposome-mediated in vivo gene transfer of the antisense K-*ras* RNA expression vector inhibits pancreatic tumor dissemination in the murine peritoneal cavity *(18,19)* and also reported that the Ad-mediated in vivo gene transfer of the antisense K-*ras* construct suppressed the growth of colorectal cancer in a murine sc tumor model *(20)*.

2. Materials

2.1. Construction of Antisense K-ras-Expressing Plasmids

1. Human placental mRNA (TOYOBO, Tokyo, Japan).
2. Oligonucleotide primers.
3. LNSX vector *(21)*.
4. Reverse transcriptase (Rt), *Taq* polymerase, restriction enzymes, T7 DNA polymerase, and T4 DNA ligase.
5. Competent cell, *Escherichia coli* strain DH5α (TaKaRa, Tokyo, Japan).
6. Luria–Bertani (LB) plate and medium.
7. Ampicillin.
8. DNA-sequencing equipment.

2.2. Construction and Purification of Recombinant Ad Vector

1. pAxCAwt vector and Ad5-dlX DNA-TPC (TaKaRa).
2. 293 cells (human embryonic primary kidney cells transformed with sheared Ad5 DNA) (American Type Culture Collection [ATCC] CRL-1573).
3. Calcium phosphate transfection reagents (Amersham Pharmacia Biotech, Piscataway, NJ).
4. Dulbecco's modified Eagle's medium (DMEM) (with 4500 mg/L of glucose, L-glutamin, and sodium phosphate) (Sigma, St. Louis, MO).
   a. Penicillin/streptomycin solution (Gibco-BRL, Gaithersburg, MD); store at –20°C.
   b. Serum (fetal calf) heat inactivated at 56°C for 30–45 min. Store at –20°C.
5. Cell propagation medium: DMEM (500 mL), penicillin/streptomycin solution (5 mL), fetal calf serum (FCS) (50 mL).
6. Infection medium: DMEM (500 mL), penicillin/streptomycin solution (5 mL), FCS (10 mL).
7. CsCl-gradient solutions: For a light CsCl (density: 1.20 g/mL), dissolve 22.39 g of CsCl in 77.61 mL of 10 m*M* Tris-HCl, pH 8.1; for a heavy CsCl (density: 1.45 g/mL), dissolve 42.33 g of CsCl in 57.77 mL of 10 m*M* Tris-HCl, pH 8.1.
8. Ultra centrifuge tubes: 50 Ultra-Clear tubes, 1 × 3.5 in. (25 × 89 mm) and 9/16 × 3. 5 in. (14 × 89 mm) (Beckman, Palo Alto, CA).
9. Bio-Gel 6-P DG chromatography column (Bio-Rad, Hercules, CA).
10. 65% Glycerol in phosphate-buffered saline (PBS); store at 4°C.
11. Tissue culture dishes (96-well, 60-mm, and 150-mm).

2.3. Expression of Antisense K-ras RNA in Transduced Cells

1. Cell lines (ATCC, Rockville):
   a. AsPC-1, a human pancreatic cancer cell line.
   b. HCT-15, a human colon cancer cell line.
2. Tissue culture dishes (10-mm).
3. RPMI-1640 medium (Sigma) (500 mL) with 5 mL of penicillin/streptomycin solution and 50 mL of FCS.
4. Antisense-K-*ras*-expressing Ad vector; store at –80°C (from **Subheading 3.2.**).
5. Nitrocellulose membrane (NitroPlus; MSI, Westboro, MA).
6. Agarose gel electrophoresis and filter transfer equipment.
7. pGEM vector (Riboprobe systems; Promega, Madison, WI).
8. T7 and T3 bacteriophage DNA-dependent RNA polymerase.

2.4. Downregulation of K-ras p21 Protein

1. Cells and viruses (same as for **Subheading 2.3.**).
2. RIPA buffer: 10 m*M* Tris-HCl, pH 7.4; 1% deoxycholate; 1% Nonidet P-40; 150 m*M* NaCl; 0.1% sodium dodecyl sulfate (SDS); 0.2 m*M* phenylmethyl-sulfonyl fluoride, 1 µg/mL of aprotinin; 1 µg/mL of leupeptin.
3. 2X Laemmli's buffer: 0.125 *M* Tris-HCl, pH 6.8, 4% SDS, 20% Glycerol, 10% 2-mercaptoethanol. 2-Mercaptoethanol should be added just before use.
4. Protein assay dye concentrate (Bio-Rad) and microplate reader.
5. SDS-polyacrylamide gel, electrophoresis and transfer equipment.
6. Polyvinylidene difluoride membrane (NEM Life Science, Boston, MA).
7. PBS containing 10% skim milk (Becton Dickinson, Sparks, MD) and 1% bovine serum albumin (BSA) (Sigma).
8. K-*ras*-specific p21 rabbit monoclonal antibody (MAb) (Calbiochem, San Diego, CA).
9. Horseradish peroxidase (HRP)-conjugated goat anti-rabbit IgG antibody (Zymed, South San Francisco, CA).
10. PBS containing 0.01% Tween-20.
11. Enhanced chemiluminescence system (Amersham Pharmacia Biotech).

2.5. Growth Suppression of AxCA-AS-K-ras-Transduced Cells In Vitro

1. Cells and viruses (same as for **Subheading 2.3.**).
2. 96-well plates and microplate reader.
3. Water-soluble tetrazolium salt (Tetracolor One; Seikagaku, Tokyo, Japan).

2.6. Growth Suppression of Antisense K-ras Vector–Transduced Cells In Vivo

1. BALB/c nude mice and severe combined immunodeficient (SCID) mice (Charles River Japan, Kanagawa, Japan).
2. Cells (same as for **Subheading 2.3.**).
3. Hank's balanced salt (HBSS) solution.

4. 26G needles and 1-mL syringes.
5. Adenoviruses:
   a. Antisense or sense-K-*ras*-expressing virus (from **Subheading 3.2.**).
   b. LacZ-expressing virus.
6. Antisense or sense K-*ras* expression plasmids (from **Subheading 3.1.**).
7. Dioctadecylamidoglycylspermine (DOGS) (Biosepra).
8. 0.15 *M* NaCl.
9. OCT compound (Miles, Elkhart, IN) and slide glasses.
10. 0.25% Glutaraldehyde in PBS.
11. 5-Bromo-4-chloro-3-indoyl-β-D-galactopyranoside (X-gal) substrate solution: 5 m*M* K₃Fe(CN)₆, 5 m*M* K₄Fe(CN)₆, 2 m*M* MgCl₂, and 1 mg/mL of X-gal.

3. Methods

3.1. Construction of Antisense K-ras-Expressing Plasmids

Dr. Dusty Miller (Fred Hutchinson Cancer Research Center, Seattle, WA) generously provided the retroviral vector plasmid LNSX. LNSX contains the selectable neomycin phosphotransferase gene, which is expressed from the retroviral long terminal repeat (LTR), and an inserted DNA is placed in the downstream of an internal SV40 early promoter *(21)*. In this experiment, the LNSX is used as an expression plasmid vector (*see* **Note 1**). A K-*ras* cDNA fragment spanning from nucleotide (nt) 171 in the first exon to nt 517 in the third exon *(22)* is amplified by RT-polymerase chain reaction from normal human placental mRNA (Toyobo). The *Cla*I site of cloned K-*ras* fragment is digested at the 5' end of the upstream primer and at the *Avr*II site at the 3' end of the downstream primer to obtain a K-*ras* cDNA fragment in antisense orientation, whereas the engineered restriction sites are reversed for the sense construct. LNSX is digested with *Cla*I and *Avr*II, and after ligation of the cloned K-*ras* fragment into LNSX, the DNA is used to transform *E. coli* DH5α cells by the standard methods *(23)*. The *E. coli* DH5α cells are then plated onto LB plates containing ampicillin (50–100 µg/mL) and incubated overnight at 37°C. Single colonies are selected and grown overnight in LB with ampicillin. The plasmid DNA is then isolated *(23)* and checked for the presence of the insert and for the correct orientation using restriction enzyme digestions and DNA sequencing. The recombinant plasmid expressing antisense K-*ras* RNA or sense K-*ras* RNA is designated AS-K-*ras*-LNSX or S-K-*ras*-LNSX, respectively (**Fig. 1**).

3.2. Construction and Purification of Recombinant Ad Vector

The cassette cosmid for constructing recombinant Ad of the E1-substitution type, pAxCAwt is an 11-kb charomid vector bearing an Ad5 genome spanning map units (mu) 0–99.3 with deletions of E1 (mu 1.3–9.3) and E3 (mu 79.6–84.8) *(24)*. The vector harbors the CAG promoter, which is a chicken β-actin

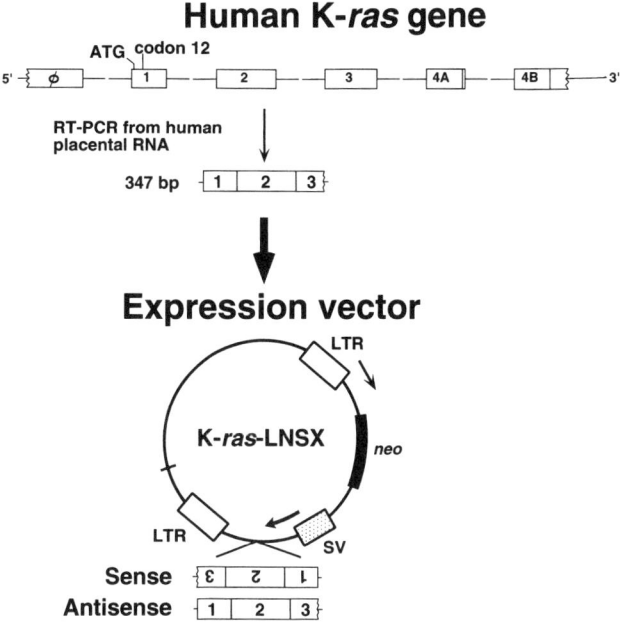

Fig. 1. K-*ras* expression plasmids. A 347-bp K-*ras* cDNA fragment containing exons 1 and 2 and part of exon 3 is subcloned into the LNSX plasmid in antisense and sense orientation. LTR, long terminal repeat; SV, SV40 early promoter and enhancer; neo, bacterial neomycin phosphotransferase.

promoter fused to a cytomegalovirus enhancer *(25,26)*, a unique *Swa*I site, and a rabbit β-globin poly (A) sequence in the leftward orientation at the AdE1-deleted position (Adenovirus expression vector kit; TaKaRa).

3.2.1. Construction of Ad Vector

1. Digest and insert the *Swa*I sites of the cloned K-*ras* fragment (spanning from nt 171 to 517) at the ends of the upstream and downstream primers into pAxCAwt (TaKaRa) in antisense (pAxCA-AS-K-*ras*) or sense orientation (pAxCA-S-K-*ras*) as described in **Subheading 3.1.**
2. Prepare the *Eco*T22I-digested adenoviral DNA-terminal protein complex (DNA-TPC) from the parent Ad5-dlX Ad, which has an E3 deletion (mu 79.6–84.8) (TakaRa) *(27)*.
3. To construct the recombinant Ad, cotransfect 1 μg of Ad5-dlX DNA-TPC digested with *Eco*T22I into 293 cells in a 60-mm dish together with 8 μg of pAxCA-AS-K-*ras* or pAxCA-S-K-*ras* by the calcium phosphate method (Cell-Phect Transfection Kit; Amersham Pharmacia Biotech). The desired recombinant Ads are generated from overlapping recombination in the 293 cells.

4. One day later, spread the cells in three 96-well plates at a 10-fold serial dilution mixed with untransfected 293 cells.

5. After maintaining in culture for 10–15 d, further isolate and propagate the virus clones to assess restriction analysis. The resulting Ad vectors expressing the K-*ras* cDNA fragments in the antisense or sense orientation are designated AxCA-AS-K-*ras* and AxCA-S-K-*ras*, respectively.

3.2.2. Purification and Storage of Recombinant Ads

1. Use viral lysates of AxCA-AS-K-*ras* or AxCA-S-K-*ras* to infect 20–30 tissue culture dishes (150 mm) of 293 cells.

2. After 3–5 d, when they show a cytopathic effect, harvest the cells and then centrifuge at 220*g* for 5 min. Remove the medium and add 20–30 mL of 10 m*M* Tris-HCl, pH 8.0. After four cycles of freezing and thawing, spin the viral lysates at 2500*g* for 7 min and collect the Ad-containing supernatant.

3. Overlay the supernatant on the first CsCl gradient (10 mL of light CsCl and 10 mL of heavy CsCl solutions) in 1×3.5 in. Ultra-Clear tubes, and centrifuge at 20,000 rpm for 2 h at 4°C in an SW28 rotor (Beckman).

4. Collect the visible virus band and then overlay on the second CsCl gradient (4 mL of heavy CsCl and 4 mL of light CsCl solutions) in $^9/_{16} \times 3.5$ in. Ultra-Clear tubes, and centrifuged at 20,000 rpm for 12–18 h at 4°C in an SW41 rotor (Beckman). Collect a sharp band consisting of virus particles.

5. Purify the viruses through a Bio-Gel P-6 DG chromatography column (EconoPac 10DG; Bio-Rad), and store at –80°C in PBS containing 13% glycerol.

3.2.3. Determination of Virus Titer

The infectivity of recombinant viruses, or titers of virus stocks, is determined by an end-point cytopathic effect assay *(24)*.

1. Dispense 50 µL of DMEM with 10% FCS into each well of a 96-well plate, and then prepare eight rows of threefold serial dilution of the virus starting from a 10^{-4} dilution.

2. Add 3×10^5 293 cells in 50 µL of DMEM with 5% FCS to each well.

3. Incubate the plates at 37°C in 5% CO_2 in air, and add 50 µL of DMEM with 10% FCS to each well every 3 d.

4. Twelve days later, determine by microscopy, the highest dilution (end point) showing the cytopathic effect, and calculate a 50% tissue culture infectious dose ($TCID_{50}$). The $TCID_{50}$/mL corresponds approx to 1 plaque-forming unit (PFU)/mL.

3.3. Expression of Antisense K-ras RNA in Transduced Cells

AsPC-1 and HCT-15 cells are obtained from ATCC and maintained in RPMI-1640 medium with 10% FCS.

1. Infect the cells with AxCA-AS-K-*ras* or AxCA-S-K-*ras* at an moi of 30 in a minimal amount of culture medium (0.3 mL/100-mm plate) in 5% CO_2 for 2 h at 37°C with the plates being rocked every 15 min.

2. Twenty-four hours later, harvest the cells and extract poly(A)$^+$ RNA from the cells.
3. Prepare strand-specific RNA probes by the K-*ras* cDNA fragment (exons 1 and 2 and part of exon 3) subcloned into a pGEM vector. Synthesize sense and antisense RNA probes using a T7 and T3 bacteriophage DNA-dependent RNA polymerase (Riboprobe systems; Promega).
4. Size-fractionate 2 μg of poly(A)$^+$ RNA on a 1.0% denaturing agarose gel, transfer onto a nitrocellulose membrane (NitroPlus; MSI), and hybridize with the strand-specific RNA probe. Perform hybridization in 50% formamide, 5X Denhardt's solution, 0.1% SDS, 5X SSPE, and 100 μg/mL of salmon testis DNA at 42°C for 16 h. Then wash the filters in 0.1X SSPE and 0.1% SDS at 65°C for 30 min and expose to X-ray film.

3.4. Downregulation of K-ras p21 Protein by Antisense K-ras-Expressing Ad

3.4.1. Protein Extraction

1. Seed the cells at 1×10^5 in 100-mm tissue culture dishes.
2. Twenty-four hours later, infect the cells with AxCA-AS-K-*ras* or Ax-CA-S-K-*ras* at an moi of 30 in a minimal amount of culture medium (0.3 mL/100-mm plate) in 5% CO_2 for 2 h at 37°C with the plates being rocked every 15 min.
3. Forty-eight hours after the infection, wash the cells two times with cold (4°C) PBS, and then lyse in 400–500 μL of RIPA buffer.
4. Measure the concentrations of cell lysates using Bio-Rad protein assay reagent: Mix 1 μL of cell lysate with 150 μL of protein assay dye concentrate (1:5 dilution in H_2O), and incubate at room temperature for 15 min. Measure the absorbance by using a microplate reader with a wavelength of 595 nm, and calculate the protein concentration as compared with references. Store the cell lysates at –20°C.

3.4.2. Western Blotting

1. Mix 80 μL of cell lysates with 2X Laemmli's buffer containing 10% fresh 2-mercaptoethanol, and heat at 95°C for 5 min (*see* **Note 2**).
2. Size-fractionate the cell lysates by 8–16% SDS-polyacrylamide gel (TEFCO, Tokyo, Japan), and electroblot onto polyvinylidene difluoride membranes (NEM Life Science).
3. Block the membranes using PBS containing 10% skim milk (Becton Dickinson) and 1% BSA (Sigma) at 4°C for 12–16 h with rocking.
4. Probe the membranes with a K-*ras* specific MAb (Calbiochem) (10X dilution in blocking buffer) for 3 to 4 h at room temperature with rocking (*see* **Note 2**).
5. Wash the membranes with PBS containing 0.01% Tween-20 three times, for 5 min each time, and then probe with an HRP-conjugated anti–rabbit IgG (Zymed) (10,000X dilution in blocking buffer) for 1 h at room temperature with rocking.
6. Wash the membranes with PBS containing 0.01% Tween-20 three times, for 5 min each time, and then detect the K-*ras* protein using an enhanced chemiluminescence system (Amersham Pharmacia Biotech).

3.5. Growth Suppression of AxCA-AS-K-ras-Transduced Cells In Vitro

1. Seed the cells at 2×10^3/well in 96-well tissue culture plates.
2. One day later, infect the cells with AxCA-AS-K-*ras* or AxCA-S-K-*ras* at an moi of 10, 30, and 100 in a minimal amount of culture medium (50 µL/well) in 5% CO_2 for 2 h at 37°C with the plates being rocked every 15 min.
3. Assay the cell numbers by a colorimetric cell viability assay using a water-soluble tetrazolium salt (Seikagaku) for 7–9 d after the infection. Determine the absorbance using a microplate reader at a wavelength of 450 nm with 595 nm as a reference. Repeat the assays (carried out in eight wells) a minimum of two times.

3.6. Growth Suppression of Antisense K-ras Vector–Transduced Cells In Vivo

Five-week-old male BALB/c nude mice and SCID mice are obtained from Charles River and kept in a specific pathogen-free environment.

3.6.1. Suppression of Tumor Dissemination in Murine Peritoneal Cavity

1. Harvest AsPC-1 cells with trypsin and resuspend in HBSS. Inject the AsPC-1 cell suspensions (6×10^6 cells/50 µL) into the ip cavity of the mice, which will result in the peritoneal dissemination and formation of tumor nodules in pancreatic regions within 28 d.
2. Three days after ip inoculation of the AsPC-1 cells, inject the BALB/c nude mice intraperitoneally with the AS-K-*ras*-LNSX plasmid complexed with DOGS lipopolyamine *(28)* at 12 h intervals for a total of three times (*see* **Note 3**). As a control, use the S-K-*ras*-LNSX plasmid complexed with DOGS. Prepare DOGS complexes as follows: Dilute 100 µg of plasmid in 300 µL of 0.15 M NaCl, and dilute 400 nmol of DOGS in 300 µL of 0.15 M NaCl; then add the DOGS solution to the plasmid solution. After incubating for 5–10 min at room temperature, inject the resulting mixture. Alternatively, one can inject 1×10^9 PFU of the AxCA-AS-K-*ras* Ad vector intraperitoneally at 12 h intervals for a total of three times, and as a control, use the AxCA-S-K-*ras* Ad vector.
3. Twenty-eight days after the injection of AsPC-1, sacrifice and examine the mice under a stereomicroscope to determine whether they have developed peritoneal dissemination and tumors in the pancreas.

3.6.2. Suppression of sc Tumor

1. Harvest HCT-15 or AsPC-1 cells with trypsin and resuspend in HBSS. Inject the cell suspensions (5×10^6 cells/50 µL) subcutaneously into the left flank of SCID or nude mice.
2. When the sc tumor nodule reaches approx 4 mm in diameter, inject 0.5×10^9 PFU of AxCA-AS-K-*ras* intratumorally three times every 24 h. As a control, use AxCA-S-K-*ras*.
3. Sacrifice one mouse from each group for the histological examination of the sc tumor using hematoxylin and eosin (H&E) and for further analyses, and observe

the remaining animals for tumor growth. Measure the short (r) and long (l) diameters of the tumors for 3 to 4 wk, and calculate the tumor volume of each as $r^2l/2$.
4. To evaluate the gene transduction efficiency, inject an Ad vector expressing lacZ (AxCA-LacZ) into the sc tumor. Forty-eight hours later, remove the sc tumors, freeze with OCT compound (Miles), and store at −80°C. Fix the cryostat sections (5–10 μm) of the specimens with 0.25% glutaraldehyde in PBS for 5 min at room temperature, wash with PBS two times, and develop in X-gal reaction solution at 37°C for 3 h. Then counterstain the X-gal-stained tissues with H&E.

4. Notes

1. The LNSX expression plasmid was employed as an expression vector in the murine peritoneal dissemination model. Northern blot analysis showed that the vector expressed a significant amount of the read-through transcript from the 5' LTR, in addition to the internal SV40 promoter–driven antisense K-*ras* RNA *(19)*. It is possible that the use of vectors, which can produce a higher amount of the proper antisense transcript, may enhance the inhibitory effect on the growth of pancreatic cancer.
2. In Western blot analysis, a K-*ras*-specific MAb (Calbiochem) should be used at a 1:10 dilution in the blocking buffer to obtain a clear K-*ras* band, and more than 80 μg of cell lysates should be loaded into SDS-polyacrylamide gel, since the expression level of K-*ras* p21 protein is not high. Blocking condition (PBS containing 10% skim milk and 1% BSA at 4°C for 12–16 h) seems to be important to reduce the membrane background.
3. Although the AS-K-*ras*-LNSX plasmid complexed with DOGS was intraperitoneally injected in a murine peritoneal dissemination model, we recently demonstrated that a linear form of polyethylenimine, which is the organic macromolecule with the highest cationic charge density potential *(29,30)*, was more efficient than several cationic lipids containing DOGS in transducing the transgene into the peritoneal dissemination of mice *(31)*. When the transfection mixture is prepared, the DOGS or polyethylenimine solutions should be added into DNA solutions. After incubation for 5–10 min, the transfection mixture becomes slightly cloudy. If the large aggregates are visible in the mixture, the total volume should be increased, because the formation of aggregates reduces gene transfer efficiency.

Acknowledgments

We thank Dr. Dusty Miller for providing the retrovirus vector LNSX, Dr. Izumu Saito (Institute of Medical Science, University of Tokyo) for providing the Ad vector expressing lacZ (AxCA-LacZ), and Kimiko Yoshida for technical help. This work was supported in part by a grant-in-aid from the second-term Comprehensive 10-Year Strategy for Cancer Control, and Health Sciences Research Grants from the Ministry of Health, Welfare and Labour of Japan.

References

1. Sugimura, T. (1992) Multistep carcinogenesis: a 1992 perspective. *Science* **258,** 603–607.
2. Howe, J. R. and Conlon, K. C. (1997) The molecular genetics of pancreatic cancer. *Surg. Oncol.* **6,** 1–18.
3. Almoguera, C., Shibata, D., Forrester, K., Martin, J., Arnheim, N., and Perucho, M. (1988) Most human carcinomas of the exocrine pancreas contain mutant c-K-ras genes. *Cell* **53,** 549–554.
4. Bos, J. L. (1989) ras oncogenes in human cancer: a review. *Cancer Res.* **49,** 4682–4689.
5. Mariyama, M., Kishi, K., Nakamura, K., Obata, H., and Nishimura, S. (1989) Frequency and types of point mutation at the 12th codon of the c-Ki-ras gene found in pancreatic cancers from Japanese patients. *Jpn. J. Cancer Res.* **80,** 622–626.
6. Motojima, K., Urano, T., Nagata, Y., Shiku, H., Tsunoda, T., and Kanematsu, T. (1991) Mutations in the Kirsten-ras oncogene are common but lack correlation with prognosis and tumor stage in human pancreatic carcinoma. *Am. J. Gastroenterol.* **86,** 1784–1788.
7. Lee, K. H., Lee, J. S., Suh, C., Kim, S. W., Kim, S. B., Lee, J. H., Lee, M. S., Park, M. Y., Sun, H. S., and Kim, S. H. (1995) Clinicopathologic significance of the K-ras gene codon 12 point mutation in stomach cancer: an analysis of 140 cases. *Cancer* **75,** 2794–2801.
8. Nakano, H., Yamamoto, F., Neville, C., Evans, D., Mizuno, T., and Perucho, M. (1984) Isolation of transforming sequences of two human lung carcinomas: structural and functional analysis of the activated c-K-ras oncogenes. *Proc. Natl. Acad. Sci. USA* **81,** 71–75.
9. Prassolov, V. S., Sakamoto, H., Nishimura, S., Terada, M., and Sugimura, T. Activation of c-Ki-ras gene in human pancreatic cancer. *Jpn. J. Cancer Res.* **76,** 792–795.
10. Lu, G. R., Wu, C. J., Ke, Y., et al. (1991) Inhibited neoplastic phenotype by the c-Ha-ras antisense RNA. *Sci. China B.* **34,** 1485–1491.
11. Daaka, Y. and Wickstrom, E. Target dependence of antisense oligodeoxy-nucleotide inhibition of c-Ha-ras p21 expression and focus formation in T24-transformed NIH3T3 cells. *Oncogene Res.* **5,** 267–275.
12. Saison-Behmoaras, T., Tocque, B., Rey, I., Chassignol, M., Thuong, N. T., and Helene, C. (1991) Short modified antisense oligonucleotides directed against Ha-ras point mutation induce selective cleavage of the mRNA and inhibit T24 cells proliferation. *EMBO J.* **10,** 1111–1118.
13. Kashani-Sabet, M., Funato, T., Florenes, V. A., Fodstad, O., and Scanlon, K. J. (1994) Suppression of the neoplastic phenotype in vivo by an anti-ras ribozyme. *Cancer Res.* **54,** 900–902.
14. Monia, B. P., Johnston, J. F., Ecker, D. J., Zounes, M. A., Lima, W. F., and Freier, S. M. (1992) Selective inhibition of mutant Ha-ras mRNA expression by antisense oligonucleotides. *J. Biol. Chem.* **267,** 19,954–19,962.

15. Georges, R. N., Mukhopadhyay, T., Zhang, Y., Yen, N., and Roth, J. A. (1993) Prevention of orthotopic human lung cancer growth by intratracheal instillation of a retroviral antisense K-ras construct. *Cancer Res.* **53,** 1743–1746.

16. Zhang, Y., Mukhopadhyay, T., Donehower, L. A., Georges, R. N., and Roth, J. A. (1993) Retroviral vector-mediated transduction of K-ras antisense RNA into human lung cancer cells inhibits expression of the malignant phenotype. *Hum. Gene Ther.* **4,** 451–460.

17. Alemany, R., Ruan, S., Kataoka, M., et al. (1996) Growth inhibitory effect of anti-K-ras adenovirus on lung cancer cells. *Cancer Gene Ther.* **3,** 296–301.

18. Aoki, K., Yoshida, T., Matsumoto, N., Ide, H., Sugimura, T., and Terada, M. (1997) Suppression of Ki-ras p21 levels leading to growth inhibition of pancreatic cancer cell lines with Ki-ras mutation but not those without Ki-ras mutation. *Mol. Carcinog.* **20,** 251–258.

19. Aoki, K., Yoshida, T., Sugimura, T., and Terada, M. (1995) Liposome-mediated in vivo gene transfer of antisense K-ras construct inhibits pancreatic tumor dissemination in the murine peritoneal cavity. *Cancer Res.* **55,** 3810–3816.

20. Nakano, M., Aoki, K., Matsumoto, N., et al. (2001) Suppression of colorectal cancer growth using an adenovirus vector expressing an antisense K-ras RNA. *Mol. Ther.* **3,** 491–499.

21. Miller, A. D., Miller, D. G., Garcia, J. V., and Lynch, C. M. (1993) Use of retroviral vectors for gene transfer and expression. *Methods Enzymol.* **217,** 581–599.

22. Capon, D. J., Seeburg, P. H., McGrath, J. P., et al. (1983) Activation of Ki-ras2 gene in human colon and lung carcinomas by two different point mutations. *Nature* **304,** 507–513.

23. Sambrook, J., Fritsch, E. F., and Maniatis, T. (1989) *Molecular Cloning: A Laboratory Manual*, 2nd ed. Cold Spring Harbor Laboratory Press, Cold Spring Harbor, New York.

24. Miyake, S., Makimura, M., Kanegae, Y., et al. (1996) Efficient generation of recombinant adenoviruses using adenovirus DNA-terminal protein complex and a cosmid bearing the full-length virus genome. *Proc. Natl. Acad. Sci. USA* **93,** 1320–1324.

25. Miyazaki, J., Takaki, S., Araki, K., et al. (1989) Expression vector system based on the chicken beta-actin promoter directs efficient production of interleukin-5. *Gene* **79,** 269–277.

26. Niwa, H., Yamamura, K., and Miyazaki, J. (1991) Efficient selection for high-expression transfectants with a novel eukaryotic vector. *Gene* **108,** 193–199.

27. Kanegae, Y., Lee, G., Sato, Y., et al. (1995) Efficient gene activation in mammalian cells by using recombinant adenovirus expressing site-specific Cre recombinase. *Nucleic Acids Res.* **23,** 3816–3821.

28. Behr, J. P., Demeneix, B., Loeffler, J. P., and Perez-Mutul, J. (1989) Efficient gene transfer into mammalian primary endocrine cells with lipopolyamine-coated DNA. *Proc. Natl. Acad. Sci. USA* **86,** 6982–6986.

29. Boussif, O., Lezoualc'h, F., Zanta, M. A., et al. (1995) A versatile vector for gene and oligonucleotide transfer into cells in culture and in vivo: polyethylenimine. *Proc. Natl. Acad. Sci. USA* **92,** 7297–7301.

30. Abdallah, B., Hassan, A., Benoist, C., Goula, D., Behr, J. P., and Demeneix, B. A. (1996) A powerful nonviral vector for in vivo gene transfer into the adult mammalian brain: polyethylenimine. *Hum. Gene Ther.* **7,** 1947–1954.
31. Aoki, K., Furuhata, S., Hatanaka, K., et al. (2001) Polyethylenimine-mediated gene transfer into pancreatic tumor dissemination in the murine peritoneal cavity. *Gene Ther.* **8,** 508–514.

10

Induction of Tumor Cell Apoptosis and Chemosensitization by Antisense Strategies

Manuel Rieber and Mary Strasberg-Rieber

1. Introduction

Radiation and traditional chemotherapy kill both normal and tumor cells owing to lack of target selectivity. In this chapter, we review methods for the use of antisense oligodeoxynucleotides (AS-DONs) directed against gene products overexpressed in tumor cells in synergy with other anticancer agents. In contrast to the low basal levels of the mitogen-dependent cyclin D1 and of the DNA-damage-inducible p53 tumor suppressor protein, tumors with greater genetic instability and poor therapeutic response show a mutant-stabilized p53 protein, a mitogen-independent constitutive cyclin D1, and overexpression of antiapoptotic genes such as *bcl-2* and *bcl-xL*. Since overexpression of any one of these gene products decreases apoptosis in malignant cells, we propose the use of subtoxic antisense technology specifically directed against some of these tumor-associated targets prior to treatment with sublethal chemotherapy as a strategy to diminish damage to normal cells and the emergence of cancer cells resistant to conventional therapy. AS-ODN technology capable of antagonizing gene sequences preferentially expressed in tumors combined with standard anticancer therapy offers an alternative approach to improve target selectivity, diminish anticancer toxicity, and lower drug resistance.

The Human Genome Project and new high-throughput technologies such as cDNA differential display or DNA and protein microarrays are allowing the rapid acquisition of information about multiple changes that occur in response to signaling events capable of activating some biochemical pathways and able to cause the reciprocal inactivation of other pathways. This alternation in specific gene activation and silencing not only depends on a particular signaling stimuli

From: *Methods in Molecular Medicine, Vol. 106: Antisense Therapeutics, Second Edition*
Edited by: I. Phillips © Humana Press Inc., Totowa, NJ

such as a cytotoxic drug or ionizing radiation, but may be influenced by tissue specificity, heterotypic or homotypic cell–cell interaction, degree of differentiation, association with particular extracellular matrix proteins, extent of malignant progression, and so on. Growth of K1735 melanoma in vivo and in vitro can be suppressed by the introduction of an antisense *cyclin D1* *(1)*. This gene product is inducible in normal cells but constitutive in tumors *(2,3)*; its targeting may produce a selective effect against highly proliferating cells *(1,2)*. In the particular case of cancer, the second leading cause of death worldwide, it has been possible to identify a group of genes like p53, which is mutated and overexpressed in more than 50% of human cancers in contrast to its low expression in normal cells unless exposed to genotoxic agents *(4,5)*. Another example of genes expressed only in certain stages of tumor progression is Bcl-2, which showed low expression in melanoma cells of stages I, II, and III, but high in normal melanocytes, whereas bcl-xL expression was high in all cell types tested *(6)*. To separate cause from effect, it is important to study specific mammalian gene function, leading to precise and targeted downregulation of gene expression. Moreover, in addition to investigating the mechanistic effect of knocking out a specific gene, it is important to use antisense technology against genes not only because they are preferentially expressed in cancer *(1–4)*, but also because they also are major determinants of resistance to a variety of anticancer agents, such as mutant *p53* *(4,5)*, *bcl-xL* *(6)*, or *cyclin D1* *(1–3)*. Moreover, cyclin D1 overexpression is associated with p53 mutation and overexpression *(7,8)*, and bcl-2 can induce cyclin D1 expression *(9)*, indicating that this group of genes show an interrelationship that makes them important targets in antitumor strategies.

1.1. Antisense Fundamentals

AS-ODNs are short stretches of 16–18 chemically modified nucleotides designed to hybridize with specific complementary mRNA to block production of proteins encoded by the targeted mRNA transcripts. AS-ODNs use Watson–Crick base pairing to specifically bind to complementary mRNAs. For this binding to occur, the targeted mRNA region must be accessible to the oligonucleotide. In addition, the oligonucleotide must be taken up efficiently by the target cells and must arrive intact to the site of the target mRNA with sufficient affinity and specificity. The likely mechanism of AS-ODN action involves RNase H recognition of an RNA-DNA duplex with selective cleavage of the RNA strand.

1.2. AS-ODN Design

Unmodified DNA oligonucleotides are either poorly taken up or rapidly degraded by serum and cell-associated nucleases. To prevent this, phosphorothioate oligonucleotides (P-ODNs) approx 16–18 bases long in which one of the nonbridging oxygen atoms in the nucleotide is replaced with sulfur provide protection from nuclease degradation without significantly interfering with RNase H activity. 2'-Methoxy-ethoxy modifications in the first and last four or five nucleotides flanking the central region, in combination with the phosphorothioate backbone, further increase resistance to nucleases and affinity of the oligonucleotide for an accessible target sequence in mRNA *(5)*. In addition to the methoxy-ethoxy and phosphorothioate modification, which decreases susceptibility to degradation, specificity is a major issue. The limited length of the AS-ODN is chosen to have both selectivity and potency. A short oligonucleotide of <15 bases may be more selective but show low efficacy.

1.3. Sequence Selection

A recent review describes computational ways to select the best candidate regions to target from known cloned genes *(10)*. If known genes are being targeted in different cell types, a useful strategy is to search antisense oligo databases, such as that from the University of Utah (http://antisense. genetics.utah.edu). For new mRNA targets, identification of the best sequence for antisense activity is not always predictable by computational methods. Some potentially useful sequences in an mRNA may not be accessible in vivo owing to conformational restriction associated with secondary structure and/or protein binding. One approach involves synthesizing 8–10 oligonucleotides that target differently located regions of the mRNA, to evaluate their activity compared with proper controls against cultured cells. To help in the design of potential antisense sequences, software used for designing polymerase chain reaction (PCR) primers can help in identifying sequence motifs providing stable hybridization with target mRNA sequences *(10)*.

1.4. Experimental Controls

Antisense experiments should be specifically targeted against a chosen complementary sequence. Anionic phosphorothioate-containing oligonucleotides and cationic agents used to increase their uptake may be toxic in a nonspecific manner. Hence, for specificity tests, it is necessary to use dose-dependent controls of cells treated only with cationic agents and those using the same cationic agent plus a scrambled control phosphorothioate-containing oligonucleotide with the same base composition but in an order different from that chosen for the putatively active AS-ODN. This scrambled control oligonucleotide should be inactive and lacking homology to any other gene. Moreover, the active AS-ODN should primarily reduce expression of the complementary mRNA and preferentially block the corresponding pathway controlled by expression of this gene *(11,12)*.

2. Materials

1. Antisense and scrambled P-ODNs: These are currently the most popular in the antisense field. It is important to use these nucleotides following purification by reverse-phase high-performance liquid chromatography, to remove sources of nonspecific toxicity contaminants. Lyophilized AS-ODNs are stable during shipment and can be stored at –20°C as such. Whenever they are dissolved in aqueous form, they should be kept at –70°C. A reliable commercial source for antisense P-ODNs can be found at www.proligo.com or www.gensetoligos.com.
2. Lipofectin (Life Technologies c/o Invitrogen.com): This is stable at 4°C and appears optimal for in vitro uptake of antisense and scrambled oligonucleotides.
3. Alamar Blue (BioSource): This dye is highly recommended for monitoring whether there is a differential morphological change or growth inhibition in cells treated with antisense or scrambled oligonucleotides. When cells are seeded in 96- or 24-well tissue culture plates, and Alamar Blue is added to a final concentration of 10%, it allows photographic evidence of changes seen under an inverted microscope; quantitation of growth inhibition is done when fluorescence at 544 nm/590 nm is read 4 h after addition of this reagent *(13)*. Because cells are not damaged by this reagent after fluorometric quantitation or photomicroscopy, they can be harvested from the tissue culture plates by trypsinization or other suitable means, for biochemical analysis of specific changes in expression of genes targeted by the AS-ODN.

3. Methods

Because the uptake of plasmids or oligonucleotides is cell type specific, transfection conditions need to be optimized. The number of hours for uptake, the dose of vehicle (preferably cationic liposomes such as Lipofectin) needed to optimize introduction of the nucleic acid sequence, and the subsequent time after transfection to allow expression of the desired function require standardization for particular cell types. Protocols useful for established adherent cell lines may be too harsh for primary diploid cells or tissues, and lymphoid and other nonadherent cells require protocols unlike those used for adherent cells. In addition, agents that favor uptake of short or long nucleic acid sequences may also increase excessively the uptake of antibiotics such as penicillin, streptomycin, and neomycin, which are frequently used in tissue culture. Hence, the presence of antibiotics should be avoided when gene transfer is being attempted together with cationic agents such as positively charged liposomes or other agents such as calcium chloride, which are commonly used during transfection *(6,10–12)*.

3.1. Transfection Protocol for AS-ODNS

1. On d 1, resuspend cells in medium with serum but without antibiotics, and seed at a concentration of 1.5×10^5 cells/6-well tissue culture plates or 3×10^3 cells/ 96-well tissue culture plates. Prior to use, we suggest covering the plates over-

night with 10 μg/mL of fibronectin, and subsequently removing it followed by blocking with 2% bovine serum albumin fraction V. This markedly increases cell adhesion and prevents detachment when cells are transfected.

2. On d 2, remove serum-containing medium and check the plates to verify that the cells are subconfluent and healthy. Oligonucleotides are delivered as 1:1 complexes with the lipofectin transfection reagent into recipient cells for intervals between 8 and 16 h in serum- and antibiotic-free medium. The following amount of mixture was used for one well of a six-well plate or four wells of a 24-well plate: Solution A contains 181.2 μL medium without serum and antibiotic plus 4.8 μL oligonucleotide stock (250 mM), prewarmed to room temperature, to give a 6.45 μM oligonucleotide. Solution B contains the lipofectin diluted 1:10 into 186 μL of serum- and antibiotic-free medium, prewarmed to room temperature, and then further warmed for 30 min at room temperature. Solutions A and B are gently mixed and incubated for 15 min at room temperature to allow formation of the nucleic acid–lipofectin complexes. then 628 μL of serum- and antibiotic-free medium are added to get an oligonucleotide concentration of 1.2 μM. Finally, this is added to the cells kept in 1 mL of serum- and antibiotic-free medium for intervals between 6 and 24 h. Final concentration of oligonucleotide is 600 nM.

3.2. Assay of Antisense Effects

Introduction of an antisense sequence either as a cDNA cloned into an expression vector or by use of AS-ODNs may be tested by *in situ* hybridization, Northern blots, or reverse transcriptase (RT)-PCR to show downregulation of the corresponding mRNA. However, even if the transcription of the desired gene is downregulated, in some instances there is no correlation between transcriptional and translational changes. It is important to show not only that the corresponding protein is not produced, but to learn whether the knockdown of a particular gene affects cellular physiology and behavior, because this not only may provide insight into the function of the gene or its ability to direct or be part of a defined pathway, but, more important, can provide clues about its potential therapeutic importance. Again, knocking out or interfering with the expression of a gene in a particular cell type may not cause an identical effect in other histotypes, unless the particular gene is not subject to alternative control in other cells. For example, if a death-inducing gene is cloned in the same vector under the control of the same promoter, but introduced into different target cells, the outcome of survival may differ in mammalian cells in which expression of proapoptotic and anti apoptotic genes is stress dependent or cell specific. In some instances, a putative AS-ODN directed against exon 10 of the p53 gene was growth inhibitory in a nonspecific manner, because it did not inhibit expression of the target gene *(14)*. This clearly prevents its use as a truly specific antisense reagent. An example of the desired effect of a bispecific AS-ODN is shown in **Fig. 1**: #4625 directed against a common

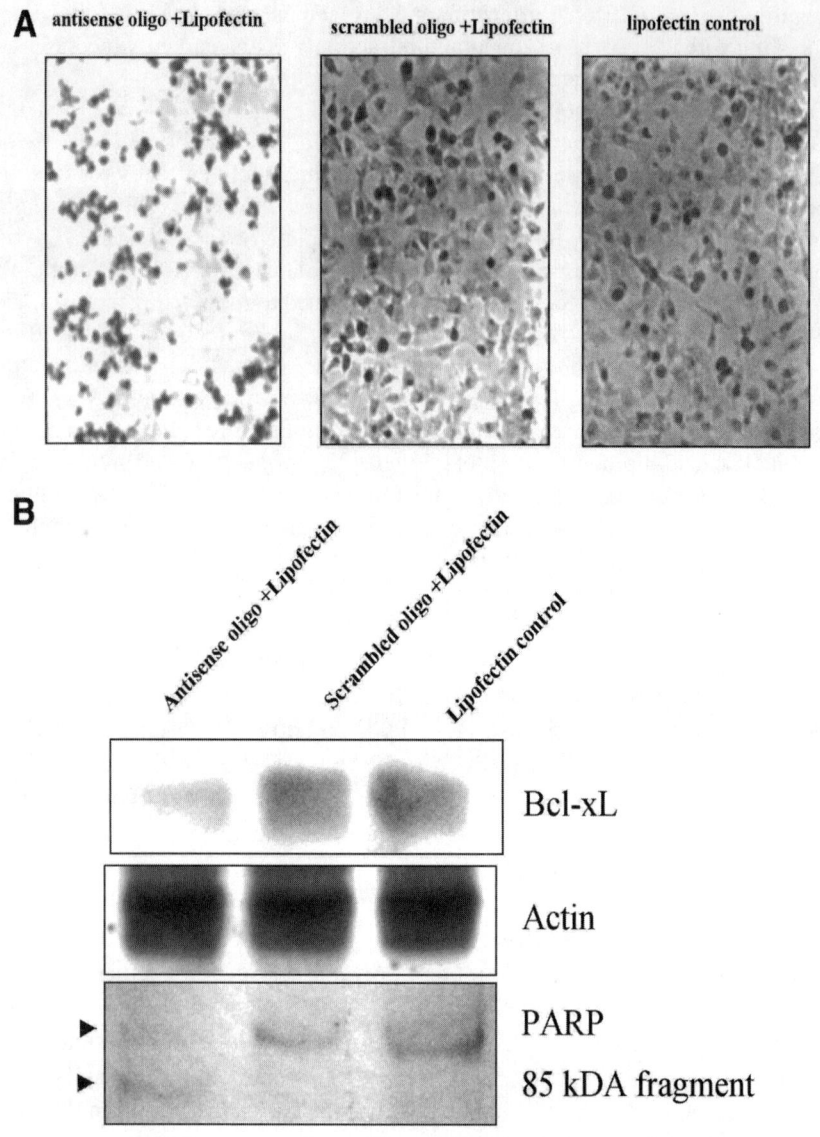

Fig. 1. Effects of a bcl2/bcl-xL bispecific AS-ODN on human MelJuso melanoma. (**A**) Morphological damage is induced only by antisense oligo plus lipofectin, with no comparable effect of the scrambled oligonucleotide plus lipofectin or the lipofectin vehicle. (**B**) A decrease in antiapoptotic bcl-xL and apoptosis-associated PARP fragmentation is induced only by antisense oligo plus lipofectin, with no comparable effect of the scrambled oligonucleotide plus lipofectin or the lipofectin vehicle. Similar results were obtained with human C8161 melanoma.

region of homology shared by antiapoptotic bcl-2 and bcl-xL mRNAs compared with that of a scrambled #4626 oligonucleotide with the same base composition but in a different order against human metastatic MelJuso or C8161 melanoma cells *(12)*. A specific cytotoxic effect of the #4625 oligonucleotide antisense sequence is evident, since no cytotoxicity is seen with the scrambled sequence or the transfection vehicle, lipofectin (**Fig. 1A**). These changes are paralleled by an apoptosis-associated PARP fragmentation *(12)* and a selective down-regulation in bcl-xL, without a comparable change in actin protein (**Fig. 1B**).

4. Conclusion

Drug resistance during cancer therapy can occur during exposure to subtoxic concentrations of death-inducing drugs. On the other hand, excessive cytotoxicity is damaging for both tumor and normal cells *(13)*. In contrast to the permanent drug resistance through constitutive, ectopic overexpression of antiapoptotic genes such as bcl-2 and bcl-xL, which may not be relevant to cellular conditions in naturally occurring tumors, the duration of antisense downregulation by a typical AS-ODN is usually transient and may produce recovery within 3 d *(15)*. Although it may sound disadvantageous, this can allow the transient knockout of specific antiapoptotic genes such as bcl-xL by antisense strategies. Such treatment simultaneously with or prior to exposure to otherwise subtoxic levels of cytotoxic drugs directed toward alternative targets, such as topoisomerase-directed inhibitors, may switch a tumor response from growth arrest or drug resistance toward apoptosis. In addition to their therapeutic potential when used in conjunction with subtoxic levels of cytotoxic drugs, AS-ODNs may allow functional dissection of specific gene functions or identification of apoptotic signaling pathways, which is useful in counteracting resistance to cancer therapy. Although not foolproof or totally predictable, antisense technology has advanced significantly in recent years *(6,10–12)*. When proper controls are used, antisense technology is a very useful research tool hepful in defining the function of specific genes in species such as humans in which knockout organisms cannot be made because of ethical and scientific reasons. Despite the use of traditional radiation, chemotherapy, or immunotherapy, the survival rate of adult cancer patients has only moderately increased. Tumorigenesis due to mutation or lack of p53 tumor suppressor function *(4,5)* and overexpression of genes such as *cyclin D1 (2,3)*, or diminished therapeutic efficacy due to overexpression of antiapoptotic proteins such as bcl-xl *(6,12)*, is involved in therapeutic resistance. AS-ODNs can be used to specifically inhibit tumor-specific unwanted gene expression and hence target the molecular basis of genetic diseases *(2–4)*. The interrelationship between cyclin D1 and bcl-2 was demonstrated in a recent report showing that bcl-2 overexpression induces cyclin D1 promoter activity and expression in the human breast epithelial cell line MCF10A independent of cell anchorage *(9)*. Because cyclin D1 overexpression is a frequent event in clinically invasive carcinomas *(2,3)* that overexpress epidermal growth factor receptor (EGF-R) *(7,8)*, our results suggest that constitutive cyclin D1 expression may prevent apoptosis even in the absence of antiapoptotic EGF signaling *(7)*. In addition, cyclin D1 overexpression results in EGF-R and p53 abnormalities in a transgenic model *(8)*. Morever, a correlation was found between cyclin D1 amplification and poor radiation response in murine carcinomas *(16)*, emphasizing the therapeutic importance of the antisense-mediated knockout of cyclin D1 overexpressed in many tumors *(1,9)*.

Antisense therapeutics also presents a novel alternative to counteract unwanted side effects of chemotherapy. A recent evaluation of 39,451 breast cancer patients diagnosed from 1980 through 2000 who were initially treated with tamoxifen found that the overall risk of subsequent uterine corpus cancer was increased more than twofold *(17)*. Because tamoxifen may induce apoptosis in breast cancer cells, partly by downregulating antiapoptotic bcl-2 *(18)* and bcl-xL *(19)*, known to be involved in drug resistance *(20)*, a potentially useful approach to overcome unwanted side effects of tamoxifen *(17)* may be to use it at a lower dosage together with a chemosensitizing AS-ODN *(21)* such as a bcl-2/bcl-xL bispecific AS-ODN *(12)*.

Since ionizing radiation and traditional chemotherapy are toxic because of either poor targeting specificity or side effects, we envisage a promising potential in the joint use of subtoxic levels of traditional anticancer strategies synergistically with antisense technology directed against tumor-associated gene products that promote drug resistance such as cyclin D1 *(7,8)*, mutant p53 *(5)*, or bcl-xL *(6)* together with AS-ODNs directed against sequences common for bcl-2/bcl-xL *(11,12)*. This antisense approach has been shown to be effective even against melanoma tumors harboring mutant p53, which usually are poorly responsive to a number of other treatments *(12)*. The antisense strategies just discussed should be used in conjunction with subtoxic levels of chemotherapy, to increase chemosensitization of the more rapidly proliferating tumor populations and diminish toxicity toward resting normal cells *(20,21)*.

References

1. Rieber, M. and Strasberg-Rieber, M. (1999) Tumor suppression without differentiation or apoptosis by antisense cyclin D1 gene transfer in K1735 melanoma involves induction of p53,p21WAF1 and superoxide dismutases. *Cell Death Differ.* **6**, 1209–1215.
2. Steeg, P. and Zhou, Q. (1998) Cyclins and breast cancer. *Breast Cancer Res. Treat.* **52**, 17–28.

3. Hall, M. and Peters, G. (1996) Genetic alterations of cyclins, cyclin-dependent kinases, and Cdk inhibitors in human cancer. *Adv. Cancer Res.* **68,** 67–108.

4. Sigal, A. and Rotter, V. (2000) Oncogenic mutations of the p53 tumor suppressor, the demons of the guardian of the genome. *Cancer Res.* **60,** 6788–6793.

5. Esteve, A., Lehman, T., Jiang, W., et al. (1993) Correlation of p53 mutation with epidermal growth factor receptor overexpression and mdm2 amplification in human esophageal carcinoma. *Mol. Carcinog.* **8,** 306–311.

6. Olie R. A., Hafner, C., Kuttel, R., et al. (2002) Bcl-2 and bcl-xL antisense oligonucleotides induce apoptosis in melanoma cells of different clinical stages. *J. Invest. Dermatol.* **118,** 505–512.

7. Mineta, H., Borg, A., Dictor, M., Wahlbert, P., and Wennerberg, J. (1997) Correlation between p53 mutation and cyclin D1 amplification in head and neck squamous cell carcinoma. *Oral Oncol.* **33,** 42–46.

8. Mueller, A., Odze, R., Jenkins, T. D., et al. (1997) A transgenic mouse model with cyclin D1 overexpression results in cell cycle, epidermal growth factor receptor, and p53 abnormalities. *Cancer Res.* **57,** 5542–5549.

9. Lin, H. M., Lee, Y. J., Li, G., Pestell, R. G., and Kim, H. R. (2001) Bcl-2 induces cyclin D1 promoter activity in human breast epithelial cells independent of cell anchorage. *Cell Death Differ.* **8,** 44–50.

10. Smith, L., Andersen, K. B., Hovgaard, L., and Jaroszewski, J. W. (2000) Rational selection of antisense oligonucleotide sequences. *Eur. J. Pharm. Sci.* **11,** 191–198.

11. Olie, R. A., Hall, J., Natt, F., Stahel, R. A., and Zangemeister-Wittke, U. (2002) Analysis of ribosyl-modified, mixed backbone analogs of a bcl-2/bcl-xL antisense oligonucleotide. *Biochim. Biophys. Acta* **1576,** 101–109.

12. Strasberg-Rieber, M., Zangemeister-Wittke, U., and Rieber, M. (2001) p53-independent induction of apoptosis in human melanoma cells by a bcl2/bcl-xL bispecific antisense oligonucleotide. *Clinical Cancer Res.* **7,** 1446–1451.

13. O'Brien, J., Wilson, I., Orton, T., and Pognan, F. (2000) Investigation of the Alamar Blue (resazurin) fluorescent dyefor the assesment of mammalian cell cytotoxicity. *Eur. J. Biochem.* 267, 5421–5426.

14. Barton, C. M. and Lemoine, N. R. (1995) Antisense oligonucleotides directed against p53 have antiproliferative effects unrelated to effects on p53 expression. *British J. Cancer* 71, 429–437.

15. Vilenchik, M., Raffo, A. J., Benimetskaya, L., Shames, D., and Stein, C. A. (2002) Antisense RNA down-regulation of bcl-xL expression in prostate cancer cells leads to diminished rates of cellular proliferation and resistance to cytotoxic therapeutic agents. *Cancer Res.* **62,** 2175–2183.

16. Milas, L., Akimoto, T , Hunter, N. R., et al. (2002) Relationship between cyclin D1 expression and poor radioresponse of murine carcinomas. *Int. J. Radiat. Oncol. Biol. Phys.* **52,** 514–521

17. Curtis, R. E., Freedman, D. M., Sherman, M. E., Fraumeni, J. F., Jr. (2004) Risk of malignant mixed mullerian tumors after tamoxifen therapy for breast cancer. *J. Natl. Cancer Inst.* **96(1),** 70–74.

18. Zhang, G.-J., Kimijima, I., Onda, M., Sato, H., Watanabe, T., Tsuchiya, A., Abe, R., and Takenoshita,S. (1999) Tamoxifen-induced apoptosis in breast cancer cells

relates to down-regulation of bcl-2 but not bax and bcl-xl, without alteration of p53 protein levels. *Clin. Cancer Res.* **5,** 2971–2977.

19. Simoes-Wust, A. P., Olie, R. A., Gautschi, O., et al. (2000) Bcl-xl antisense treatment induces apoptosis in breast carcinoma cells. *Int. J. Cancer.* **87,** 582–590.

20. Heere-Ress, E., Thallinger, C., Lucas, T., et al. (2002) Bcl-X(L) is a chemoresistance factor in human melanoma cells that can be inhibited by antisense therapy. *Int. J. Cancer* **99,** 29–34.

21. Wacheck, V., Heere-Ress, E., Halaschek-Wiener, J., et al. (2001) Bcl-2 antisense oligonucleotides chemosensitize human gastric cancer in a SCID mouse xenotransplantation model. *J. Mol. Med.* **79,** 587–93.

11

Utility of Antioncogene Ribozymes and Antisense Oligonucleotides in Reversing Drug Resistance

Tadao Funato

1. Introduction

The development of new anticancer drugs and the identification of novel targets are a major focus for pharmaceutical and biotech companies, universities, and research institutes worldwide *(1)*. However, the therapeutic efficacy of anticancer drugs against malignant diseases is limited because of the selection and regrowth of drug-resistant cells. The development of approaches to overcome and/or circumvent drug resistance will depend on a precise understanding of the mechanisms of resistance not only at the target tumor cell level but also in vivo. Resistance to treatment with anticancer drugs results from a variety of factors including individual variations in patients and genetic differences in somatic cells in tumors. We have focused on two anticancer drugs: cisplatin, which is exceptionally effective against testicular cancer, ovarian cancer, and others *(2)*; and Ara-C (1-β-D-arabinofuranosyl cytidine, cytosine arabinoside), now a standard for the treatment of acute and chronic leukemia *(3)*.

One of the underlying mechanisms of multidrug resistance (MDR) is cellular overproduction of P-glycoprotein, which acts as an efflux pump for various anticancer drugs *(4)*. However, drug resistance to cisplatin and Ara-C is not related to the MDR phenotype. We have searched for targets based on the differentiation of gene expression in drug-resistant cells compared with wild-type cells (sensitive). For genetic alterations, we suggested that certain oncogenes were candidates for novel indicators of drug resistance. Other mechanisms may play an important role in acquired anticancer drug resistance.

From: *Methods in Molecular Medicine, Vol. 106: Antisense Therapeutics, Second Edition*
Edited by: I. Phillips © Humana Press Inc., Totowa, NJ

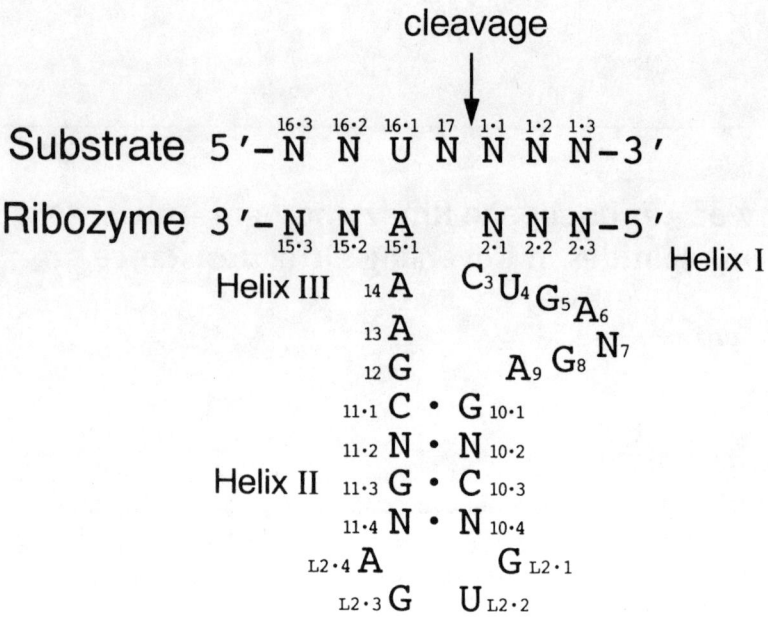

Fig. 1. Conserved sequences of the hammerhead ribozyme according to Haseloff and Gerlach *(54)*. The numbering system for the ribozyme and target sequences are according to Hertel *(55)*.

Antisense oligonucleotides (AS-ODNs) are short modified DNA or RNA molecules designed to bind selectively to mRNA and inhibit the synthesis of the encoded protein. Ribozymes are catalytic RNAs that can cleave specific RNA sequences and do not require an external energy source (**Fig. 1**). In the last 20 yr, antisense technology has emerged as an exciting and promising strategy, especially for treating cancer. Current preclinical antisense strategies in cancer research include the inhibition of proliferation and induction of tumor cell differentiation, reversal of immunosuppression by tumor-secreted molecules, and induction of apoptosis *(5)*. The use of phosphorothioate oligonucleotides as antisense agents has shown promising results in various preclinical cancer models. Thus, downregulation of the expression of related genes by ribozyme or AS-ODNs may help to circumvent the sensitivity to anticancer drugs.

2. Cisplatin Resistance

Cancer chemotherapeutic agents primarily act by damaging cellular DNA directly or indirectly. Tumor cells, in contrast to normal cells, respond to cisplatin with transient gene expression to protect and/or repair their chromosomes *(6)*. Repeated cisplatin treatment results in a stable resistant cell line with enhanced gene expression but lacking gene amplification for the proteins that limit cisplatin cytotoxicity. Recently, several new human cell lines have been characterized for cisplatin resistance, which has led to a better understanding of the molecular and biochemical basis of cisplatin resistance. Recently, gene expression profiles were analyzed by cDNA microarray for a cisplatin-sensitive cell line and a cisplatin-resistant cell line *(7)*. We have found somatic genetic differences in the cisplatin-resistant cells *(8)*.

2.1. c-*fos*

The c-*fos* proto-onocogene, a master switch for turning on other genes in response to a wide range of stimuli, has been shown to play an important role in cisplatin resistance both in vitro and in patients *(9)*. The involvement of c-*fos* in DNA synthesis has been supported by studies utilizing antisense fos RNA to inhibit 3T3 cell proliferation and the G_0-G_1 transition, and Fos-specific antibodies to inhibit DNA synthesis *(10)*. However, the question still remains: How does the gene-regulating function of Fos contribute to DNA synthesis?

A ribozyme was constructed to site specifically cleave fos RNA to investigate further the role of c-*fos* in resistance to antineoplastic agents and in the signal transduction response. **Figure 2** depicts the sequence of the ribozyme including its hammerhead structure and the complementary sequence of c-*fos* RNA containing the target GUC site. DNA encoding fos ribozyme was cloned into a pMAMneo plasmid containing the murine mammary tumor virus dexamethasone-inducible promoter to regulate expression of the ribozyme. A plasmid with the ribozyme in the opposite orientation or mutated sequences was constructed and showed no activity. The resistant cells transfected with *fos*-ribozyme were screened for changes in cisplatin cytotoxicity (**Table 1**). The results support the causal role that c-*fos* plays in the development of resistance. Furthermore, we provide evidence linking the gene expression of c-*fos* with that of integral components of the cellular machinery for DNA synthesis and repair such as dTMP synthase, DNA polymerase β, and topoisomerase I *(11)*. These studies establish a role for c-*fos* in drug resistance and in mediating DNA synthesis and repair processes by modulating the expression of genes.

2.2. K-*ras*

Oncogenic mutations in the *ras* gene are present in approx 30% of all human cancers. Somatic mutations of K-*ras* occur frequently at hot spots in non–small cell lung, colorectal, and pancreatic carcinomas *(12)*. The identification of molecular markers, such as *p53*, thymidylate synthase, and K-*ras*, may provide medical oncologists an important tool for defining subsets of patients with gastrointestinal cancers more likely to benefit from chemotherapy *(13)*. The *ras*-

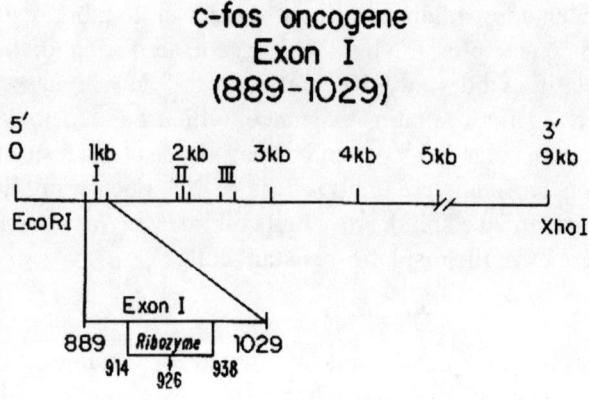

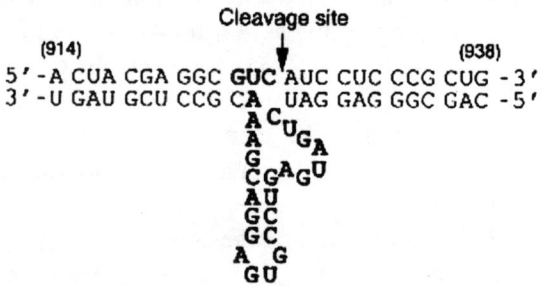

Fig. 2. Structure of fos ribozyme with conserved sequence in boldface type (hammerhead ribozyme). The complementary c-*fos* RNA (bp 914–938) is also shown with the GUC cleavage site on c-*fos* mRNA in boldface type.

signaling pathway has attracted considerable attention as a target for anticancer therapy because of its important role in carcinogenesis.

Colon cancer is an incurable carcinoma because of its poor sensitivity to chemotherapy and radiotherapy, despite great efforts to develop an effective therapy. Antisense therapeutics, which exploit the high degree of specificity offered by genetic information, may have promising therapeutic value in cancers, including colon cancers. This therapy has the ability to suppress neoplastic phenotypes in cancer cells, but successful results in antisense technology have been demonstrated only in vitro or in culture systems. An important factor in the strategy for cancer gene therapy is to select the genes for targeting. Thus, we showed that transfer of the K-*ras* ribozyme may be a useful therapeutic strategy for colon cancer in vivo as well as in vitro *(14)*. The hammerhead ribozyme was constructed with the same stem loop region and distinct flanking

Table 1
Cisplatin Cytotoxicity (IC$_{50}$) in A2780 Cell Lines Using
Fos-Ribozyme[a]

Cell line	IC$_{50}$ (mM)
1. A2780S	9.9 ± 2.0
2. A2780S pMMV-fos+pKoneo	72.5 ± 5.0
3. A2780DDP	104.5 ± 5.0
4. A2780DDP+Dex	95.5 ± 5.0
5. A2780DDPfosR-2	71.0 ± 5.0
6. A2780DDPfosR-2+Dex	37.0 ± 4.0
7. A2780DDPfosR-3	95.0 ± 5.0
8. A2780DDPfosR-3+Dex	6.0 ± 2.0
9. A2780DDPfosR-6	87.0 ± 9.0
10. A2780DDPfosR-6+Dex	86.0 ± 9.0
11. A2780DDPfosR-8	96.5 ± 5.0
12. A2780DDPfosR-8+Dex	43.0 ± 4.5

[a]Data are means the ± SD. IC$_{50}$ represents half the number of colonies formed in comparison to the untreated A2780S cells. Dexamethasone (Dex) was administrered to the A2780DDP cells for 24 h prior to cisplatin treatment.

sequences complementary to the substrate K-*ras* RNA and was cloned into the plasmid pLNCX (**Fig. 3**). The ribozyme against activated K-*ras* oncogene (K-*ras* ribozyme) has the ability to cleave the targeted RNA in vitro. Similar results were obtained in the in vivo system; namely, the tumors were smaller in athymic mice transfected with the K-*ras* ribozyme than in the controls. We also evaluated the chemosensitivity of the cells to a variety of agents. The sensitivity of tumors with K-*ras* ribozyme was increased, as shown in **Table 2**. We have demonstrated the effectiveness of in vivo transfection of ribozyme for inhibiting growth as well as for increasing the sensitivity to anticancer agents. Therefore, altered K-*ras* genes may be an important target for strategies to enhance chemosensitivity in colon cancers.

2.3. c-myc

Along this line, we have tried to identify the genes responsible for drug resistance to cisplatin in cancers and revealed that the *fos* and K-*ras* genes are implicated in cellular resistance to cisplatin. Isonishi et al. *(15)* also demonstrated cisplatin resistance in NIH3T3 cells transfected with a mutated H-*ras* gene. However, the mechanism of acquired resistance to cisplatin based on DNA damage is complex and other factors may be involved, making it important to clearly identify the effectors of cisplatin resistance.

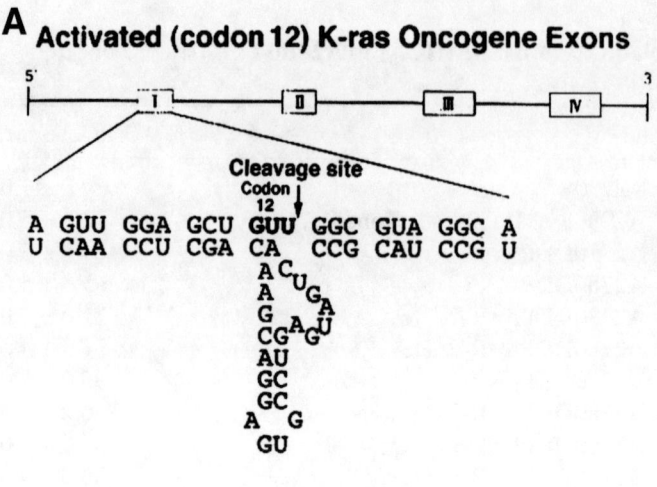

LNCX-*Kras* Rib.

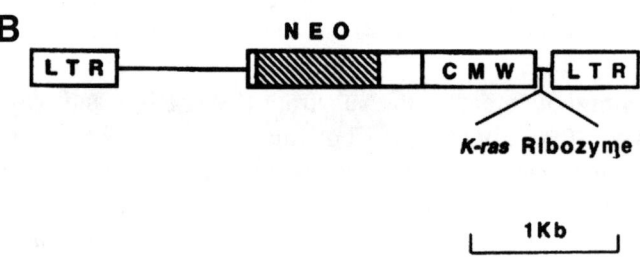

Fig. 3. Structure of K-*ras* ribozyme (**A**) and schematic representation of retroviral vector pLNCX (**B**). The conserved hammerhead sequence includes a base alteration in the catalytic core. Target codon-12 is bold at cleavage site. The complementary K-*ras* RNA is also shown with the GUC cleavage site in mutated K-*ras* RNA. The position of the ribozyme and a synthetic RNA substrate used to test the ribozyme are shown with respect to exon I of K-*ras*. This vector contains the neo gene (NEO) and the promoter of the cytomegalovirus gene (CMV). The K-*ras* ribozyme as a double-stranded DNA was inserted into the *Hin*dIII site. LTR; long terminal repeat.

The c-*myc* oncogene, a regulator of cell growth associated with cell division, is highly expressed in more than 70% of human colon cancers, and colon cancer is often resistant to cisplatin treatment *(16)*. The role of c-*myc* protein in cellular susceptibility to anticancer drugs is controversial. In fact, overexpression of the protein has been reported to enhance tumor cell sensitiv-

Table 2
Sensitivity (IC$_{50}$) of Cells Excised From Tumors by K-ras Ribozyme to Drugs In Vivo

Group[a]	5-FU (IC$_{50}$, nM)	ADR (IC$_{50}$, nM)	VCR (IC$_{50}$, nM)	CDDP (IC$_{50}$, μM)	VP16 (IC$_{50}$, nM)
1. SW620S	12.41	2.36	4.57	25.9	16.4
2. SW620V	10.60	3.62	5.22	30.5	14.1
3. KRZM1	9.52	0.12	1.06	9.7	8.6
4. KRZM2	4.16	0.24	0.85	10.8	5.7
5. KRZM3	5.89	0.11	1.64	6.5	7.3
6. KRZM4	6.23	0.37	1.28	5.5	10.4

[a]The cells from each group were exposed to drugs. SW620V was transfected with a vector not containing the ribozyme, and KRZM1–4 were transfected with a vector containing the ribozyme. A total of 3000 cells in each group were incubated for 24 h and subsequently treated with seven different concentrations of the various drugs for 2 h. 5-FU, 5-fluorouracil.

ity and to induce resistance in response to antineoplastic agents *(17)*. This prompted us to investigate the role of the c-*myc* gene in cisplatin resistance in colon cancer.

We used AS-ODNs and transfection technology to determine whether the c-*myc* gene regulates cisplatin sensitivity in colon cancer. Expression of the c-*myc* gene was partially inhibited after incubation with c-*myc* AS-ODNs but was not inhibited with c-*myc* sense oligonucleotides. Regarding cell cytotoxicity, after incubation with c-*myc* AS-ODNs, cells displayed increased resistance to cisplatin, but resistance to adriamycin (ADR) or etoposide (VP16) was unchanged with 3-(4,5-dimethylthiazolyl-2)-2,5-diphenyl tetrazolium bromide (MTT) assay (**Table 3**).

For overexpression of c-*myc*, SW480DDP and SW620DDP cells in culture were transfected with pLNCX including c-*myc* cDNA. These cells indicated reversed resistance to cisplatin but not to other agents (**Table 3**). The c-*myc*-overexpressing cells also exhibited increased sensitivity to cisplatin, but not to other agents. We have shown here the potential of the c-*myc* gene to confer cisplatin resistance in human colon cancer. We used AS-ODNs to c-*myc* to determine whether targeting c-*myc* modifies drug resistance to cisplatin in colon cancer cells. AS-ODNs for c *myc* mRNA exhibited the potential to block specific c-*myc* expression. Uses of c-*myc* AS-ODNs have been demonstrated in several reports, including investigations of the inhibitory effects of growth and proliferative modulation *(18)*, induced apoptosis *(19)*, and inhibition of telomerase activity *(20)* in cancer cells. However, previous reports relied on the use of c-*myc* AS-ODNs to enhance chemosensitivity in combination with

Table 3
Drug Sensitivity (IC_{50}) of Cells After Incubation with c-*myc* Oligonucleotides

Cell line	Cisplatin (IC_{50}, μM)	ADR (IC_{50}, nM)	VP16 (IC_{50}, nM)
1. SW480	18.3 ± 0.8 ⎤	0.073 ± 0.019	5.60 ± 0.44
2. SW480+*myc*-AS[a]	40.8 ± 2.5 ⎥[b] [c]	0.054 ± 0.043	5.02 ± 0.72
3. SW480+*myc*-S	21.3 ± 1.5 ⎦	0.092 ± 0.035	6.21 ± 0.84
4. SW620	15.1 ± 1.7 ⎤	0.034 ± 0.024	6.85 ± 0.35
5. SW620+*myc*-AS	54.2 ± 3.4 ⎥[b] [c]	0.029 ± 0.047	5.76 ± 5.76
6. SW620+*myc*-S	20.7 ± 2.9 ⎦	0.044 ± 0.085	6.24 ± 1.24

[a]*myc*-AS and *myc*-S: oligonucleotides for the c-*myc* gene. SW480 and SW620 cells were treated with 5 nmol/m of myc-AS or myc-S for 2 d, then with one of seven different concentrations of various drugs for 2 h. After being washed with phosphate-buffered saline (PBS) once, the cells were incubated for 5 d, and the cytotoxicity of the drugs was determined by MTT assay. The results represent the mean ±2 SDs for three separate experiments.
[b]$p < 0.05$
[c]$p < 0.01$

cisplatin in cases of human melanoma (*21*). The mechanisms of acquired cisplatin resistance to c-*myc* are not yet known. Therefore, we directly assessed the efficacy of c-*myc* AS-ODNs in drug sensitivity to cisplatin. The results indicate that c-*myc* antisense sequences confer resistance to cisplatin, but not other drugs, in drug-sensitive cell lines. Thus, we postulate that c-*myc* is one of the factors related to drug sensitivity and the acquisition of cisplatin resistance.

Furthermore, we examined whether the potential of the c-*myc* gene to enhance sensitivity to cisplatin can be modified by transfected drug-resistant colon cancer cells with c-*myc* cDNA. We demonstrated that expression of the c-*myc* gene reversed resistance to cisplatin but not other drugs. Thus, the induction of apoptosis by cisplatin is related to cisplatin sensitivity and the expression of c-*myc*, because the c-*myc* gene is a well-known factor in the induction of apoptosis (*22*). High c-*myc* levels are necessary for cisplatin-induced apoptosis in resistant cells. Thus, the present results suggest that c-*myc* confers sensitivity and resistance to cisplatin in human colon cancer. Modulation of c-*myc* activity employing antisense sequence techniques or gene transfer may provide a means to circumvent cisplatin resistance in colon cancers. Such therapeutic modulation has become a powerful tool for selectively inhibiting the expression of target genes in vitro and is of increasing interest in the development of in vivo therapeutic strategies. In fact, it is important to clarify how c-*myc* is expressed in vivo for more effective cisplatin therapy in colon cancer.

Therefore, further investigation of why the c-*myc* gene is related to the sensitivity to cisplatin alone but not to ADR and VP16 in other colon cancer cell lines is needed.

2.4. c-myb

The development of human hematopoietic cells frequently upregulates expression of the c-*myb* gene, but it is unclear whether this is a cause or a consequence of the neoplastic state *(23)*. The c-*myc* and Myb proteins are transcription factors that regulate cell proliferation and differentiation and can be seen as specific partner proteins in the regulation of gene expression. Expression of c-*myb* has been observed in numerous human cancer cells. In particular, the c-*myb* expression level is higher in the neoplastic mucosa of colon cancer patients than in normal mucosa *(24)*. These findings led us to question whether c-*myb* gene expression was related to drug sensitivity in colon cancer. The mechanism of acquired resistance to cisplatin based on DNA damage is complex and other factors may be involved, making it important to identify clearly the effectors of cisplatin resistance. This prompted us to investigate the role of the c-*myb* gene in cisplatin resistance in colon cancer. In this study, we used AS-ODN technology to determine whether the c-*myb* gene regulated cisplatin sensitivity in colon cancer.

Expression of c-*myb* mRNA was markedly increased in SW480DDP and SW620DDP compared with the parent SW480 or SW620 cell lines. Expression of the c-*myb* gene was partly inhibited after incubation with c-*myb* AS-ODNs (MYB-2) at the optimal concentration and time, but it was not inhibited with c-*myb* sense oligonucleotides (MYB-1). To shed more light on the association between the expression of c-*myb* activity and increased resistance to cisplatin, we then tested the effect of AS-ODNs on cisplatin sensitivity. Assessment of IC_{50} indicated that resistance to cisplatin was reversed in MYB-2-treated SW480DDP or SW620DDP cells, but not in MYB-1-treated cells (**Table 4**). c-*myb* antisense or sense oligonucleotides had no effect on the numbers or growth rates of SW480DDP or SW620DDP cells in terms of cell cytotoxicity.

We suggest that the c-*myb* gene is able to confer cisplatin resistance in human colon cancer. It has been reported that c-*myb* mRNA can be detected at low levels in all normal mucosa, but that detection levels increase in the neoplastic mucosa in colon cancer. In other reports, c-*myb* expression has not been detected in inflammatory mucosa or colon cancer. We used AS-ODN technology to determine whether targeting c-*myb* cells would modify drug resistance to cisplatin in colon cancer. AS-ODNs for c-*myb* exhibited the potential to block specific c-*myb* expression. The use of c-*myb* AS-ODNs has been demonstrated in several reports, including investigations of the role of hematopoiesis *(25)*, cell-cycle progression such as restenosis *(26)*, and promoter regulation *(27)*.

Table 4
Sensitivity (IC$_{50}$) of Cells to Cisplatin After Incubation with c-*myb* Oligonucleotides

Cell line	Cisplatin (IC$_{50}$, µM)
1. SW480DDP	18.5 ± 2.5
2. SW480DDP+MYB-1[a]	20.1 ± 3.8
3. SW480DDP+MYB-2[a]	10.3 ± 3.5
4. SW620DDP	24.9 ± 2.1
5. SW620DDP+MYB-1	25.3 ± 3.4
6. SW620DDP+MYB-2	10.4 ± 4.2

[a]MYB-1, sense oligonucleotides for c-*myb*; MYB-2, AS-ODNs for c-*myb*. SW480DDP and SW620DDP cells were treated with 5 nmol/m of MYB-1 or MYB-2 for 48 h, then with one of seven different concentrations of cisplatin for 2 h. After being washed with PBS once, the cells were incubated for 5 d, and the cytotoxicity of cisplatin was then determined by MTT assay. The results represent the mean ± 2 SDs for three separate experiments.
[b]$p < 0.01$.

The role of c-*myb* as an oncogene has not yet been analyzed by antisense strategies in cancer. The mechanisms of acquired cisplatin resistance related to c-*myb* are also not yet known. We investigated the effects of c-*myb* AS-ODNs on cisplatin resistance. Therefore, we directly assessed the efficacy of c-*myb* AS-ODNs in terms of sensitivity to cisplatin. Our results indicated that c-*myb* AS-ODNs increase sensitivity to cisplatin. The modulation of c-*myb* activity by AS-ODN techniques may be a way of circumventing cisplatin resistance in colon cancer.

2.5. c-erbB-2

The HER-2/*neu* (c-*erb*B-2) oncogene encodes a transmembrane tyrosine kinase receptor with extensive homology to the epidermal growth factor (EGF) receptor *(28)*. Overexpression of the c-*erb*B-2 gene in human adenocarcinomas such as breast cancer or non–small cell lung cancer is thought to be associated with prognosis *(29)*. The relationship between expression of *erb*B-2 and drug resistance in cancer has not been elucidated.

The KATOIII and MKN7 cell lines expressed the c-erbB-2 gene but not SW480. The introduction of AS-ODNs (*erb*B-2-AS) into KATOIII and MKN7 cells decreased the expression of c-*erb*B-2, but that of sense oligonucleotides (*erb*B-2-S) had no effect. An antisense and a control were transferred to the cells. The intensity of the fluorescent peaks in the cells treated with antisense was shifted to the left. This shift was observed in the KATOIII and MKN7 lines. We then tested the effect of AS-ODNs for c-*erb*B-2 on cisplatin sensitivity. All of four independent tests showed a severalfold increase in resistance to

Table 5
Inhibition of Drug Sensitivity (IC_{50}) by c-erbB-2 AS-ODNs on Cells[a]

Cells	Cisplatin (IC_{50}, μM)	VP16 (IC_{50}, nM)	ADR (IC_{50}, nM)	CR (IC_{50}, nM)
1. KATOIII 2	8.5 ± 0.9	34.7 ± 1.8	242.1 ± 3.4	157.5 ± 14.8
2. KATOIII+AS	15.6 ± 1.8	31.8 ± 2.7	221.9 ± 5.6	137.5 ± 22.4
3. KATOIII+S	26.2 ± 1.8	32.0 ± 5.7	219.7 ± 4.8	145.1 ± 20.2
4. MKN-7	34.1 ± 0.9	28.4 ± 3.4	187.2 ± 6.0	121.5 ± 7.5
5. MKN-7+AS	14.2 ± 0.9	26.8 ± 2.4	166.9 ± 5.2	113.1 ± 6.7
6. MKN-7+S	32.4 ± 0.9	25.6 ± 4.1	176.5 ± 4.8	121.5 ± 5.1
7. SW480	23.7 ± 0.9	24.7 ± 8.5	157.9 ± 4.8	131.2 ± 6.6

[a]Cells were pretreated with AS-ODNs (AS) or sense oligonucleotides (S) for 48 h, and then the concentrations that inhibited 50% of cell growth (IC_{50} values) were determined for seven different concentrations of various drugs after 5 d by MTT assay. The results represent the mean ±2 SDs for three separate experiments. SW480 cells were used as a control.

cisplatin after treatment of *erb*B-2-AS but not erbB-2-S (**Table 5**). KATOIII and MKN7 cells treated with *erb*B-2-AS did not show crossreactivity to ADR, vincristine (VCR), and VP16. The degree of sensitivity to cisplatin depends on the increase in the concentration of erbB-2-AS.

In recent years, many strategies for the manipulation of gene expression have been established to circumvent drug resistance, such as the use of AS-ODNs for *bcl*-2 *(30)* or c-*myc*. Kondo et al. *(31)* reported that inhibition of telomerase with cisplatin in glioblastoma cells induced apoptosis. AS-ODNs for EGF have already been extensively used to abrogate cisplatin-induced apoptosis in a breast cancer model *(32)*. In our study, therefore, we focused on the c-*erb*B-2 gene and used antisense when the gene regulates cell proliferation but not drug sensitivity to cisplatin in gastric cancer cell lines.

We have studied the role of c-*erb*B-2 expression in cisplatin-mediated cyto-toxicity and found that AS-ODNs for c-*erb*B-2 inhibited expression of c-*erb*B-2 and cell growth in gastric cell lines. The inhibitory effects on c-*erb*B-2 amplification and overexpression by *erb*B-2 AS-ODNs in various cancers were mainly obtained during cancer cell proliferation *(33)*. It was indicated that overexpression of *erb*B-2 is important for the proliferation of cancer cells that have been seleced for *erb*B-2 gene amplification. The effects by *erb*B-2 AS-ODNs for overexpressing *erb*B-2 are consistent with a similar inhibition of proliferation observed with *erb*B-2 antibodies *(34)*. Similar strategies were reported with a vector system in breast cancer *(35)*, which certainly improved efficiency but not sensitivity to nucleases. The inhibitory effects by the ribozyme for *erb*B-2 were as good as an antisense effect in ovarian cancer *(36)*.

It was reported that anchorage-independent growth and tumor take were reduced by *erb*B-2 antisense in lung, ovarian, and breast cancer; however, we could not examine the effects on tumorigenicity in vivo.

The focus of our study was whether the expression of c-*erb*B-2 was related to drug sensitivity, particularly cisplatin in gastric cancer. We demonstrated that expression of c-*erb*B-2 was associated with sensitivity to cisplatin using AS-ODN technology. A previous article on the relationship between *erb*B-2 expression and sensitivity to cisplatin reported that a combination of p185HER2/neu monoclonal antibody with cisplatin was more effective than cisplatin alone in breast cancer *(37)*. In another article, it was reported that cisplatin-resistant ovarian cancer cells expressed p185HER-2 and HER-2 mRNA *(38)*. In addition, HER-2/*neu* overexpression in lung cancer is known to induce chemoresistance, and tyrosine kinase inhibitors such as emodin improved the sensitivity to cisplatin or VP16 for antiproliferative effects. These reports did not directly demonstrate that expression of *erb*B-2 is related to sensitivity to cisplatin. Finally, there is no report about inhibition of *erb*B-2 expression by AS-ODNs and sensitivity to cisplatin in gastric cancer. We conclude that the inhibitory effect on c-*erb*B-2 should be useful by causing increased sensitivity to cisplatin in gastric cancer. Therefore, the c-*erb*B-2 oncogene is an attractive target for tumor-specific gene therapy.

3. Ara-C Resistance

Ara-C is widely used to treat hematological malignancies but has become ineffective because of increased resistance to the drug *(39)*. The acquisition of drug resistance may lead to a poorer prognosis. It is important, therefore, to understand the mechanism responsible for drug resistance and to develop new therapies to overcome the resistance. The involvement of MDR-1 or MRP in the mechanism of Ara-C resistance in leukemia has not, however, been established. Some patients with clinical resistance to Ara-C have a reduced expression of deoxycytidine kinase (dCK) mRNA in their leukemia cells *(40)*. Nevertheless, data on the molecular events leading to Ara-C resistance are still sparse. We have demonstrated a model of Ara-C-resistant cells, and a new cell line with Ara-C resistance, developed to facilitate studies of the mechanisms of drug resistance *(41)*. Other target genes likely to play an important role in Ara-C sensitivity have been investigated.

3.1. c-raf-1

The c-*raf* genes encode a family of cytoplasmic proteins (A-*raf*, B-*raf*, and c-*raf*-1) with associated serine/threonine kinase activities *(42)*. The proto-oncogene c-*raf*-1 is an important mediator of signal transduction pathways involving cell growth, transformation, and differentiation. c-*raf*-1 is expressed in many human tissues, suggesting that c-*raf*-1 may be activated as an oncogene in carcinogenesis *(43)*. However, it is not yet known whether c-*raf*-1 is expressed in leukemia, since a variant of this gene was found in lymphoma patients *(44)*. Furthermore, in terms of the relationship between c-*raf* and drug resistance, inhibition of RNA synthesis prevents Raf-1 activation and *bcl*-2 phosphorylation, suggesting that an intermediate protein acts upstream of Raf-1 in the microtubule damage-activating pathway *(45)*. It has also been reported that expression of the c-*raf*-1 gene significantly enhanced the activity of the MDR1 promoter *(46)*. Another report indicated that although treatment of human U937 myeloid leukemia cells with phorbol ester (TPA) is associated with activation of the Raf-1 kinase, there was no detectable decrease in cells resistant to TPA *(47)*. However, to our knowledge, it has not been demonstrated that c-*raf*-1 is associated with drug sensitivity.

AS-ODNs targeted against c-*raf*-1 kinase resulted in potent antiproliferative and antitumor effects *(48)*. Downregulation of c-*raf*-1 expression by AS-ODNs inhibited BCR/ABL-dependent growth of chronic myelogenous leukemia cells and growth factor–dependent proliferation of normal hematopoietic progenitors, as did inhibition of c-*raf*-1 activity by its dominant-negative mutants *(49)*. Furthermore, sensitizing effects of these AS-ODNs have also been reported in radioresistant tumors *(50)*. To define the role of c-*raf*-1 in the mechanisms of drug sensitivity, we investigated whether AS-ODNs targeted against c-*raf*-1 kinase could reverse drug resistance in leukemia.

We had previously determined the expression of the c-*raf*-1 gene in several human leukemia cell lines; however, almost all the lines showed lower levels of expression compared with the U937 cell line. First, treatment of U937 cells with AS-ODNs (ISIS 5132) caused decreased expression of c-*raf*-1. Treatment with the AS-ODNs was compared in a kinetic analysis at the mRNA level to determine the relative potency and duration of the effect. Second, similar treatment of K562AC cells with AS-ODNs (ISIS 5132) also resulted in decreased expression of c-*raf*-1, which again was not changed by mismatched control oligonucleotides (mismatches).

We then examined the effect of AS-ODNs against c-*raf*-1 on sensitivity to Ara-C. The results of four independent tests showed a two- to threefold increase in sensitivity to Ara-C after treatment with AS-ODNs, but not after treatment with mismatched control oligonucleotides (**Table 6**). However, the sensitivity to other drugs such as ADR, VCR, or VP16 was not changed in any cell lines after treatment.

In this study, we examined the role of c-*raf*-1 expression in growth- and Ara-C-mediated cytotoxicity in human leukemia cells. We found that AS-ODNs against c-*raf*-1 markedly inhibited the expression of c-*raf*-1 mRNA in leukemia cells but did not affect cell growth. The inhibitory effects of AS-ODNs for

Table 6
Increased Sensitivity (IC$_{50}$) to Ara-C in Human Leukemia Caused by c-*raf*-1 AS-ODNs[a]

Cell line	Ara-C (IC$_{50}$, μM)	ADR (IC$_{50}$, nM)	VCR (IC$_{50}$, nM)	VP16 (IC$_{50}$, μM)
1. U937	2.36 ± 1.68	0.67 ± 0.21	1.56 ± 0.16	5.8 ± 2.1
2. U937 with antisense	1.54 ± 0.85 [b][b]	0.57 ± 0.33	1.41 ± 0.25	5.2 ± 1.3
3. U937 with control	2.44 ± 0.60	0.53 ± 0.26	1.55 ± 0.32	6.1 ± 0.9
4. K562	1.96 ± 0.48	0.34 ± 0.27	0.88 ± 0.55	4.8 ± 2.2
5. K562AC	42.63 ± 3.14	0.57 ± 0.41	1.25 ± 0.34	5.1 ± 1.5
6. K562AC with antisense	16.42 ± 8.32 [b][b]	0.51 ± 1.02	1.17 ± 0.82	6.4 ± 3.0
7. K562AC with control	40.31 ± 5.73	0.91 ± 0.84	1.38 ± 0.71	5.6 ± 2.9

[a]Values are given as the mean ± SEM ($n = 5$).
[b]Statistical significance at the $p < 0.05$ level.

c-*raf*-1 on overexpression of c-*raf*-1 in various diseases were mainly obtained in cellular signal transduction pathways. However, c-*raf*-1 also has survival-promoting functions, indicating a possible role for raf antisense in the management of radioresistant malignancies, as demonstrated by its radiotherapeutic efficacy. However, although c-*raf*-1 plays important roles in cell growth and proliferation, its role in drug sensitivity is still unclear.

The focus of our study was whether the expression of c-*raf*-1 is related to drug sensitivity, particularly to Ara-C in leukemia. We demonstrated by using AS-ODN technology that c-*raf*-1 expression is associated with drug sensitivity to Ara-C. A study of sensitivity to Ara-C in leukemia revealed that clinical resistance to Ara-C is associated with reduced expression of dCK mRNA in leukemia cells. Nevertheless, data on the molecular events leading to Ara-C resistance are still sparse. Although expression of c-*raf*-1 enhanced the activity of the MDR1 promoter, in the present study modification of c-*raf*-1 expression did not affect sensitivity to drugs such as ADR or VCR that are related to the MDR phenotype. To our knowledge, there are no previous reports directly demonstrating that c-*raf*-1 is related to drug sensitivity for Ara-C or that link inhibition of c-*raf*-1 expression by AS-ODNs and drug sensitivity to Ara-C in human leukemia. We conclude that the inhibitory effect on c-*raf*-1 should be useful by causing increased sensitivity to Ara-C in leukemia. Therefore, the c-*raf*-1 oncogene expressed in leukemia may be an attractive target for tumor-specific therapy. Further, clinical trials of AS-ODNs (ISIS 5132) for c-*raf*-1 have recently been reported in refractory malignancies for phase I *(51)*, in advanced cancer *(52)*, and in ovarian cancer with mutated p53 *(53)*. The preferential targeting of drug resistance may develop for antisense therapeutics.

The c-*raf*-1 oncogene plays an important role in cellular signal transduction pathways that mediate cell growth and proliferation. To evaluate the effect on the drug sensitivity of a well-characterized raf AS-ODN, we examined whether expression of c-*raf*-1 is related to resistance to Ara-C in human leukemia. Modulation of gene expression by AS-ODNs altered drug sensitivity to Ara-C in leukemia cells.

4. Conclusion

Drug resistance in cancer is a major obstacle to successful chemotherapy. However, the genetic basis of cellular resistance to an anticancer drug is not well understood. The genetic mechanisms of drug resistance have been elucidated. The delineation of gene function has always been a subject of intense investigation. Recent advances in the synthesis and chemistry of AS-ODNs have provided important molecular tools to study and identify gene functions and regulation. These oligonucleotides were used to elucidate cellular signaling pathways. The key in this pathway is oncogenes, which have important growth and tumorigenicity. Thus, we have attempted to identify targets of anticancer drugs using the antisense strategy and support the need to develop anticancer therapeutics based on oncogene-targeted AS-ODN technology.

References

1. Broxterman, H. J. and Georgopapadakou, N. (2001) Cancer research 2001: drug resistance, new targets and drug combinations. *Drug Resist. Updat.* **4,** 197–209.
2. Go, R. S. and Adjei, A. A. (1999) Review of the comparative pharmacology and clinical activity of cisplatin and carboplatin. *J. Clin. Oncol.* **17,** 409–422.
3. Grant, S. (1998) Ara-C: cellular and molecular pharmacology. *Adv. Cancer Res.* **72,** 197–233.
4. Scheffer, G. L. and Scheper, R. J. (2002) Drug resistance molecules: lessons from oncology. *Novartis Found. Symp.* **243,** 19–31.
5. Gutierrez-Puente, Y., Zapata-Benavides, P., Tari, A. M., and Lopez-Berestein, G. (2002) *Semin. Oncol.* **29(Suppl. 11),** 71–76.
6. Trimmer, E. E. and Essigmann, J. M. (1999) Cisplatin. *Essays in Biochem.* **34,** 191–211.
7. Sakamoto, M., Kondo, A., Kawasaki, K., et al. (2001) Analysis of gene expression profiles associated with cisplatin resistance in human ovarian cancer cell lines and tissues using cDNA microarray. *Hum. Cell* **14,** 305–315.
8. Scanlon, K. J., Kashani-Sabet, M., Miyachi, H., Sowers, L., and Rossi, J. J. (1989) Molecular basis of cisplatin resistant human carcinomas: model systems and patients. *Anticancer Res.* **9,** 1301–1312.
9. Tulchinsky, E. (2000) Fos family members: regulation, structure and role in oncogenic transformation. *Histol. Histopathol.* **15,** 921–928.

10. Holt, J. T., Gopal, T. V., Moulton, A. D., and Nienhuis, A. W. (1986) Inducible production of *c-fos* antisense RNA inhibits 3T3 cell proliferation. *Proc. Natl. Acad. Sci. USA* **83,** 4794–4798.

11. Funato, T., Yoshida. E., Jiao, L., Tone, T., Kashani-Sabet. M., and Scanlon, K. J. (1992) The utility of an anti-*fos* ribozyme in reversing cisplatin resistance in human carcinomas. *Advan. Enzyme Regul.* **32,** 195–209.

12. Minamoto, T., Mai, M., and Ronai, Z. (2000) K-*ras* mutation: early detection in molecular diagnosis and risk assessment of colorectal, pancreas, and lung cancers-a review. *Cancer Detect. Prev.* **24,** 1–12.

13. Catalano, V., Baldelli, A. M., Giordani, P., and Cascinu, S. (2001) Molecular markers predictive of response to chemotherapy in gastrointestinal tumors. *Crit. Rev. Oncol.-Hematol.* **38,** 93–104.

14. Funato, T., Ishii, T., Kambe, M., Scanlon, K. J., and Sasaki, T. (2000) Anti-K-*ras* ribozyme induces growth inhibition and increased chemosensitivity in human colon cancer cells. *Cancer Gene Ther.* **7,** 495–500.

15. Isonishi, S., Hom, D. K., Thiebaut, F. B., et al. (1991) Expression of the c-Ha-*ras* oncogene in mouse NIH 3T3 cells induces resistance to cisplatin. *Cancer Res.* **51,** 5903–5909.

16. Rigas, B. (1990) Oncogenes and suppressor genes: their involvement in colon cancer. *J. Clin. Gastroenterol.* **12,** 494–499.

17. Sklar, M. D. and Prochownik, E. V. (1991) Modulation of cis-platinum resistance in Friend erythroleukemia cells by c-*myc*. *Cancer Res.* **51,** 2118–2123.

18. Cerutti, J., Trapasso, F., Battaglia, C., et al. (1996) Block of c-*myc* expression by antisense oligonucleotides inhibits proliferation of human thyroid carcinoma cell lines. *Clin. Cancer Res.* **2,** 119–126.

19. Hashiramoto, A., Sano, H., Maekawa, T., et al. (1999) C-*myc* antisense oligonucleotides can induce apoptosis and down-regulate Fas expression in rheumatoid synoviocytes. *Arthrith. Rheumat.* **42,** 954–962

20. Fujimoto, K. and Takahashi, M. (1997) Telomerase activity in human leukemia cell lines is inhibited by antisense pentadecadeoxynucleotides targeted against c-*myc* mRNA. *Biochem. Biophys. Res. Commun.* **241,** 775–781.

21. Citro, G., D'Agnano, I., Leonetti, C., et al. (1998) C-*myc* antisense oligodeoxynucleotides enhance the efficacy of cisplatin in melanoma chemotherapy in vitro and in nude mice. *Cancer Res.* **58,** 283–289.

22. Thompson, E. B. (1998) The many roles of c-Myc in apoptosis. *Ann. Rev. Phys.* **60,** 575–600.

23. Sumner, R., Crawford, A., Mucenski, M., and Frampton, J. (2000) Initiation of adult myelopoiesis can occur in the absence of c-*Myb* whereas subsequent development is strictly dependent on the transcription factor. *Oncogene* **19,** 3335–3342.

24. Torelli, G., Venturelli, D., Colo, A., et al. (1987) Expression of c-*myb* protooncogene and other cell cycle-related genes in normal and neoplastic human colonic mucosa. *Cancer Res.* **47,** 5266–5269.

25. Gilmore, M. M. and Bishop, T. R. (1999) The role of c-*myb* during the maturation of murine CFU-E. *Blood Cells Mol. Dis.* **25,** 68–77.

26. Lee, M., Simon, A. D., Stein, C. A., and Rabbani, L. E. (1999) Antisense strategies to inhibit restenosis. *Antisense Nucleic Acid Drug Dev.* **9,** 487–492.
27. Salomoni, P., Perrotti, D., Marinez, R., Franceschi, C., and Calabretta, B. (1997) Resistance to apoptosis in CTLL-2 cells constitutively expressing c-*Myb* is associated with induction of *BCL-2* expression and *Myb*-dependent regulation of *bcl-2* promoter activity. *Proc. Natl. Acad. Sci. USA* **94,** 3296–3301.
28. Akiyama, T., Sudo, C., Ogawara, H., Toyoshima, K., and Yamamoto, T. (1986) The product of the human c-*erb*B-2 gene: a 185-kilodalton glycoprotein with tyrosine kinase activity. *Science* **232,** 1644–1646.
29. Ramanathan, R. K. and Belani, C. P. (1997) Chemotherapy for advanced non-small cell lung cancer: past, present, and future. *Sem. Oncol.* **24,** 440–454.
30. Zangemeister-Wittke, U., Schenker, T., Luedke, G. H., and Stahel, R. A. (1998) Synergistic cytotoxicity of *bcl*-2 antisense oligodeoxynucleotides and etoposide, doxorubicin and cisplatin on small-cell lung cancer cell lines. *Br. J. Cancer* **78,** 1035–1042.
31. Kondo, Y., Kondo, S., Tanaka, Y., Haqqi, T., Barna, B. P., and Cowell, J. K. (1998) Inhibition of telomerase increases the susceptibility of human malignant glioblastoma cells to cisplatin-induced apoptosis. *Oncogene* **16,** 2243–2248.
32. Dixit, M., Yang, J. L., Poirier, M. C., Price, J. O., Andrews, P. A., and Arteaga, C. L. (1997) Abrogation of cisplatin-induced programmed cell death in human breast cancer cells by epidermal growth factor antisense RNA. *J. Natl. Cancer Inst.* **89,** 365–373.
33. Colomer, R., Lupu, R., Bacus, S. S., and Gelmann, E. P. (1994) *erb*B-2 antisense oligonucleotides inhibit the proliferation of breast carcinoma cells with *erb*B-2 oncogene amplification. *Br. J. Cancer* **70,** 819–825.
34. Hudziak, R. M., Lewis, G. D., Winget, M., Fendly, B. M., Shepard, H. M., and Ullrich, A. (1989) p185HER2 monoclonal antibody has antiproliferative effects in vitro and sensitizes human breast tumour cells to tumour necrosis factor. *Mol. Cell. Biol.* **9,** 1165–1172.
35. Allal, C., Sixou, S., Kravzoff, R., Soulet, N., Soula, G., and Favre, G. (1998) SupraMolecular BioVectors (SMBV) improve antisense inhibition of *erb*B-2 expression. *Br. J. Cancer* **77,** 1448–1453.
36. Wiechen, K., Zimmer, C., and Dietel, M. (1998) Selection of a high activity c-*erb*B-2 ribozyme using a fusion gene of c-*erb*B-2 and the enhanced green fluorescent protein. *Cancer Gene Ther.* **5,** 45–51.
37. Pegram, M. D., Lipton, A., Hayes, D. F., et al. (1998) Phase II study of receptor-enhanced chemosentivity using recombinant humanized anti-p185IIER2/*neu* monoclonal antibody plus cisplatin in patients with HER2/*neu*-overexpressing metastatic breast cancer refractory to chemotherapy treatment. *J. Clin. Oncol.* **16,** 2659–2671.
38. Marth, C., Widschwendter, M., Kaern, J., et al. (1997) Cisplatin resistance is associated with reduced interferon-gamma-sensitivity and increased HER-2 expression in cultured ovarian carcinoma cells. *Br. J. Cancer* **76,** 1328–1332.

39. Norgaard, J. M., Langkjer, S. T., Palshof, T., Pedersen, B., and Hokland, P. (2001) Pretreatment leukemia cell drug resistance is correlated to clinical outcome in acute myeloid leukaemia. *Eur. J. Haematol.* **66,** 160–167.
40. Flasshove, M., Srumberg, D., Ayscue, L., et al. (1994) Structure analysis of the deoxycytidine kinase gene in patients with acute myeloid leukemia and resistance to cytosine arabinoside. *Leukemia* **8,** 780–785.
41. Funato, T., Satou, J., Nishiyama, Y., et al. (2000) In vivo leukemia cell models of Ara-C resistance. *Leukemia Res.* **24,** 535–541.
42. Williams, N. G. and Roberts, T. M. (1994) Signal transduction pathways involving the Raf proto-oncogene. *Cancer Met. Rev.* **13,** 105–116.
43. Storm, S. M., Brennscheidt, U., Sithanandam, G., and Rapp, U. R. (1990) *Raf* proto-oncogenes in carcinogenesis. *Crit. Rev. Oncol.* **2,** 1–8.
44. Trench, G. C., Southall, M., Smith, P., and Kidson, C. (1989) Allelic variation of the c-*raf*-1 proto-oncogene in human lymphoma and leukemia. *Oncogene* **4,** 507–510.
45. Blagosklonny, M. V., Giannakakou, P., El-Deiry, W., et al. (1997) *Raf*-1/*bcl*-2 phosphorylation: a step from microtubule damage to cell death. *Cancer Res.* **57,** 130–135.
46. Cornwell, M. M. and Smith, D. E. (1993) A signal transduction pathway for activation of the mdr1 promoter involves the proto-oncogene c-*raf* kinase. *J. Biol. Chem.* **268,** 15,347–15,350.
47. Hass, R., Hirano, M., Kharbanda, S., Rubin, E., Meinhardt, G., and Kufe, D. (1993) Resistance to phorbol ester-induced differentiation of a U-937 myeloid leukemia cell variant with a signaling defect upstream to *Raf*-1 kinase. *Cell Growth Differ.* **4,** 657–663.
48. Monia, B. P., Johnston, J. F., Geiger, T., Muller, M., and Fabbro, D. (1996) Antitumor activity of a phosphorothioate antisense oligonucleotide targeted against c-*raf* kinase. *Nature Med.* **2,** 668–675.
49. Skorski, T., Nieborowska-Skorska, M., Szczylik, C., et al. (1995) C-*RAF*-1 serine/ threonine kinase is required in BCR/ABL-dependent and normal hematopoiesis. *Cancer Res.* **55,** 2275–2278.
50. Gokhale, P. C., McRae, D., Monia, B.P., et al. (1999) Antisense *raf* oligodeoxyribonucleotide is a radiosensitizer in vivo. *Antisense Nucleic Acid Drug Dev.* **9,** 191–201.
51. Stevenson, J. P., Yao, K. S., Gallagher, M., et al. (1999) Phase I clinical/pharmacokinetic and pharmacodynamic trial of the c-*raf*-1 antisense oligonucleotide ISIS 5132 (CGP 69846A). *J. Clin. Oncol.* **17,** 2227–2236.
52. Cunningham, C. C., Holmlund, J. T., Schiller, J. H., et al. (2000) A phase I trial of c-*Raf* kinase antisense oligonucleotide ISIS 5132 administered as a continous intravenous infusion in patients with advanced cancer. *Clin. Cancer Res.* **6,** 1626–631.
53. Britten, R. A., Perdue, S., Eshpeter, A., and Merriam, D. (2000) *Raf*-1 kinase activity predicts for paclitaxel resistance in TP53mut, but not TP53wt human ovarian cancer cells. *Oncol. Rep.* **7,** 821–825.

54. Haseloff, J. and Gerlach, W. L. (1988) Simple RNA enzymes with new and highly specific endoribonuclease activity. *Nature* **334,** 585–591.
55. Hertel, K. J., Pardi, A., Uhlenbeck, O. C., et al. (1992) Numbering system for the hammerhead. *Nucleic Acids Res.* **20,** 3252.

IV

BLOOD–BRAIN BARRIER

12

Transport of Antisense Across the Blood–Brain Barrier

Laura B. Jaeger and William A. Banks

1. Introduction

Antisense-mediated therapy holds great promise for the treatment of central nervous system (CNS) diseases in which neurodegeneration is linked to over-production of endogenous protein. Administration of antisense therapy could be difficult, however, because peripherally administered antisense would have to cross the blood–brain barrier (BBB) in effective quantities. Several studies have investigated various modifications that can be introduced into antisense molecules to improve their ability to cross the BBB. Recently, two types of unmodified antisense analogs have been shown to effectively cross the BBB and affect CNS function. One of the analogs, a phosphorothioate oligodeoxy-nucleotide (P-ODN) directed against the amyloid β protein, was peripherally administered via iv injection. This analog crosses the BBB via a saturable transport mechanism termed oligonucleotide transport system-1 (OTS-1) *(1)*. Another unmodified P-ODN was also shown to cross the BBB. This P-ODN was directed at methionine enkephalin (Met-Enk), an opiate peptide associated with alcohol-ism (unpublished results). The second type of unmodified analog shown to cross the BBB is a peptide nucleic acid (PNA) that was directed against the neurotensin receptor NTS-1 in rats and was administered via ip injection *(2)*. Antisense molecules that can cross the BBB would be potential therapeutic agents and useful tools in the elucidation of CNS pathophysiology.

2. The BBB: What Is It?

The endothelia of the CNS regulate the composition of the interstitial fluid that surrounds the neurons and glia in the brain. The endothelial cells forming brain capillaries differ anatomically and biochemically from general systemic

From: *Methods in Molecular Medicine, Vol. 106: Antisense Therapeutics, Second Edition*
Edited by: I. Phillips © Humana Press Inc., Totowa, NJ

capillaries. These differences allow the capillary wall of the CNS, also known as the BBB, to maintain a composition of cerebrospinal fluid (CSF) that is different from the composition of the plasma.

2.1. History of the BBB

The first studies elucidating the BBB were conducted in the late nineteenth century. When Paul Ehrlich and others injected dyes into the periphery, they found that the dye was able to stain almost all body tissues. The brain and spinal cord were the major exceptions (3). Later studies conducted by Stern and Gautier demonstrated the selectivity of the BBB. While recording the movement of substances from the peripheral circulation into the CSF of the brain, they found that after a short incubation period, substances such as bromide and morphine could be found in the CSF, whereas others such as epinephrine and fluorescein remained in the blood (4). Stern and Gautier (5) also experimentally demonstrated that chemicals that were able to enter the CSF could affect CNS functioning, whereas those that remained in the blood showed no effect on CNS function. The anatomical basis for the BBB was finally demonstrated by studies conducted by Brightman, Reese, and Karnovsky in the late 1960s. These experiments showed that the tight junctions between capillary endothelial cells provided the barrier between the blood and the CNS (6–8).

2.2. Characteristics of the Brain Capillary Endothelia

The term *barrier* does not fully explain the functions and roles of the BBB. This monolayer of endothelial cells has features that allow selective entry of molecules into the CNS that are required for brain growth and function, while excluding or effluxing molecules and substances that may be harmful. Unlike the cells of the systemic capillaries, the endothelial cells of the brain capillaries are not fenestrated but, instead, are sealed together by intercellular tight junctions (*see* **Fig. 1**). These cells also have fewer pinocytotic vesicles than cells comprising systemic capillary walls. The endothelial cells of the BBB contain a higher number of mitochondria for the support of energy-dependent transport across the capillary walls (9).

Because of the tight junctions connecting the endothelial cells of the BBB, paracellular movement of substances is greatly reduced. Most molecules must cross by either passive diffusion (nonsaturable system) or saturable transport (saturable system) across the membrane.

The passive diffusion of blood-borne substances across the BBB correlates with the physiochemical properties of the molecule attempting to cross. Properties that are related to the lipid solubility of the molecule are especially important in this respect. Rate of entry into the CNS by a nonsaturable system is directly related to lipid solubility/hydrogen bonding and inversely related to the square

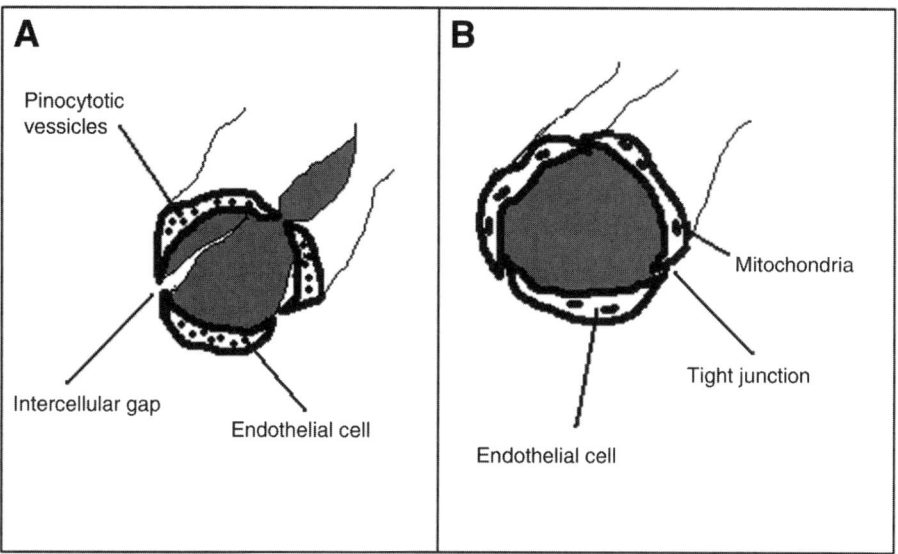

Fig. 1. Peripheral capillaries vs capillaries of the CNS. (**A**) In peripheral capillaries, fluid and solutes move into their target tissue through the clefts between the endothelial cells of the capillary bed. (**B**) The capillaries of the CNS, however, lack fenestra. Instead, these endothelial cells are joined by tight junctions to create a barrier that will aid in the regulation of substances entering and leaving the CNS. (Adapted from **refs. 3** and **62**).

root of the molecular weight *(10–12)*. Many small, lipid-soluble molecules can enter the brain by passively dissolving into the lipid bilayer of the cells of the BBB. Lipid solubility can be experimentally determined by calculating the fraction of substance that partitions into the two layers of an aqueous/octanol mixture. Other, more complex measures can improve agreement with in vivo results *(13)*. Hydrophobic substances, such as urea and sucrose, demonstrate a large partition coefficient and are readily absorbed across the endothelial cells of the BBB. Molecules that are hydrophilic, however, do not passively diffuse well across the BBB. These molecules are associated with a small partition coefficient and limited brain uptake unless a saturable transport mechanism exists.

For a molecule that utilizes a carrier-mediated system to cross the BBB, the rate of transport cannot be readily predicted by the physiochemical characteristics of the molecule. Carrier-mediated transport across the BBB is saturable because the rate of uptake depends on a limited number of transporters and, in the presence of excess ligand, these transporters can become saturated. Carrier-mediated transport mechanisms ensure that the brain receives an adequate supply of the nutrients that are too large to cross the BBB via passive diffusion. For example, glucose enters the brain via the GLUT-1 transporter that is present in the endothelial cells of the BBB *(14,15)*. Amino acids are also delivered to the CNS by various saturable transport systems *(16,17)*.

3. Antisense Oligonucleotides
3.1. Therapeutic Possibilities

Antisense oligodeoxynucleotides (AS-ODNs) are potential neuropharmaceuticals that would be useful for the treatment of CNS diseases, such as Huntington's disease, Alzheimer's disease, and alcoholism, in which overproduction or underproduction of endogenous protein causes disease. Either the protein's mRNA can be directly inhibited or antisense can be directed at the protein's receptor, synthetic enzymes, catabolic enzymes, or brain efflux systems. Antisense directed at the last two categories would allow for the treatment of conditions in which protein levels are low, not high. AS-ODNs are advantageous for therapeutic use because they can be easily synthesized with equipment found in most laboratories. However, AS-ODNs are large, highly polar molecules that are readily digested in blood and tissues by nucleases—the body's defense against stray DNA. Two general approaches have been developed to overcome these issues: (1) enzymatically resistant analogs have been produced to overcome the problem of stability, and (2) AS-ODNs have been modified with protecting groups or carrier systems to increase lipid solubility and thus enhance BBB permeability.

3.2. Production of Enzymatically Resistant Analogs

Binding of an AS-ODN to its complementary mRNA will inhibit protein expression either by sterically blocking the translation machinery or by promoting recognition by degrading enzymes, such as RNase H *(18)*. The method of antisense-mediated inhibition of protein expression varies depending on where on the target mRNA strand the AS-ODN binds. For example, AS-ODNs that are directed at the 5' region of the target mRNA will form a DNA–RNA complex that blocks protein expression by directly inhibiting the binding of the ribosome to the mRNA. Conversely, AS-ODNs that target regions farther downstream of the 5' terminus can block either elongation or splicing of the mRNA. AS-ODNs can also aid in the destruction of their target mRNA. AS-ODNs containing a stretch of six or more nucleotide bases can form DNA–RNA complexes that are recognized by RNase H *(19,20)*. This enzyme cleaves AS-ODN–mRNA hybrids and thus aids in antisense-mediated decreases in protein expression *(21)*.

Modifications are often introduced into the AS-ODN backbone to increase chemical stability, reduce sensitivity to nuclease digestion, and prevent non-specific interactions with unintended targets. These modifications allow for the administration of smaller quantities of antisense to achieve efficacy, which reduces the risk of toxicity. Three types of backbone modifications are discussed below along with their chemical properties, that contribute to AS-ODN stability and promote specific interactions with target mRNA.

One of the first modifications developed to reduce AS-ODN sensitivity to nuclease digestion involved replacing a nonbridging oxygen atom at both ends of the AS-ODN chain with a methyl group (*see* **Fig. 2A**). The result of this modification was called a methylphosphonate oligodeoxynucleotide *(22)*. This type of modification renders the antisense nuclease resistant. However, because the backbone is now uncharged, this modification reduces antisense activity by decreasing RNase H competency *(23)*.

Another modification developed to increase stability involves replacing a nonbridging oxygen atom at each phosphorus at the two ends of the AS-ODN chain with a sulfur (*see* **Fig. 2A**), producing a P-ODN *(24)*. This type of antisense is not completely resistant to nuclease digestion; however, it is still charged, so it is water soluble. P-ODNs have been extensively studied in vitro. These antisense analogs can specifically downregulate many protein targets in cell culture, including protein kinase-α *(25,26)* intercellular adhesion molecule-1 *(27)*, and bcl-2 *(28,29)*. The introduction of a sulfur atom makes P-ODNs more water soluble, but it also increases nonspecific effects, such as binding to heparin-binding proteins, some of which (e.g., fibronectin, laminin 30) have cell adhesion properties *(30)*. P-ODNs are RNase H competent *(31)*. Activation of this enzyme may help to mediate the antisense effects of P-ODNs.

The PNAs are another form of antisense that were designed to enhance the affinity of binding to target mRNA. PNAs are created by replacing the phosphate deoxyribose backbone with uncharged *N*-(2-aminoethyl)-glycine linkages (*see* **Fig. 2B**). Bases are attached to the glycine amino group via methylene carbonyl linkages *(32,33)*. Because PNAs contain a neutral backbone, this enhances hybridization with target mRNA by decreasing the repulsion between strands *(34)*. However, because PNA–RNA complexes do not act as substrates for RNase H, PNAs must block protein translation sterically to exhibit their antisense effects *(35)*.

3.3. Carrier Systems Designed to Increase Permeation Across the BBB

Because AS-ODNs are often large, highly polar molecules, their transport across the BBB by way of passive diffusion is likely to be low. Many studies assumed that AS-ODNs would not cross the BBB without the aid of either a carrier protein or modification of the BBB itself. Carrier systems designed to

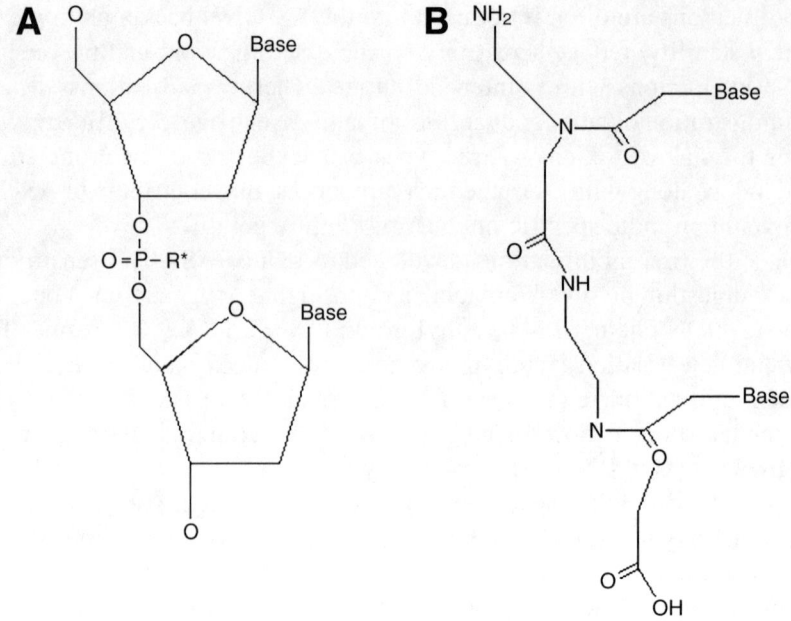

Fig. 2. Various AS-ODN analogs. (**A**) R^* represents a variable group. If R^* = oxygen atom (O), then this would depict an endogenous phosphodiester AS-ODN. If R^* = sulfur atom (S), then this would depict a phosphorothioate antisense analog. If R^* = methyl group (CH_3), then this would depict a methylphosphonate antisense analog. (**B**) Structure of a PNA.

enhance transport across the BBB include antisense complexed with biotin, glycosylated polycations, or antibodies directed at receptors present within the BBB endothelium *(36,37)*.

For example, in a study by Wu et al. *(37)*, antisense molecules injected intravenously into rats were delivered to the brain via attachment to biotin. The modified antisense formed a complex with streptavidin molecules attached to an antibody directed at the rat transferrin receptor, which is present in the BBB under normal circumstances. On binding to the receptor, the antisense/ antibody complex would be transported across the BBB and delivered to the target tissue.

The BBB itself has also been a target for modifications aimed at increasing antisense delivery to the CNS. The BBB's permeability can be increased when certain vasoactive agents are introduced *(38–40)* . For example, bradykinin has been shown to increase permeability through a mechanism that involves B2 receptors and nitric oxide *(41,42)*. Koga et al. *(43)* showed that intracarotid infusion of bradykinin increased delivery of AS-ODN into brain tumors. This type of delivery system allows therapeutic agents to be infused systemically and then delivered to malignant tumors of the brain without the level of disruption that occurs with other approaches to opening the BBB.

4. Unmodified Antisense and the BBB: Are Modifications Necessary?

4.1. Antisense Therapy in Pathophysiology Involving Increased Protein Production

Although many studies have investigated various modification strategies designed to enhance penetration of antisense across the BBB, few studies have considered if unmodified antisense can cross the BBB. It is possible that enzymatically resistant analogs of unmodified AS-ODNs enter the CNS by binding to transporters/receptors present in the BBB for the import of endogenous substances. By analogy, viruses often use this method to enter cells. For example, human immunodeficiency virus-1 binds to the CD4 receptor to promote fusion and entry of the virus into T-cells, dendritic cells, and macrophages *(44)*. Currently, studies examining the ability of unmodified antisense to cross the BBB have shown that two AS-ODN analogs are capable of permeation into the CNS: PNAs and phosphorothioate analogs *(1,2)*. To illustrate the points above, we now consider studies that have focused on treatment of pathophysiology through the use of unmodified AS-ODN analogs to target the proteins or receptors associated with disease states such as schizophrenia, pain perception, Alzheimer's disease, and alcoholism.

4.1.1. Schizophrenia and Pain Perception

Neurotensin (NT) is an endogenous peptide that exerts potent effects on the CNS. Because NT is rapidly degraded in the blood, direct injection into the brain is necessary for NT to exert its CNS effects of hypothermia, antinociception (decreased perception of pain), modulation of dopamine (DA) neurotransmission, and increased locomotor activity. Because the pharmacological activity of NT is similar to that of known neuroleptic agents, NT agonists have implications in the treatment of schizophrenia.

Schizophrenia is a disorder that is often associated with increased levels of the neurotransmitter DA. Many of the drugs designed to treat schizophrenia are DA receptor agonists. Because administration of NT is associated with increased DA turnover, this peptide could provide a useful tool in the treatment of schizophrenia. Peripheral administration of NT will not produce the effects seen on direct administration of NT into the CNS *(45)*. Peripheral administration of NT analogs has been shown to induce the CNS responses associated with the direct administration of NT into the brain *(46,47)*. NT69L, an NT analog developed by Tyler et al. *(47)*, has been shown to cross the BBB at levels sufficient to cause CNS activity, as assessed by hypothermic and antinociceptive effects. Like NT, the NT69L analog increases DA turnover rates, as demonstrated by increased DA metabolite (DOPAC) *(48)*. The effects of NT are mediated through binding to its receptors, NTR-1 and NTR-2 *(49,50)*. Both are G-protein-coupled receptors, but they are distributed differentially throughout the brain *(51)*. NT knockout animals would be useful in the elucidation of the specific roles of these receptors in the maintenance of critical DA concentrations within the brain. However, these animals may produce misleading results. Some knockout animals are not compatible with life, and others may develop mechanisms to compensate for the lack of the deleted protein. Antisense technology would provide a useful tool to better understand the interaction between the NT and DA systems in the CNS.

A study by Tyler et al. *(47)* investigated an unmodified 12mer PNA antisense directed at the NT receptor NTR-1. The normal physiological response to an increase in NT is hypothermia and antinociception. Using a gel-shift assay, Tyler et al. *(2)* demonstrated that the unmodified PNA could cross the BBB and affect CNS function after a single ip injection in rats. The effects of NT microinjected into the brain were inhibited within 24 h of ip administration of the PNA directed at NTR-1. The effects of the antisense PNA were completely reversible. The normal physiological response to NT returned 48 h postinjection. Radioligand binding assays showed that administration of antisense PNA was accompanied by a reduction in receptor sites; however, there was no change in the levels of mRNA, possibly because of the nature of inhibition caused by antisense PNAs. Unlike AS-ODN–RNA complexes, PNA–RNA complexes do not activate RNase H, so protein levels are reduced by sterically blocking translation machinery *(35,52)*.

4.1.2. Alzheimer's Disease

Alzheimer's disease is characterized by the presence of amyloid plaques in the extracellular space of the CNS. The amyloid plaques consist mainly of a 40- to 42-amino-acid peptide known as amyloid β protein (Aβ). These aggregates are toxic and could accumulate around blood vessels, leading to apoptosis of the vascular smooth muscle cells *(53)*. Aβ is a cleavage product of the amyloid precursor protein (APP). Increased brain levels of Aβ in Alzheimer's patients could result from many factors including increased production of APP. Another possibility is altered cleavage of APP, resulting in the formation of a mutant form of Aβ that is more likely to aggregate. Regardless of the cause, levels of Aβ could be reduced with antisense therapy.

Kumar and colleagues *(54,55)* tested a series of P-ODNs directed against Aβ with the above in mind. They tested the P-ODNs in a strain of mice that spontaneously overexpresses Aβ peptide as it ages, the senescence accelerated mouse (SAMP8). By 12 mo of age, this strain demonstrates a twofold increase in the brain levels of Aβ and severe learning and memory deficits *(54,55)*. This increase is similar to that which occurs in patients with Alzheimer's disease. Kumar and colleagues *(54,55)* found that either antibody or a phosphorothioate antisense directed against Aβ reversed the cognitive impairments present in mature SAMP8 mice after injection directly into the brain.

To determine whether the phosphorothioate antisense directed at Aβ could cross the BBB, it was radioactively labeled with ^{32}P and administered intravenously *(1)*. Purification techniques using gel electrophoresis and autoradiography revealed that the AS-ODN was transported intact across the BBB. Because of its large molecular weight and poor lipid solubility (partition coefficient = log[–3.52]), passive transmembrane diffusion was an unlikely means of transport for the antisense. Penetration of this antisense into the CNS was inhibited by an excess of unlabeled antisense directed against Aβ, thus demonstrating saturable transport. This system was named OTS-1 *(1)*.

Transport of AS-ODN across the BBB was also verified behaviorally. Intravenous administration of the antisense against Aβ reversed the learning and memory deficits normally seen in the mature SAMP8 mouse.

4.2. Antisense Therapy in Pathophysiology Involving Lowered Protein Production (Alcoholism)

Alcoholism is associated with major changes in neurochemistry. One peptide that is especially affected is Met-Enk, an opiate peptide. In ethanol naïve animals, administration of Met-Enk is correlated with a decrease in the amount of ethanol drank *(57)*. Conversely, in animals addicted to ethanol, endogenous Met-Enk levels are low *(58–61)*. Both Met-Enk and Tyr-MIF-1 (a peptide with opiate-like properties) are transported out of the brain across the BBB by the saturable efflux peptide transport system (PTS-1). During alcohol dependency and physical withdrawal from alcohol, Met-Enk levels in the brain are independent of preproenkephalin mRNA levels *(62)*. Because the mRNA levels do not correlate with the decreased brain levels of Met-Enk, one can assume that control is no longer at the level of transcription. Met-Enk levels at this point, therefore, are largely determined by posttranscriptional processes, including PTS-1 activity.

In mice, Met-Enk has been shown to increase when an animal is initially exposed to ethanol. Chronic exposure, however, is correlated with a decrease in Met-Enk levels in the brain *(60,61)*. On withdrawal from alcohol, Met-Enk levels remain low or decrease even further *(62,63)*. This condition is associated with seizures in mice. Low levels of or resistance to Met-Enk is correlated with a genetic predisposition to abuse alcohol and also with increased seizure susceptibility *(64)*. Enzymes such as neutral endopeptidase, aminopeptidase, angiotensin-converting enzyme, and dipeptidlyaminopeptidase have been implicated in the metabolism of the enkephalins in vitro *(65,66)*. Administration of AS-ODNs directed at the metabolic enzymes that degrade enkephalin peptides would provide a new method for the treatment of alcoholism and the seizures associated with alcohol withdrawal.

As a first step in determining whether antisense could be effective in the treatment of alcoholism, we have demonstrated that phosphorothioate AS-ODNs directed at Met-Enk are able to cross the BBB. Three different analogs (10mer, 17mer, and 19mer) of AS-ODNs were radioactively labeled with ^{32}P and administered individually via iv injection. The 10mer and 17mer AS-ODNs crossed at a higher rate than the 19mer AS-ODN (unpublished results). Expected behavioral effects associated with low levels of Met-Enk were confirmed in mice that were subcutaneously administered a cocktail of the three different antisense oligomers. As predicted, these mice drank more alcohol than the control mice (unpublished results).

4.3. Other Strategies for Antisense Therapy (Metabolic Disorders)

Although we have primarily considered AS-ODNs as a therapeutic option for conditions associated with abnormal protein production, these potential neuropharmaceuticals would also be useful in the treatment of metabolic disorders, in which abnormal lipid accumulation often results in severe neurodegeneration. The glycosphingolipid (GSL) lysosomal storage diseases are a cluster of metabolic disorders each caused by mutations in the genes that encode for the glycohydrolases *(67)*. This class of enzymes is essential for the metabolism of GSL within lysosomes. Because the defective glycohydrolase is unable to metabolize its substrate properly, the GSL will accumulate within the lysosome, causing cell dysfunction and eventually organ failure. Tay-Sachs disease (TSD) is a lysosomal storage disease characterized by progressive neurodegeneration due to accumulation of GSL in neurons of the CNS.

Children born with TSD appear healthy at birth, but by about 2 yr of age, they begin to experience the mental and physical deterioration associated with the accumulation of a fatty substance called the GM2 ganglioside. This accumulation is caused by the deficient activity of the enzyme, β-hexosaminidase A (Hex-A) *(68)*. The substrate for Hex-A, the GM2 ganglioside, accumulates abnormally in cells, especially the neurons of the CNS. The ongoing accumulation progressively destroys nerve cells until the nervous system is so badly damaged that it can no longer sustain life. Current therapeutic options available for the treatment of TSD are limited in their effectiveness. Enzyme replacement for neuronal lysosomes would require a system that would not only target the enzyme to the lysosme, but also allow the enzyme to cross the BBB.

A strategy for lysosomal delivery of Hex-A was examined in feline cerebral cortex cell culture; however, only a minimal effect on the lysosomal storage of GM2 was observed in that study *(69)*. AS-ODNs directed against enzymes responsible for GSL synthesis would be another option for the treatment of children afflicted with TSD.

5. Conclusion

AS-ODNs are potential neuropharmaceuticals for the treatment of CNS diseases in which overproduction or underproduction of endogenous protein causes pathology. However, because unmodified AS-ODNs are large, highly polar molecules, their passage across the BBB may be limited unless an endogenous transporter is available to promote passage across the BBB. To reach therapeutic concentrations in the CNS, peripherally administered AS-ODNs must be able to cross the BBB in quantities sufficient to affect brain function. Currently, it has been shown that an unmodified phosphorothioate antisense molecule directed at Aβ is able to cross the BBB via an active transport mechanism, termed OTS-1 *(1)*. Other unmodified AS-ODNs have been shown to cross the BBB: a P-ODN directed at Met-Enk (unpublished results) and a PNA directed at NTR-1 *(2)*. Future studies of the transport mechanisms that allow antisense molecules to cross the BBB will aid in the elucidation of CNS pathophysiology and also provide a better understanding of the dynamics of the BBB.

References

1. Banks, W. A., Farr, S. A., Butt, W., Kumar, V. B., Franko, M. W., and Morley, J. E. (2001) Delivery across the blood-brain barrier of antisense directed against amyloid b : reversal of learning and memory deficits in mice overexpressing amyloid precursor protein. *J. Pharmacol. Exp. Ther.* **297,** 1113–1121.
2. Tyler, B. M., Jansen, K., McCormick, D. J., et al. (1999) Peptide nucleic acids targeted to the neurotensin receptor and administered i.p. cross the blood-brain barrier and specifically reduce gene expression. *Proc. Natl. Acad. Sci. USA* **96,** 7053–7058.
3. Ehrlich, P. (1906) *Uber die Beziehungen von Chemische Constituion Vertheilung, und Pharmakologischer Wirkung: Collected Studies in Immunity,* Wiley, New York.
4. Stern, L. and Gautier, R. (1921) Rapports entre le liquide céphalo-rachidien et la circulation sanguine. *Arch. Int. Physiol.* **17,** 138–192.
5. Stern, L. and Gautier, R. (1922) Les rapports entre le liquide céphalo-rachidien et les éléments nerveux de l'axe cérébrospinal. *Arch. Int. Physiol.* **17,** 391–448.
6. Brightman, M. W. and Reese, T. S. (1969) Junctions between intimately apposed cell membranes in the vertebrate brain. *J. Cell Biol.* **40,** 648–677.
7. Reese, T. S. and Karnovsky, M. J. (1967) Fine structural localization of a blood-brain barrier to exogenous peroxidase. *J. Cell Biol.* **34,** 207–217.

8. Reese, T. S. and Brightman, M. W. (1968) Similarity in structure and permeability to peroxidase of epithelia overlying fenestrated cerebral capillaries. *Anat. Rec.* **160,** 414.

9. Oldendorf, W. H., Cornford, M. E., and Brown, W. J. (1977) The large apparent work capability of the blood-brain barrier: a study of the mitochondrial content of capillary endothelial cells in brain and other tissues of the rat. *Ann. Neurol.* **5,** 409–417.

10. Chikhale, E. G., Ng, K. Y., Burton, P. S., and Borchardt, R. T. (1994) Hydrogen bonding potential as a determinant of the in vitro and in situ blood-brain barrier permeability of peptides. *Pharm. Res.* **11,** 412–419.

11. Cornford, E. M., Braun, L. D., Oldendorf, W. H., and Hill, M. A. (1982) Comparison of lipid-mediated blood-brain barrier penetrability in neonates and adults. *Am. J. Physiol.* **243,** C161–C168.

12. Oldendorf, W. H. (1973) Lipid-solubility and drug penetration of the blood-brain barrier. *Proc. Soc. Exp. Biol. Med.* **147,** 813–816.

13. Fischer, H., Gottschlich, R., and Seelig, A. (1998) Blood-brain barrier permeation: molecular parameters governing passive diffusion. *J. Memb. Biol.* **165,** 201–211.

14. Pardridge, W. M., Boado, R. J., and Farrell, C. R. (1990) Brain-type glucose transporter (GLUT-1) is selectively localized to the blood-brain barrier. Studies with quantitative western blotting and in situ hybridization. *J. Biol. Chem.* **265,** 18,035–18,040.

15. Kalaria, R. N., Gravina, S. A., Schmidley, J. W., Perry, G., and Harik, S. I. (1988) The glucose transporter of the human brain and blood-brain barrier. *Ann. Neurol.* **24,** 757–764.

16. Cornford, E. M. (1985) The blood-brain barrier, a dynamic regulatory interface. *Mol. Physiol.* **7,** 219–260.

17. Christensen, H. N. (1990) Role of amino acid transport and countertransport in nutrition and metabolism. *Physiol. Rev.* **70,** 43–77.

18. Bonham, M. A., Brown, S., Boyd, A. L., et al. (1995) An assessment of the antisense properties of RNase H-competent and steric-blocking oligomers. *Nucleic Acids Res.* **23,** 1197–1203.

19. Monia, B. P., Lesnik, E. A., Gonzalez, C., et al. (1993) Evaluation of 2'-modified oligonucleotides containing 2'-deoxy gaps as antisense inhibitors of gene expression. *J. Biol. Chem.* **268,** 14,514–14,522.

20. Agrawal, S., Jiang, Z., Zhao, Q., et al. (1997) Mixed-backbone oligonucleotides as second generation antisense oligonucleotides: in vitro and in vivo studies. *Proc. Natl. Acad. Sci. USA* **94,** 2620–2625.

21. Walder, R. Y. and Walder, J. A. (1988) Role of RNase H in hybrid-arrested translation by antisense oligonucleotides. *Proc. Natl. Acad. Sci. USA* **85,** 5011–5015.

22. Miller, P. S. and Ts'o, P. O. (1987) A new approach to chemotherapy based on molecular biology and nucleic acid chemistry : Matagen (masking tape for gene expression). *Anticancer Drug Des.* **2,** 117–128.

23. Crooke, S. T., Lemonidis, K. M., Neilson, L., Griffey, R., Lesnik, E. A., and Monia, B. P. (1995) Kinetic characteristics of Escherichia coli RNase H1: cleavage of various antisense oligonucleotide-RNA duplexes. *Biochem. J.* **312,** 599–608.

24. Stec, W. J., Zon, G., Egan, W., and Stec, B. (1984) Automated solid-phase synthesis, separation, and stereochemistry of phosphorothioate analogues of oligodeoxyribonucleotides. *J. Am. Chem .Soc.* **106,** 6077–6080.

25. Geiger, T., Muller, M., Dean, N. M., and Fabbro, D. (1998) Antitumor activity of a PKC-alpha antisense oligonucleotide in combination with standard chemotherapeutic agents against various human tumors transplanted into nude mice. *Anticancer Drug Des.* **13,** 35–45.

26. Dean, N. M. and McKay, R. (1994) Inhibition of protein kinase C-alpha expression in mice after systemic administration of phosphorothioate antisense oligodeoxynucleotides. *Proc. Natl. Acad. Sci. USA* **91,** 11,762–11,766.

27. Chiang, M. Y., Chan, H., Zounes, M. A., Freier, S. M., Lima, W. F., et al (1991) Antisense oligonucleotides inhibit intercellular adhesion molecule 1 expression by two distinct mechanisms. *J. Biol. Chem.* **266,** 18,162–18,171.

28. Ziegler, A., Luedke, G. H., Fabbro, D., Altmann, K. H., and Stahel, R. A. (1997) Induction of apoptosis in small-cell lung cancer cells by an antisense oligodeoxynucleotide targeting the Bcl-2 coding sequence. *J. Natl. Cancer Inst.* **89,** 1027–1036.

29. Miayake, H., Tolcher, A., and Gleave, M. E. (2000) Chemosensitization and delayed androgen-independent recurrence of prostate cancer with the use of antisense Bcl-2 oligodeoxynucleotides. *J. Natl. Cancer Inst.* **92,** 34–41.

30. Khaled, Z., Benimetskaya, L., Zeltser, R., Khan, T., Sharma, H. W., et al (1996) Multiple mechanisms may contribute to the cellular anti-adhesive effects of phosphorothioate oligodeoxynucleotides. *Nucleic Acids Res.* **24,** 737–745.

31. Stein, C. A., Subasinghe, C., Shinozuka, K., and Cohen, J. S. (1988) Physiochemical properties of phosphorothioate oligodeoxynucleotides. *Nucleic Acids Res.* **16,** 3209–3221.

32. Nielsen, P. E., Egholm, M., Berg, R. H., and Buchardt, O. (1991) Sequence-selective recognition of DNA by strand displacement with a thymine-substituted polyamide. *Science* **254,** 1497–1500.

33. Larsen, H. J., Bentin, T., and Nielsen, P. E. (1999) Antisense properties of peptide nucleic acid. *Biochim. Biophys. Acta* **1489,** 159–166.

34. Egholm, M., Buchardt, O., Christensen, L., et al. (1993) PNA hybridizes to complementary oligonucleotides obeying the Watson-Crick hydrogen bonding rules. *Nature* **365,** 566–568.

35. Knudsen, H. and Nielsen, P. E. (1996) Antisense properties of duplex- and triplex-forming PNAs. *Nucleic Acids Res.* **24,** 494–500.

36. Stewart, A. C., Pichon, C., Meunier, L., Midoux, P., Monsigny, M., and Roche, A. C. (1996) Enhanced biological activity of antisense oligonucleotides complexed with glycosylated poly-L-lysine. *Mol. Pharmacol.* **50,** 1487–1494.

37. Wu, D., Boado, R. J., and Pardridge, W. M. (1996) Pharmacokinetics and blood-brain barrier transport of [³H]-biotinylated phosphorothiolate oligodeoxynucleotide conjugated to a vector-mediated drug delivery system. *J. Pharmacol. Exp. Ther.* **276,** 206–211.

38. Inamura, T. and Black, K. L. (1994) Bradykinin selectively opens blood-tumor barrier in experimental brain tumors. *J. Cereb. Blood Flow Metab.* **14,** 862–870.

39. Inamura, T., Nomura, T., Bartus, R.T., and Black, K. L. (1994) Intracarotid infusion of RMP-7, a bradykinin analog: a method for selective drug delivery to brain tumors. *J. Neurosurg.* **81,** 752–758.
40. Matsukado, K., Inamura, T., Nakano, S., Fukui, M., Bartus, R. T., and Black, K. L. (1996) Enhanced tumor uptake of carboplatin and survival of glioma-bearing rats by intracarotid infusion of bradykinin analog, RMP-7. *Neurosurgery* **39,** 125–133.
41. Elliott, P. J., Hayward, N. J., Huff, M. R., Nagle, T. L., Black, K. L., and Bartus, R. T. (1996) Unlocking the blood-brain barrier : a role for RMP-7 in brain tumor therapy. *Exp. Neurol.* **141,** 214–224.
42. Nakano, S., Matsukado, K., and Black, K. L. (1996) Increased brain tumor microvessel permeability after intracarotid bradykinin infusion is mediated by nitric oxide. *Cancer Res.* **56,** 4027–4031.
43. Koga, H., Inamura, T., Ikezaki, K., Samoto, K., Matsukado, K., and Fukui, M. (1999) Selective transvascular delivery of oligodeoxynucleotides to experimental brain tumors. *J. Neurooncol.* **43,** 143–151.
44. Willey, R. L., Martin, M. A., and Peden, K. W. C. (1994) Increase in soluble CD4 binding to and CD4-induced dissociation of gp120 from virions correlates with infectivity of human immunodeficiency virus type 1. *J. Virol.* **68,** 1029–1039.
45. Nemeroff, C. B., Bissette, G., Prange, A. J., Jr., Loosen, P. T., Barlow, T. S., and Lipton, M. A. (1977) Neurotensin: CNS effects of a hypothalamic peptide. *Brain Res.* **128,** 485–496.
46. Banks, W. A., Wustrow, D. J., Cody, W. L., Davis, M. D., and Kastin, A. J. (1995) Permeability of the blood-brain barrier to the neurotensin 8-13 analog NT1. *Brain Res.* **695,** 59–63.
47. Tyler, B. M., Douglas, C. L., Fauq, A. H., et al. (1999) In vitro binding and CNS effects of novel neurotensin agonists that cross the blood brain barrier. *Neuropharm.* **38,** 1027–1034.
48. Boules, M., Cusack, B., Zhao, L., Fauq, A., McCormick, D. J., and Richelson, E. (2000) A novel neurotensin peptide analog given extracranially decreases food intake and weight in rodents. *Brain Res.* **865,** 35–44.
49. Le, F., Cusack, B., and Richelson, E. (1996) The neurotensin receptor: is there more than one subtype? *Trends Pharmacol. Sci.* **17,** 1–3.
50. Vincent, J. P., Mazella, J., and Kitabgi, P. (1999) Neurotensin and neurotensin receptors. *Trends Pharmacol. Sci.* **20,** 302–309.
51. Walker, N., Lepee-Lorgeoux, I., Fournier, J., et al. (1998) Tissue distribution and cellular localization of the levocabastine-sensitive neurotensin receptor mRNA in adult rat brain. *Brain Res. Mol. Brain Res.* **57,** 193–200.
52. Mollegaard, N. E., Buchardt, O., Egholm, M., and Nielsen, P. E. (1994) Peptide nucleic acid: DNA strand displacement loops as artificial transcription promoters. *Proc. Natl. Acad. Sci. USA* **91,** 3892–3895.
53. Davis, J., Cribbs, D. H., Cotman, C. W., and Van Nostrand, W. E. (1999) Pathogenic amyloid-β protein induces apoptosis in cultured human cerebrovascular smooth muscle cells. *Amyloid* **6,** 157–164.

54. Kumar, V. B., Farr, S. A., Flood, J. F., et al. (2000) Site-directed antisense oligonucleotide decreases the expression of amyloid precursor protein and reverses deficits in learning and memory in aged SAMP8 mice. *Peptides* **21,** 1769–1775.
55. Morley, J. E., Kumar, V. B., Bernardo, A. F., et al. (2000) β-Amyloid precursor polypeptide in SAMP8 mice affects learning and memory. *Peptides* **21,** 1761–1767.
56. Morley, J. E., Farr, S. A., and Flood, J. F. (2002) Antibody to amyloid β protein alleviates impaired acquisition, retention, and memory processing in SAMP8 mice. *Neurobiol. Learn. Mem.* **78,** 125–138.
57. Blum, K., Briggs, A. H., Wallace, J. E., Hall, C. W., and Trachtenberg, M. A. (1987) Regional brain [Met]-enkephalin in alcohol-preferring and non-alcohol-preferring inbred strains of mice. *Experientia* **43,** 408–410.
58. Gianoulakis, C. (1989) The effect of ethanol on the biosynthesis and regulation of opioid peptides. *Experientia* **45,** 428–435.
59. Ng, G. Y. K., O'Dowd, B. F., and George, S. R. (1996) Genotypic differences in mesolimbic enkephalin gene expression in DBA/2J and C57BL/6J inbred mice. *Eur. J. Pharmacol.* **311,** 45–52.
60. Schulz, R., Wuster, M., Duka, T., and Herz, A. (1980) Acute and chronic ethanol treatment changes endorphin levels in brain and pituitary. *Psychopharmacology (Berlin),* **68,** 221–227.
61. Seizinger, B. R., Bovermann, K., Mayslinger, D., Holtt, V., and Herz, A. (1983) Differential effects of acute and chronic ethanol treatment on particular opioid peptide systems in discrete regions of rat brain and pituitary. *Pharmacol. Biochem. Behav.* **18,** 361–369.
62. Plotkin, S. R., Banks, W. A., Waguespack, P. J., and Kastin, A. J. (1997) Ethanol alters the concentration of Met-enkephalin in brain by affecting peptide transport system-1 independent of preproenkephalin mRNA. *J. Neurosci. Res.* **48,** 273–280.
63. Hong, J. S., Majchrowicz, E., Hunt, W. A., and Gillin, J. C. (1981) Reduction in cerebral methionine-enkephalin content during the ethanol withdrawal syndrome. *Subst. Alcohol Actions Misuse* **2,** 233–240.
64. Plotkin, S. R., Banks, W. A., Cohn, C. S., and Kastin, A. J. (2001) Withdrawal from alcohol in withdrawal seizure-prone and -resistant mice: evidence for enkephalin resistance. *Pharmacol. Biochem. Behav.* **68,** 379–387.
65. Gros, C., Giros, B., Llorens, C., et al. (1985) Enkephalin metabolism and its inhibition. *Biochem. Soc. Trans.* **13,** 47–50.
66. Melzig, M. F. and Heder, G. (1995) Dexamethasone induced enhanced enkephalin degradation by angiotensin-converting enzyme (ACE) of endothelial cells. *Pharmazie* **50,** 139–141.
67. Friedman, B. (1971) Tay-Sachs and other lipid storage diseases. *HSMHA Health Rep.* **86,** 769–774.
68. Rattazzi, M. C., Brown, J. A., Davidson, R. G., and Shows, T. B. (1976) Studies on complementation of beta hexosaminidase deficiency in human GM2 gangliosidosis. *Am. J. Hum. Genet.* **28,** 143–154.
69. Dobrenis, K., Joseph, A., and Rattazzi, M. C. (1992) Neuronal lysosomal enzyme replacement using fragment C of tetanus toxin. *Proc. Natl. Acad. Sci. USA* **89,** 2297–2301.

V

DERMAL

13

Transdermal Delivery of Antisense Oligonucleotides

Rhonda M. Brand and Patrick L. Iversen

1. Introduction

Transdermal delivery is an appealing method of introducing therapeutic agents because it allows medication to bypass the gastrointestinal (GI) tract. This reduces degradation by the acid and proteolytic enzymes in the gastric environment *(1)*, as well hepatic first-pass elimination *(2)* and incomplete absorption due to GI motility disorders *(3)*. Transdermal delivery also provides steady-state drug levels and improves patient compliance because of its extended duration. Another beneficial use of this technique is the treatment of skin disorders using local delivery of chemicals. Careful chemical design and formulation can modulate whether topically applied medications will reside within the skin or penetrate transdermally to achieve therapeutic systemic levels. The disadvantages of transdermal delivery include limited numbers of potential drug candidates due to their inability to penetrate the skin in sufficient levels, and potential irritation at the application site *(4)*.

2. Skin

The skin is the largest organ of the human body, covering nearly $2m^2$ of surface area on an average adult and receiving about one-third of all blood circulating through the body. The primary function of the skin is to act as the body's major barrier. It maintains the fluid homeostasis within the body while preventing compounds from entering from the external environment. To do this, the skin has developed two layers, the epidermis and the dermis. The epidermis can be subdivided into the viable epidermis and the stratum corneum *(5)*.

The outermost layer of the skin, the stratum corneum, is approx 15–20 μm thick in human skin and consists of terminally differentiated keratinocytes

From: *Methods in Molecular Medicine, Vol. 106: Antisense Therapeutics, Second Edition*
Edited by: I. Phillips © Humana Press Inc., Totowa, NJ

(corneocytes) that are embedded in a matrix of lipid bilayers *(6)*. It is the skin's principle barrier and can be thought of as a brick wall, with corneocytes serving as the bricks and lipids surrounding them as mortar *(7)*. The viable epidermis is located just below the stratum corneum, and its principal function is the production of stratum corneum *(8)*. The dermis is the innermost layer, consisting mainly of collagen fibers in an aqueous gel matrix that imparts elastic properties to the skin. It contains blood vessels, lymphatics, and nerve endings and is the physiological support mechanism for the epidermis. The skin also contains hair follicles and sebaceous glands that can act as shunts for transdermal penetration of chemicals. The surface area of these appendages, however, is quite small when compared with the total surface area of skin, and, therefore, they have a minor role in penetration *(9)*.

Keratinocytes are formed at the epidermal/dermal junction. As they begin the process of terminal differentiation, they are replaced with newer keratinocytes. As differentiation occurs, cells move away from the epidermal/dermal barrier toward the stratum corneum. By the time that they have reached the stratum corneum, they have elongated and flattened, lost their nuclei and other organelles, and are surrounded by a thick band of protein forming a cornified envelope. After approx 2 wk in the stratum corneum, the corneocytes reach the outside of the stratum corneum and are sloughed off in a process called desquamation *(10)*.

The lipids associated with keratinocytes change as the cells differentiate. Unlike most biological tissues, the stratum corneum contains no phospholipids but, instead, ceramines, cholesterol, fatty acids, sterol esters, and cholesteryl sulfate *(8,11)*. These lipids form the bilayer that becomes the major barrier to water and water-soluble chemicals. For appreciable quantities of chemicals to cross the skin, therefore, a permeant must pass through the stratum corneum's brick wall. Evidence has shown that for a molecule to transverse this barrier successfully, it must rely on partitioning into the lipid mortar surrounding the keratinocytes. It then remains in the lipids and diffuses through the thickness of the stratum corneum. On successfully crossing the stratum corneum, the permeant must then diffuse into the more hydrophilic viable epidermis before proceeding to the blood vessels in the dermis. The need to partition into stratum corneum lipids and then into the hydrophilic bloodstream favors molecules that are moderately lipophilic *(12)*.

Movement through the skin is different from many other tissues in that it relies on passive diffusion instead of active transport. Flux across the skin can be described by Fick's law, which demonstrates a linear relationship among penetration, donor concentration, and exposed area. Furthermore, smaller and uncharged chemicals tend to penetrate better than comparable larger and charged molecules. Several additional factors affect the feasibility of the therapeutic delivery of drugs through the skin, including the surface area of the delivery patch and dose of drug needed. A transdermal patch cannot be too large; generally, it is assumed that 50 cm^2 should be the largest coverage area. Based on Fick's law, chemical flux is linearly related to surface area. Therefore, calculations have shown that the maximum dose feasibly delivered for chemicals that readily penetrate the skin should be in the low milligram range *(13)*.

Given the information provided above, it seems obvious that delivery of antisense oligonucleotides (AS-ODNs) through the skin is unlikely to succeed. They are large, highly charged, and hydrophilic and require doses in humans that are too large to be feasible by transdermal delivery. Thus, sufficient quantities of oligonucleotides would not be able to penetrate through the stratum corneum. Two approaches have been used to reduce these problems and make transdermal delivery more feasible. The first is to improve delivery across the skin using chemical and physical penetration enhancers, and the second is to alter the oligonucleotide chemistry to increase potency, thereby reducing the quantity that must be delivered. The latter is not discussed in this chapter but is addressed in other areas of this book.

3. Penetration Enhancement

Because the skin has evolved as such an efficient barrier, several techniques have been developed to increase the absorption of therapeutic chemicals into the skin. This section provides a brief overview of the techniques for penetration enhancement that have been used to improve oligonucleotide delivery.

3.1. Techniques for Physical Penetration Enhancement

3.1.1. Iontophoresis

Iontophoresis uses a small electrical potential applied across the skin to drive ionized drug molecules into the skin *(14,15)*. Charged molecules are repelled away from an electrode of similar polarity and move into the skin. The direct effect of the potential is on the drug itself and not on the skin. Current densities used are no greater than 0.5 mA/cm^2, so iontophoresis produces little skin damage *(16)*. Because the skin has a pH between 3 and 4, it is negatively charged at physiological pH *(17)*. Therefore, transfer of positive ions is favored, resulting in a net volume flow in the direction of the positive (anode) to negative (cathode) electrodes. This electroosmotic flow can carry uncharged molecules with it through the skin, allowing iontophoresis to even enhance the delivery of uncharged molecules *(18,19)*. Although iontophoresis is most effective for positively charged molecules, the transdermal penetration of negatively charged species, such as oligonucleotides, may also be increased *(20)*.

The ability of iontophoresis to enhance the transdermal delivery of AS-ODNs has been established. The properties of the solution in which the oligonucleotide is placed (e.g., pH, salt concentration) can influence the level of penetration enhancement *(21)*. The influence of oligonucleotide size and sequence on iontophoretic permeation has also been examined. The iontophoretic transport of phosphorothioate antisense molecules with lengths ranging from 6 to 40 bases through hairless mouse skin was found to be between 2 and 26 pmol/(cm^2·h) from a 5 µM donor solution *(22)*.

Substituting a C$_5$ propyne into a phosphorothioate oligonucleotide targeted to cytochrome p450-3A2 (CYP3A2) increased potency by a factor of at least 67 over the unmodified phosphorothioate. The transdermal penetration of this sequence was examined in vivo. A patch containing either an oligonucleotide solution or buffer control was placed on an animal's back, and an iontophoretic current of 0.5 mA/cm^2 was applied for 3.5 h. Twenty-four hours later, CYP3A2 levels were measured noninvasively using the midazolam-induced sleeping rat model. Liver and small intestinal microsomes were made on completion of sleep studies. The midazolam-treated group of animals treated with antisense to CYP3A2 slept significantly longer than the controls ($p < 0.05$). Microsomal CYP3A2 levels were significantly lower from antisense-treated animals than from either buffer control ($p < 0.001$), demonstrating that transdermally delivered phosphorothioate oligonucleotides can reach concentrations sufficient to induce changes in specific target enzymes in vivo *(23)*.

3.1.2. Electroporation

Electroporation uses a high-voltage, short duration (microseconds to milliseconds) electrical pulse to enhance the penetration of chemicals through the skin. These pulses generate a high-intensity electric field that causes the lipid bilayers in the stratum corneum to rearrange. This change in structure destabilizes the membrane and thus leads to increased permeability *(24)*. Electroporation is commonly used as a method for introducing DNA into isolated cells. Unlike iontophoresis, electroporation acts principally on the skin by creating temporary "pores" *(25)*.

Electroporation has been used to deliver oligonucleotides in a number of in vitro studies. This technique enhanced the delivery of a 24mer and 15mer phosphodiester oligonucleotide across human cadaver stratum corneum *(26)*. Electroporation increased topical delivery of a 3' end modified phosphodiester by two orders of magnitude over passive diffusion. The end modifications were successful in reducing degradation by the skin *(27)*. The amount of oligonucleotide delivered through the skin can be controlled by the components of the electrophoretic pulse (voltage, pulse time, and number of pulses) and the donor concentration *(28)*. Fluorescent and confocal microscopies were used to examine localization of a phosphorothioate oligonucleotide after electroporation and iontophoresis. Iontophoretic transport occurred principally through hair follicles, whereas electroporation led to transcellular transport. The quantity of phosphorothioate in the skin's viable tissue after iontophoresis and electroporation was equivalent; however, electroporation produced greater levels in the stratum corneum than iontophoresis *(29)*.

3.1.3. Microarray Patches

The newest method for providing medications through the skin is microarray patches. These "patches" consist of stainless steel or titanium microprojections formed in an array. The projections act as needles that penetrate through the stratum corneum, thereby creating pathways through which the drug can penetrate. The projections are so small that there is little to no pain associated with their use *(30)*.

Lin et al. *(31)* created a microarray system fabricated from a 30-μm thick piece of stainless steel containing perpendicular projections 430-μm in length. The 2-cm^2 patch had a microprojection density of 240 projections/cm^2. Patches with and without iontophoretic capabilities were used to transdermally deliver a 20-base phosphorothioate oligonucleotide in hairless guinea pigs. The passive patch successfully delivered the oligonucleotide when topical application without the microprojections did not. Adding iontophoresis to the microprojection system further increased delivery by a factor of >20 *(31)*.

3.2. Chemical Penetration Enhancers

A wide variety of chemicals can act as penetration enhancers. The mechanisms by which these chemicals alter delivery are dependent on their chemical class. First, they may act on the drug itself by altering its thermodynamic activity, leading to greater partitioning into the skin. Second, enhancers may interact with the stratum corneum lipids, causing a reduced barrier *(32)*. Effective enhancers include water, alcohols (methanol and ethanol), alkylmethyl sulfoxides, pyrrolidones, laurocapram, solvents (acetone, propylene glycol), surfactants, and fatty acids (linoleic acid) *(6)*. Enhancement has also been reported for liposomes *(33)*, transferosomes *(34)*, and ethosomes *(35)*. Additionally, chemical penetration enhancers may be used in combination with physical enhancement techniques such as iontophoresis *(36)*.

In vitro transdermal studies were performed using passively applied methylphosphonate antisense compounds from a saturated solution. Absorption of methylphosphonate AS-ODNs from 6 to 18 bases long was examined using combinations of the chemical penetration enhancers propylene glycol, ethanol, decylmethyl sulfoxide, oleic acid, and ethyl acetate. Measurement of antisense in the dermal layer resulted in levels as high as 1.0 mM. Furthermore, modifying the uncharged methylphosphonate by adding a phosphate linkage to create a single negative charge decreased percutaneous penetration by a factor of 10 *(37)*.

The influence of vehicle on the in vitro transdermal delivery of several phosphorodiamidate morpholino oligomers (PMOs) was also examined. Oligomers with different sizes, lengths, base compositions, sequences, and lipophilicities were synthesized and then delivered across hairless mouse skin in vitro using vehicles composed of 95% propylene glycol (PG), 5% linoleic acid (LA), water, 50% water: 50% PG/LA, and 75% water: 25% PG/LA. The results suggest that size, sequence, and guanine composition all influence transdermal penetration and that some oligomers and vehicles would be better suited for transdermal delivery and others for topical applications *(38)*.

In vivo studies were performed using a lotion containing a 5'-^{32}P-labeled oligonucleotide applied to mouse ear helices. Systemic availability was demonstrated by successfully recovering intact oligonucleotide from blood and pancreas *(39)*.

4. In Vitro Transdermal Delivery Methods

The feasibility of transdermal delivery studies for a given chemical is first examined in an in vitro system. Skin is placed in either a static Franz diffusion chamber *(40)*, a static side-by-side chamber *(41)*, or a flow-through Bronaugh diffusion cell system *(42)*. In all three systems, the skin is sandwiched between two pieces of glass or plastic with a hole cut out of the center. The unit is then clamped tightly shut to prevent fluid leakage. The drug of interest is placed on the epidermal side of the skin while the dermal side is bathed in a physiological receiver solution. Because the system does not allow leakage, the only way the drug can reach the solution bathing the dermis is if it has diffused through the skin. Samples of the receiver solution are collected at given time intervals and are assayed for the target drug.

The amount of drug recovered in each fraction is determined and converted into either the percentage of the donor or, more commonly, the amount penetrating per square centimeter. Cumulative penetration is then calculated by adding current values to the previous one and plotting as a function of time. Steady-state flux (amount/[cm^2·h]) is calculated by determining the slope of the line defining the linear portion of the cumulative penetration curve. This line is extended to the x-axis (time), and the x-intercept is defined as the lag time.

Penetration enhancers are easily added to the diffusion systems. In the cases of iontophoresis and electroporation, electrodes are added such that the skin is in the middle of the circuit and the same fundamental procedures are followed *(43)*. Chemical penetration enhancers can be added to the diffusion chamber either prior to or in conjunction with the drug being tested.

Human skin is the ideal barrier for these studies; however, it can be difficult to obtain on a reliable basis. Several animal models including pig, hairless mouse, rat, hairless rat, guinea pig, and rabbit have been used in its place. Of these, pig skin most closely mimics the penetration properties of human skin *(44)*. Hairless mouse skin has been used for several transdermal oligonucle-otide studies. The absorption of methyl phosphonate oligonucleotides through mouse skin was only slightly greater than through human skin. The exact differences were dependent on the oligonucleotide itself and the chemical penetration enhancer used *(37)*. Human skin was one-half as permeable as hairless mouse skin to a six-base phosphorothioate oligonucleotide delivered with iontophoresis *(20)*.

5. The Importance of Selecting the Best Penetration Enhancers

The long-term goal of most research involving the transdermal delivery of AS-ODNs is to deliver them therapeutically. A great deal of effort is exerted to determine optimal delivery conditions as well as the best oligonucleotide sequence and structure. My laboratory examined the in vitro transdermal penetration of four oligonucleotides with different structures and chemistries to determine which could potentially be active in vivo. All studies used hairless mouse skin and followed the procedure described above.

The effect of a physical enhancement technique (iontophoresis) and a chemical penetration enhancer (90% PG and 10% LA) were examined for each oligo-nucleotide studied. The first oligonucleotide was a six base phosphorothioate telomere-mimic (MW = 1847) that inhibits telomerase activity with the sequence 5'-d(TTAGGG)-3' (TAG 6) *(45)*. The next three antisense molecules examined were targeted to CYP3A2. The first was a phosphorodiamidate morpholine (MW = 8438) with the sequence 5'-(GAGCTGAAAGCAGGTCCATCCC)-3'. PMOs represent multiply modified DNA molecules in which the deoxyribose sugar is replaced with a six-member morpholine sugar. The backbone comprises nonionic phosphorodiamidate linkages. The second oligonucleotide tested was in the form of a ribozyme (MW = 13136), with the sequence 5'-fluoresceinyl-AGUGUGACUgaUgaGGCCGUGAGGCCgaaaGCUG AAAiT-3'. The capital letters represent 2'-*O*-allyl ribonucleotides. The low-ercase letters represent ribonucleotides, and the 3'-end has an inverted thymi-dine linkage to protect from degradation by 3' exonucleases. The last antisense tested was a circle (MW = 11514) with the sequence 5'-GAAGAGAAttAAGA GAAGGGGGAGAAttAAGAGGGG-3'. The lowercase letters indicate the ends of the circle. A fluorescein was conjugated to one of the "t"s.

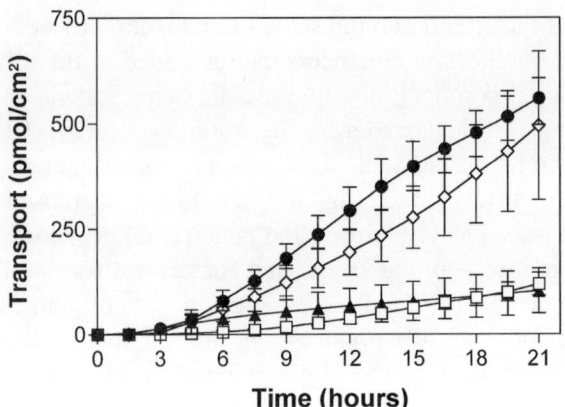

Fig. 1. In vitro cumulative absorption of four AS-ODN sequences when applied to hairless mouse skin in a donor solution containing 5 μM antisense and 90% PG, 10% LA. (●) circle; (□) TAG6; (◇) PMO; (▲) ribozyme. The steady-state flux presented in **Table 1** is calculated by obtaining the slope of the linear portion of the curve. Data are presented as the mean ± SEM.

These oligonucleotides were selected for these studies because (1) they are biologically relevant with the potential for in vivo use, and (2) they represent four very different chemistries that have distinct behaviors in vivo. The phosphorothioates are somewhat nuclease resistant and act by recruiting RNase H to cut the mRNA that is bound *(46)*. The ribozymes are catalytically active nucleic acids but are rapidly metabolized at requisite RNA linkage sites *(47)*. The circle is a phosphodiester, and because there are no ends there is no degradation in the body, but efficacy requires triple helix formation to form the so-called "RNA clamp" *(48)*. The PMOs are completely resistant to nucleases and are active via steric blockade of ribosomal assembly *(49)*.

Figure 1 shows the transdermal penetration of each of these oligonucleotides in the presence of PG and LA from a 5 μM donor solution. The passive penetration over 21 h was much greater for the circle (560 ± 33 pmol/cm^2) and phosphorodiamidate morpholine (495 ± 176 pmol/cm^2) than for the TAG6 (120 ± 27 pmol/cm^2) and ribozyme (51 ± 18 pmol/cm^2).

Table 1 compares the transdermal steady-state flux for each of these oligonucleotides when exposed to the skin using iontophoresis, PG and LA, and a buffer control. **Figure 2** shows the transdermal penetration of TAG6 under these three conditions. The comparison demonstrates the importance of choosing appropriate techniques for penetration enhancement based on the chemistry of the molecule. Passive transdermal delivery for both the phosphorothioate and the circle was lowest when applied in a buffer solution. In both cases,

Table 1
Steady State Flux from Transdermal Delivery Through Hairless Mouse Skin In Vitro Using Iontophoresis, Chemical Enhancer, or Control

Oligonucleotide	Iontophoresis (pmol/[cm^2·h])	Passive:PG/LA (pmol/[cm^2·h])	Passive: Buffer (pmol/[cm^2·h])[a]
TAG 6	17.9 ± 2.4	7.1 ± 2.0	1.6 ± 0.6
Circle	39.8 ± 0.3	37.6 ± 7.2	15.3 ± 6.9
PMO	1.0 ± 0.1	24.6 ± 13.6	ND
Ribozyme	19.5 ± 7.4	2.5 ± 1.1	ND

[a]ND, not determined.

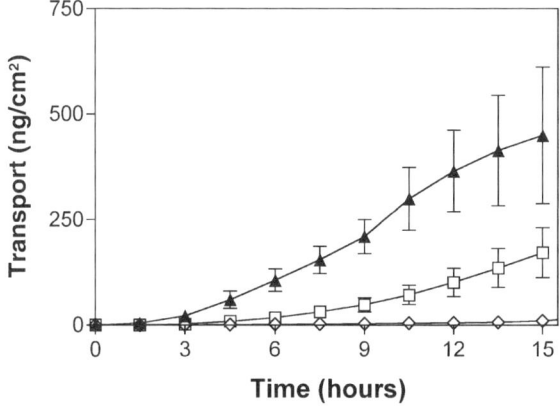

Fig. 2. In vitro cumulative absorption of phosphorothioate oligonucleotide TAG6 when applied to hairless mouse skin in a donor solution containing 5 μ*M* antisense in presence of iontophoresis (▲), 90% PG and 10% LA (□), or buffer control (◇).

altering the donor solution to PG and LA greatly enhanced penetration. Examination of the iontophoresis results demonstrates that enhancement potential is a function of the oligonucleotide chemistry. For TAG6 and the ribozyme, penetration enhancement was the greatest with iontophoresis (2.5-fold better for TAG6 and 7.8-fold better for the ribozyme). Conversely, iontophoresis did not increase the penetration of the circle when compared with the chemical enhancer and actually inhibited transport by a factor of 24.6 for the PMO. Although quite dramatic, this is not completely surprising because phosphorodiamidate

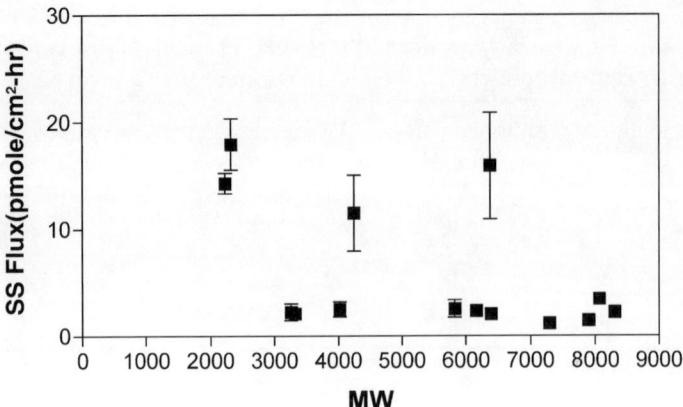

Fig. 3. Influence of molecular weight (MW) on the steady-state (SS) flux of phosphorothioate oligonucleotides iontophoretically transported across hairless mouse skin. Data are plotted as the mean ± SEM.

morpholine substitutions result in an uncharged oligonucleotide. Iontophoresis should not be as effective on an uncharged molecule as for the other charged oligonucleotides.

Figure 3 demonstrates the steady-state flux of 14 phosphorothioate oligonucleotides with different sequences as a function of molecular weight during transdermal delivery using iontophoresis in vitro *(22)*. The data demonstrate that, in general, smaller oligonucleotides have greater penetration. There are several exceptions, however, and this demonstrates that although generalities can be made regarding choosing enhancers, based on backbone properties, the individual sequence to be delivered is also important in determining penetration.

6. In Vitro Studies to Predict In Vivo Feasibility

The values obtained from in vitro experiments using the techniques described above can then be used to determine whether therapeutic levels could be feasibly delivered for animal and then human studies. The best way to make transdermal delivery feasible in humans is to find potent oligonucleotides that can be readily delivered. The uncharged antisense molecules of phosphorodiamidate morpholine oligomers are 30-fold more potent than phosphorothioate oligonucleotides. Arora et al. *(51)* have shown that intraperitoneally delivered PMOs at a dose of 0.5 mg/kg has antisense activity against the c-*myc* gene in rat liver. Converting the data in **Table 1** to mg shows that the PMOs targeted to CYP3A2 mRNA translational start had an in vitro transdermal penetration of 0.2 ± 0.1 μg/(cm^2·h) or 3.9 ± 1.3 μg/(cm^2·d) from a 38 μg/mL solution. The target dose is 0.125 mg for a 250-g rat. According to Fick's law, penetration should be linearly related to donor concentration, until saturation begins to occur, as well as to surface area. Therefore, the dose that needs to be delivered (125 μg/3.9 μg/[cm^2·d]) requires a 32-fold increase in the amount delivered. This can easily be achieved by increasing a combination of the concentration and size by a factor of 32.

Although delivery to animals is important in the experimental process, the information gained must be able to predict feasibility in humans. Freireich et al. *(50)* demonstrate how to extrapolate dose data from rats to humans:

$$(\text{dose in mg/kg}) = \frac{km_{animal}}{(km_{human})} (\text{dose in mg/kg})$$

in which $km_{rat} = 5.9$ and $km_{human} = 37$.

Extrapolation to human data using this method demonstrates that the required dose of oligonucleotide for a 70-kg person would be

$$\text{Dose (mg)} = \frac{5.9 \times 0.5 \text{ mg} \times 70 \text{ kg}}{37 \text{kg}} = 5.6 \text{ mg}$$

Given that 3.9 μg/cm can be delivered per d from a 38 μg/mL solution, an increase in the combination of donor concentration and surface area of 147-fold is necessary. The solubility of the PMO is good and solutions of 30 mg/mL are routinely prepared. Hence, a 38 mg/mL donor solution could increase the delivery to as much as 3.9 mg/(cm^2·d), and a patch of 5 cm^2 should provide sufficient delivery. This would also allow for the variability in dermal penetration between species.

Because of the potential feasibility, the antisense PMO targeted to CYP3A was applied topically to adult male rats at doses of 0.03, 0.3, and 3.0 mg/rat. CYP3A enzyme activity in the underlying skin and liver was evaluated 24 h following application, using hydroxylation of 7-benzyloxy-4-(trifluoromethyl)-coumarin as the assay (nmol of product/[100 μg of S9 protein·h]). A topical dose of 0.03 mg inhibited enzyme levels from 576 ± 17 (vehicle) and 564 ± 20 (control PMO) to 432 ± 20 in the antisense-treated liver ($p < 0.05$). Raising the dose to 0.3 mg further inhibited enzyme level to 278 ± 13 ($p < 0.005$), but an increase to 3 mg did not improve the response. In the skin, topical delivery of 0.03 mg reduced enzyme levels by half, from 171 ± 9 to 89 ± 32 ($p < 0.05$), whereas increasing the dose to 0.3 and 3.0 mg did not produce any further inhibition *(51)* .

7. Conclusion

The differences among transdermal delivery of AS-ODNs are a result of their physical properties, including size, sequence, and backbone. Uncharged

PMOs are poorly delivered by iontophoresis but can be adequately delivered passively with chemical penetration enhancers. The opposite is true of phosphorothioate and ribozymal oligonucleotides. Absorption of the circular AS-ODN is equally effective for the two techniques. Selection of AS-ODNs for in vivo studies must factor in molecular potency as well as delivery potential and methodology.

Critical considerations for evaluating transdermal antisense delivery have centered on feasibility. The studies presented indicate that this method of drug delivery is feasible. Future considerations include studies in larger populations of humans to determine whether topical and transdermal antisense therapy will be effective. Most drug therapy involves chronic treatment, so studies must examine whether local irritation will be a problem. Patient compliance will probably favor the passive delivery methods, which are easy to reproduce and allow for feasible self-administration. It remains to be seen if passive transdermal delivery will be more or less erratic than oral delivery.

References

1. Guy, R. H. and Hadgraft, J. J. (1987) Transdermal drug delivery: a perspective. *J. Control. Release* **4,** 237–251.
2. Pitt, C. G. (1990) The controlled parenteral delivery of polypeptides and proteins. *Int. J. Pharm.* **59,** 173–196.
3. Brand, R. M. and Quigley, E. M. M. (1997) Transdermal delivery of erythromycin lactobionate—implications for the therapy of gastroparesis. *Aliment. Pharmacol. Ther.* **11,** 589–592.
4. Hogan, D. J. and Cottan, J. (1996) Dermatological aspects of transdermal drug delivery systems, in *Dermatotoxicology* (Marzulli, F. N. and Maibaich, H. I., eds.), Taylor & Francis, Washington, DC, pp. 75–86.
5. Monteiro-Riviere, N. A. (1996) Anatomical factors affecting barrier function, in *Dermatotoxicology* (Marzulli, F. N. and Maibach, H. I., eds.), Taylor & Francis, Washington, DC., pp. 3–19.
6. Walters, K. A. (1989) Penetration enhancers and their use in transdermal therapeutic systems, in *Transdermal Drug Delivery*, (Hadgraft, J. and Guy, R. H., eds.), Marcel Dekker, New York, pp. 197–247.
7. Elias, P. M. (1983) Epidermal lipids, barrier function, and desquamation. *J. Invest. Dermatol.* **86,** 187–190.
8. Wertz, P. W. and Downing, D. T. (1989) Stratum corneum: biological and biochemical considerations, in *Transdermal Drug Delivery* (Hadgraft, J. and Guy, R.H., eds.), Marcel Dekker, New York, pp. 1–22.
9. Junginger, H. E., Bodde, H. E., and de Haan, F. H. N. (1990) Visualization of drug transport across human skin and the influence of penetration enhancers, in *Drug Permeation Enhancement - Theory and Applications* (Hsieh, D. S., ed.) Marcel Dekker, Malvern, PA, pp. 59–89.

10. Cullander, C. and Guy, R. H. (1992) Routes of Delivery: case studies (6). Transdermal delivery of peptides and proteins. *Adv. Drug Deliv. Rev.* **8,** 291–329.
11. Fartasch, M. (1996) The nature of the epidermal barrier: structural aspects. *Adv. Drug Deliv. Rev.* **18,** 273–282.
12. Magee, F. P. (1996) Reaffirming the complexity of transdermal transport, in *Dermatotoxicology* (Marzulli, F. N. and Maibach, H. I., eds.), Taylor & Francis, Washington, DC., pp. 61–74.
13. Guy, R. H. and Hadgraft, J. (1989) Selection of drug candidates for transdermal drug delivery, in *Transdermal Drug Delivery* (Hadgraft, J. and Guy, R. H., eds.), Marcel Dekker, New York, pp. 59–81.
14. Tyle, P. (1986) Iontophoretic devices for drug delivery. *Pharm. Res.* **3,** 318–326.
15. Singh, P. and Maibach, H. I. (1996) Iontophoresis: an alternative to the use of carriers in cutaneous drug delivery. *Adv. Drug Deliv. Rev.* **18,** 379–394.
16. Ledger, P. W. (1992) Skin biological issues in electrically enhanced transdermal delivery. *Adv. Drug Deliv. Rev.* **9,** 289–307.
17. Rosendal, T. (1942) Studies on the conducting properties of the humans skin to direct current. *Acta. Physiol. Scand.* **5,** 130–130.
18. Kim, A., Green, P. G., Rao, G., and Guy, R. H. (1993) Convective solvent flow across the skin during iontophoresis. *Pharm. Res.* **10,** 1315–1320.
19. Pikal, M. J. and Shah, S. (1990) Transport mechanisms in iontophoresis. III. An experimental study of the contributions of electroosmotic flow and permeability change in transport of low and high molecular weight solutes. *Pharm. Res.* **7,** 222–229.
20. Brand, R. M. and Iversen, P. I. (1996) Iontophoretic delivery of a telomeric oligonucleotide. *Pharm. Res.* **13,** 851–854.
21. Oldenburg, K. R., Vo, K. T., Smith, G. A., and Selick, H. E. (1995) Iontophoretic delivery of oligonucleotides across full thickness hairless mouse skin. *J. Pharm. Sci.* **84,** 915–921.
22. Brand, R. M., Wahl, A., and Iversen, P. L. (1997) Effects of size and sequence on the iontophoretic delivery of oligonucleotides. *J. Pharm. Sci.* **87,** 49–52.
23. Brand, R. M., Hannah, T. L., Norris, J., and Iversen, P. L. (2001) Transdermal delivery of antisense oligonucleotides can induce changes in gene expression *in vivo*. *Antisense Nucleic Acid Drug Dev.* **11,** 1–6.
24. Banga, A. K. and Prausnitz, M. R. (1998) Assessing the potential of skin electroporation for the delivery of protein- and gene-based drugs. *TIBTECH* **16,** 408–412.
25. Banga, A. K., Bose, S., and Ghosh, T. K. (1999) Iontophoresis and electroporation: comparisons and contrasts. *Int. J. Pharm.* **179,** 1–19.
26. Zewert, T. E., Pliquett, U. F., Langer, R., and Weaver, J. C. (1997) Transdermal transport of DNA amtosense oligonucleotides by electroporation. *Biochem. Biophys. Res. Commun.* **212,** 286–292.
27. Regnier, V., Tahiri, A., Andre, N., Lemaitre, M., Le Doan, T., and Preat, V. (2000) Electroporation-mediated delivery of 3'-protected phosphodiester oligodeoxynucleotides to the skin. *J. Control Release* **67,** 337–346.

28. Regnier, V., LeDoan, T., and Preat, V. (1998) Parameters controlling topical delivery of oligonucleotides by electroporation. *J. Drug Target* **5,** 275–289.
29. Regnier, V., De Morre, N., Jadoul, A., and Preat, V. (1999) Mechanisms of a phosphorothioate oligonucleotide delivery by skin electroporation. *Int. J. Pharm.* **184,** 147–156.
30. McAllister, D. V., Allen, M. G., and Prausnitz, M. R. (2000) Microfabricated microneedles for gene and drug delivery. *Ann. Rev. Biomed. Eng.* **2,** 289–313.
31. Lin, W., Cormier, M., Samiee, A., et al. (2001) Transdermal delivery of antisense oligonucleotides with microprojection patch (Macroflux) technology. *Pharm. Res.* **18,** 1789–1793.
32. Hadgraft, J., Walters, K. A., and Guy, R. H. (1992) Epidermal lipids and topical drug delivery. *Semin. Dermatol.* **11,** 139–144.
33. Yarosh, D. and Klein, J. (1996) The role of liposomal delivery incutaneous DNA repair. *Adv. Drug Deliv. Rev.* **18,** 325–333.
34. Cevc, G., Blume, G., Schatzlein, A., Gebauer, D., and Paul, A. (1996) The skin: a pathway for systemic treatment with patches and lipid-based agent carriers. *Adv. Drug Deliv. Rev.* **18,** 349–378.
35. Touitou, E., Dayan, N., Bergelson, L., Godin, B., and Eliaz, M. (2000) Ethosomes—novel vesicular carriers for enhanced delivery: characterization and skin penetration properties. *J. Control Release* **65,** 403–418.
36. Mitragorti, S. (2000) Synergystic effect of enhancers for transdermal drug delivery. *Pharm. Res.* **17,** 1354–1359.
37. Nolen III, H. W., Catz, P., and Friend, D. R. (1994) Percutaneous penetration of methyl phosphonate antisense oligonucleotides. *Int. J. Pharm.* **107,** 169–177.
38. Pannier A. K., Arora, V., Iversen, P. L., and Brand, R. M. (2004) Transdermal delivery of phosphorodiamidate morpholino oligomers across hairless mouse skin. *Int. J. Pharm.* **275,** 217–226.
39 Vlassov, V. V., Karamyshev, V. N., and Yakubov, L. A. (1993) Penetration of oligonucleotides into mouse organism through mucosa and skin. *FEBS Lett.* **327,** 271–274.
40. Franz, T. J. (1975) Percutaneous absorption on the relevance of in vitro data. *J. Invest. Dermatol.* **64,** 190–195.
41. Delgado-Charro, M. B. and Guy, R. H. (1995) Iontophoretic delivery of nafarelin across the skin. *Int. J. Pharm.* **117,** 165–172.
42. Bronaugh, R. L. and Stewart, R. F. (1985) Methods for in vitro percutaneous absorption studies IV: the flow-through diffusion cell. *J. Pharm. Sci.* **74,** 64–67.
43. Glikfeld, P., Cullander, C., Hinz, R. S., and Guy, R. H. (1998) A new system for in vitro studies of iontophoresis. *Pharm. Res.* **5,** 443–446.
44. Bronaugh, R. L. (2000) In vitro percutaneous absorption models. *Ann. NY Acad. Sci.* **919,** 188–191.
45. Mata, J. E., Jackson, J. D., Joshi, S. S., et al. (2000) Pharmacokinetics and in vivo effects of a six-base phosphorothioate oligodeoxynucleotide with anticancer and hematopoetic activites in swine. *J. Hematother. Stem Cell Res.* **9,** 205–214.

46. Gao, W. Y., Han, F. S., Storm, C., Egan, W., and Cheng, Y. C. (1992) Phosphoro-thioate oligonucleotides are inhibitors of human DNA polymerases and RNase H: implications for antisense technology. *Mol. Pharmacol.* **41,** 223–229.

47. Desjardins, J. P., Sproat, B. S., Beijer, B., et al. (1996) Pharmacokinetics of a synthetic, chemically modified hammerhead ribozyme against the rat cytochrome P-450 3A2 mRNA after single intravenous injections. *J. Pharmacol. Exp. Ther.* **278,** 1419–1427.

48. Rowley, P. T., Kosciolek, B. A., and Kool, E. T. (1999) Circular antisense oli-gonucleotides inhibit growth of chronic myeloid leukemia cells. *Mol. Med.* **5,** 693–700.

49. Iversen, P. L. (1902) Phosphorodiamidate morpholino oligomers: favorable prop-erties for sequence-specific gene inactivation. *Curr. Opin. Mol. Ther.* **3,** 235–238.

50. Freireich, E. J., Gehan, E. A., Rall, D. P., Schmidt, L. H., and Skipper, H. E. (1966) Quantitative comparison of toxicity of anticancer agents in mouse, rat, hamster, dog, monkey, and man. *Cancer Chemother. Rep.* **50,** 219–244.

51. Arora, V., Hannah, T. L., Iversen, P. L., and Brand, R. M. (2002) Transdermal use of phosphorodiamidate morpholino oligomer AVI-4472 inhibits cytochrome p450 3A2 activity in male rat. *Pharm. Res.* **19,** 1465–1470.

VI

DRUGS

14

Antisense Strategies for Redirection of Drug Metabolism

Using Paclitaxel as a Model

Vikram Arora

1. Introduction

The concept of metabolic redirection involves shifting the metabolism of a substrate away from one biochemical pathway to another *(1)*. There are numerous instances in biological systems of a single substrate being metabolized by more than one enzyme with differing rates and activities, resulting in substantially different products. While an extremely large number of compounds are substrates for these enzymes, the types of reactions catalyzed are finite. This chapter focuses on Taxol® (paclitaxel; **Fig. 1**), an important antineoplastic agent that is used extensively in treating several types of malignancies including ovarian, breast, and lung, as well as melanomas *(2–5)*.

Paclitaxel, which was originally isolated from the bark of pacific yew *Taxus brevifolia (6)*, causes promotion and excessive stabilization of the microtubule polymer during cell division *(7)*. It is generally accepted to be metabolized in the human liver by cytochromes P450 (CYP) 3A4 and 2C8 *(8–12)*. Although both major metabolites exhibit the excessive microtubule stabilizing properties of the parent compound, the metabolism promotes their excretion in bile in addition to reducing their potency and efficacy. It is reported that the in vitro cytotoxic activity of the CYP3A4 metabolite in cultures of leukemia cells is 10-fold lower than the parent compound *(13)*. Furthermore, it has been reported that administration of paclitaxel causes an induction in the specific content of CYP3A4 in the liver, thereby promoting its own metabolism *(14)*. This scenario presents an opportunity to improve the bioavailability and reduce clearance of paclitaxel through inhibition of CYP3A4.

From: *Methods in Molecular Medicine, Vol. 106: Antisense Therapeutics, Second Edition*
Edited by: I. Phillips © Humana Press Inc., Totowa, NJ

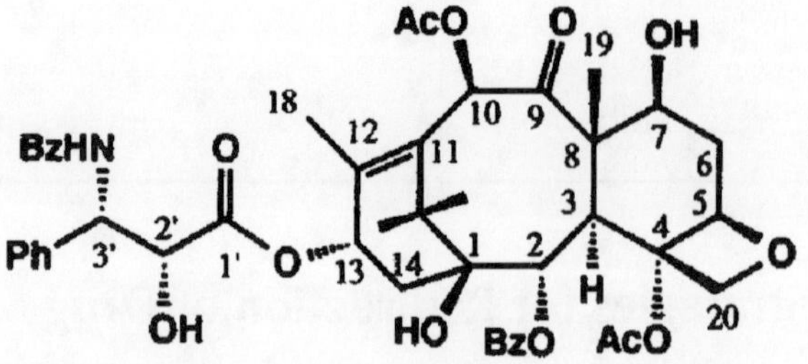

Fig. 1. Chemical structure of paclitaxel illustrating numbering system. Ac, acetyl; Bz, benzoyl; Ph, phenyl.

The studies presented in this chapter were conducted in rats and utilize the antisense approach to inhibit CYP3A2, which is the rat ortholog of human CYP3A4. Animals were pretreated with AVI-4472, a phosphorodiamidate morpholino antisense oligomer targeted to the CYP3A2 mRNA, and changes in the bioavailability of paclitaxel were examined.

1.1. Cytochrome P450 Enzymes

CYPs are a superfamily of heme-thiolate monooxygenases that were originally characterized by their absorption spectrum at 450 nm in the presence of carbon monoxide. These enzymes are located on the endoplasmic reticulum and present predominantly in the liver, but significant quantities are also found in the small intestine, kidney, and skin *(15–17)*.

This family of enzymes oxidizes a wide variety of endogenous and exogenous substrates, including xenobiotics. The term *xenobiotic* refers to molecules foreign to the body and includes pollutants, pesticides, and drugs. The importance of CYPs in pharmacology arises from their participation in the oxidative (phase I) metabolism of a wide number of medications *(18)*. The CYP monooxygenase catalytic activity can be described as $RH + NADPH + H^+ + O_2 \rightarrow ROH + NADP + H_2O$, in which RH is the drug molecule and ROH is a more hydrophilic metabolite of the drug *(19)*. CYPs biotransform drugs to more polar metabolites to enhance renal excretion. Typically, the metabolism results in less bioactivity although some medications, such as cyclophosphamide *(20)*, undergo biotransformation to pharmacologically active agents. In brief, CYP-mediated metabolism is an important determinant of the eventual disposition of most drugs *(21–25)*.

There are a particularly large number of such examples for phase I metabolism of drugs by multiple CYPs. The antiepileptic hydantoin is N-demethylated by CYP2B6 *(26)* and C4'-hydroxylated by CYP2C18 *(18)* and CYP2C19 *(27)*. The tricyclic antidepressant imipramine is C2-hydroxylated by CYP2D6 *(28)* and *N*-demethylated by CYP1A2 and CYP2C18 *(29)*. The analgesic codeine is *O*-demethylated by CYP2D6 *(30)* and *N*-demethylated by CYP3A4 *(31)*. The ability to selectively inhibit a particular metabolic pathway can be used creatively to enhance the pharmacological disposition of a drug. For example, a strategy to inhibit the CYP3A4-mediated *N*-demethylation of codeine to norcodeine is likely to redirect its metabolism to CYP2D6, which *O*-demethylates it to the highly active metabolite morphine *(32)*.

1.1.1. The CYP3A Family

The CYP3A family is the most abundant CYP in the human liver, representing about 40% of the total specific content of CYP. The entire CYP3A family is located at chromosome 7q21.1. The human CYP3A family includes CYP3A4 and its allelic variant CYP3A3, as well as CYP3A5 and CYP3A7 *(33)*. CYP3A7 is considered a fetal enzyme, and its expression is not observed in the liver after birth, but expression has been observed in the endometrium and placenta *(34)*. CYP3A5 is expressed in only 10–30% of livers but is expressed in 80% of all human kidneys.

The major isoform CYP3A4 is expressed in human liver, small intestine, and skin. This 57-kDa enzyme has also been referred to as nifedipine oxidase, NF-25, P450-PCN1, or polypeptide 4. This versatile enzyme can catalyze the following types of reactions: *O*-demethylation, oxidative deamination, *N*-hydroxylation, aliphatic oxidation, and *N*-dealkylation *(35,36)*.

Enormous interindividual variations have been reported in both enzyme content and activity in the liver and small intestines. The specific content of CYP3A4 in the liver has been reported to vary by a factor of 20-fold, and the enzyme activity, as determined by the erythromycin breath test, has been reported to vary by a factor of 10-fold *(37)*. The CYP3A4 content of the small intestine has been reported to vary by a factor of 10- to 49-fold *(38–40)*. These large differences in CYP3A4 activity, in turn, are likely responsible for interindividual variations in therapeutic efficacy and disposition of a variety of drugs *(25)*.

A variety of factors, including genetic polymorphisms, influence CYP3A4 activity. Two well-defined genetic polymorphisms are reported: (1) 12 known alternative coding variants, each containing defined alterations in amino acid sequence identified as CYP3A4*1 (wild type) through CYP3A4*13; (2) an A-to-G transition located 295 bp upstream of the coding region, A(-292)G, representing a genetic variant within the CYP3A4 promoter (nifedipine-specific element) that has been associated with lower CYP3A4 expression *(41)* and higher cancer risk *(42, 43)*. Various xenobiotics also influence CYP3A4 activity. Drugs that inhibit CYP3A4 activity include clotrimazole, ethinylestradiol, gestodene, ketaconazole, miconazole, cimetidine, and troleandomycin. On the other hand, carbamazepine, dexamethasone, phenobarbital, phenytoin, rifampin, sulfadimidine, sulfinpyrazone, troleandomycin, and paclitaxel are reported to elevate CYP3A4 activity *(44)*. Overall, the importance of CYP3A4 is underlined by the fact that it contributes to the metabolism, and eventual disposition, of more than half of all medications *(23)*.

1.1.2. Rat CYP3A2

Because the studies described in this chapter were conducted in rats, it is important to highlight important differences between human and rat CYP3A families. The rat CYP3A family includes 3A1 and 3A2, which are functionally equivalent to human isoforms 3A3 and 3A4, respectively. CYP3A1 is difficult to detect in a normal, untreated rat *(45)* whereas 3A2 constitutes approx 25% of the total specific content of CYP in male rat liver *(46)*. CYP3A2 is constitutively expressed in neonates of either sex but is present only in males on maturation. Hence, the use of male rats was employed in the current studies. The lack of the CYP3A2 expression in females may be related to the continuous female secretion pattern of growth hormone, which generally suppresses the expression of CYP3A family *(47–49)*.

AVI-4472, the antisense oligomer used in the studies presented in this chapter, targets the AUG translation initiation region of the rat CYP3A2 mRNA.

1.2. Metabolism of Paclitaxel

Most of the information about paclitaxel's metabolism is derived from incubation with human microsome fractions. Two significant metabolites are detectable (**Fig. 2**):

1. 6α-Hydroxytaxol (6HOT), resulting from CYP2C8-mediated hydrolysis, is the major metabolite detected after adult liver microsomal incubations *(8,50)* formed by a single stereospecific hydroxylation at the six position of the taxane ring. This metabolite is not observed in rats *(50)*, demonstrating important interspecies differences of the hepatic metabolism of paclitaxel.
2. 3'-(*p*-Hydroxyphenyl)taxol (3'HOT) is the result of CYP3A4 metabolism from a single hydroxylation at the para position of the C3 phenyl group on the C13 side chain of paclitaxel. Unlike 6HOT, the 3'HOT metabolite is also observed following incubation of paclitaxel with rat liver microsomes *(51)*.
3. Dihydroxylated taxol (DHOT) is formed by stepwise hydroxylations at the two previously described sites successively by CYP2C8 and CYP3A4.

The role of the CYP3A family in the metabolism of paclitaxel is likely to have been underestimated from analysis of microsome incubation studies. Isolated microsome studies are generally extremely helpful in characterizing individual chemical reactions but lack the complexity of the intact cell or organism to fully appreciate the true relative contribution of individual enzymes partici-

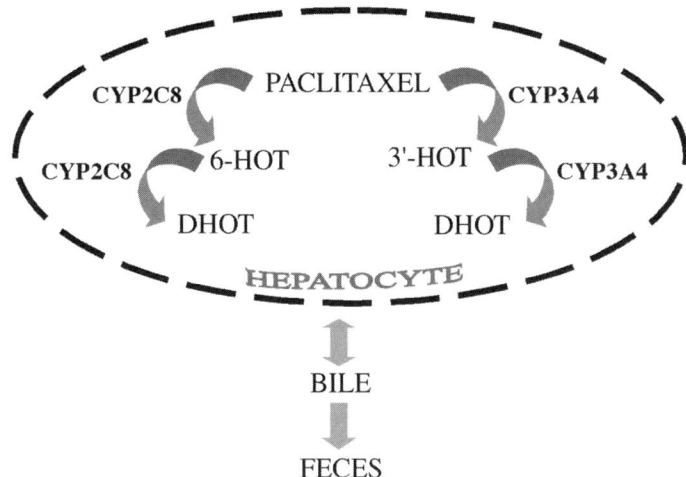

Fig. 2. Schematic overview of metabolism of paclitaxel as it applies to a human hepatocyte. (*see* **Subheading 1.2.** for details).

pating in these reactions. In contrast to other widely cited studies, Sonnischen et al. *(50)* have reported that 6HOT is not always the predominant metabolite based on observations from clinical data and 49 separate human liver microsome preparations, underlining the importance of CYP3A4 in the metabolism of paclitaxel. Furthermore, the fact that CYP3A is induced in primary hepatocytes following incubation of paclitaxel *(14)* leads to the speculation that CYP3A4 could be the major player in the metabolism of paclitaxel in vivo.

1.3. Phosphorodiamidate Morpholino Oligomers

The studies presented in this chapter utilized phosphorodiamidate morpholino oligomer (PMO) antisense agents. The chemistry of PMO overcomes some of the major shortcomings of phosphorothioate oligonucleotides (*see* Chapter 13). The PMO backbone is nonionic at physiological pH and consists of a six-membered morpholine ring instead of a deoxyribose sugar, and the charged phosphodiester internucleoside linkage is replaced by an uncharged phosphorodiamidate linkage *(52)*. PMOs are highly resistant to various nucleases and proteases *(53)*, as well as extremely efficient inhibitors of translation via a non-RNase H *(54)*, sequence-specific steric-blockade process *(55)*. PMOs also have favorable organ distribution following systemic administration at all organ sites rich in CYP activity, namely the liver *(56,57)*, kidneys *(58)*, and small intestines (unpublished data). PMOs have been successfully used to inhibit CYP3A4 in primary human hepatocytes *(59)* and CYP3A2 in rats in vivo following ip, oral *(60)*, and transdermal *(61)* administration.

2. Hypothesis and Study Plan

The hypothesis states that the inhibition of CYP3A2 by intraperitoneal antisense PMO pretreatment in rats will decrease the metabolism of paclitaxel and reduce its clearance, thereby increasing its bioavailability.

Vehicle, control, or antisense PMOs targeted to CYP3A2 were injected intraperitoneally into male rats at a dosage of 2.5 mg/kg/d for 3 d. Administration of PMOs was started 48 h prior to administration of paclitaxel, which was administered as a single bolus intravenous injection at a dose of 8.0 mg/kg. Blood samples were drawn from the jugular vein port at these time points following administration of paclitaxel: preinjection, 20 min, 40 min, 60 min, 90 min, 150 min, 180 min, 240 min, and 24 h. Plasma was analyzed for levels of paclitaxel, and CYP3A2 activity was determined in livers following termination of the study. All animal protocols conformed to the ethical guidelines of the 1975 Declaration of Helsinki.

3. Materials and Methods

3.1. Animals

Male Sprague-Dawley rats (Zivic Miller, Porterville, PA) with jugular vein cannulas and weighing between 240 and 260 g were used. The animals were maintained in a climate-controlled room with a 12-h light/dark cycle and allowed access to a commercial rat chow and tap water ad libitum.

3.2. PMO Synthesis and Sequences

PMOs were synthesized at AVI BioPharma (Corvallis, OR) as previously described *(62)*. Purity was greater than 90% full length, as determined by reverse-phase high-performance liquid chromatography and matrix-assisted laser desorption/ionization time-of-flight mass spectroscopy. Lyophilized PMO compounds were dissolved in sterile apyrogenic saline and filtered through 0.2-μm Acrodisc filters (Gelman). The antisense PMO AVI-4472 is complementary to the AUG translation initiation region of the rat CYP3A2 mRNA (Genbank accession number U09742) and has the following base sequence: 5'-GAGCTGAAAGCAGGTCCATCCC-3'. The control PMO 1-22-144 has the following base sequence: 5'-ACTGTGAGGGCGATCGCTGC-3'. Both PMOs were administered to rats intraperitoneally.

3.3. Paclitaxel

Taxol (a gift from Mead Johnson, a Bristol-Myers Squibb company, Princeton, NJ) was supplied as a sterile solution containing 6 mg of paclitaxel, 527 mg of Cremophor®EL (polyoxyethylated castor oil), and 49.7% (v/v) dehydrated alcohol (USP) per milliliter.

3.4. Analysis of Paclitaxel in Plasma

Rat plasma samples were stored at −80°C until they were analyzed at Kansas City Analytical Services, Shawnee, KS. Paclitaxel and the added internal standard cephalomannine were extracted from rat plasma by a liquid–liquid phase extraction. This extract was subjected to RP-HPLC. Paclitaxel and cephalomannine in the effluent were detected by a PE/Sciex API III+ LC/MS/ MS system in the multiple-resolution mode. Quantitation was achieved by monitoring the product ions (m/z 569 for paclitaxel and m/z 264 for cephalomannine) of precursor ions m/z 871 for paclitaxel and m/z 832 for cephalomannine, respectively. System calibration was accomplished by a weighted ($1/x$) linear regression of the peak area ratio (analyte/internal standard) vs the concentration of the analyte. The lower limit of quantitation was optimized at 10 ng/mL for paclitaxel, and the method was determined to be linear through at least 500 ng/mL. Samples were diluted as necessary to conform to the linear range of detection. A representative chromatogram is presented in **Fig. 3**.

3.5. CYP1A1/1A2 Activity

CYP1A1/1A2 activity was determined in the liver S9 fraction (postmitochondrial supernatant of tissue homogenate) by ethoxyresorufin-*O*-dealkylation assay *(63)*. A 1-mg aliquot of S9 protein, 2 µ*M* 5-ethoxyresorufin (Pierce, Rockford, IL), and 1 m*M* β-NADPH in a volume of 1 mL of potassium phosphate buffer (pH 7.4) was incubated for 10 min at 37°C. Samples were analyzed spectrofluorometrically using an excitation wavelength of 530 nm and emission wavelength of 585 nm. Concentrations of unknowns were calculated from a standard curve of resorufin (Pierce). Results were recorded in picomoles of resorufin per milligram of S9 protein per minute.

3.6. CYP3A2 Activity

The activity of CYP3A2 was measured using erythromycin *N*-demethylation *(63,64)*. Samples were prepared by mixing 1.0 mg of S9 protein, 0.4 mM erythromycin, and 1.0 m*M* β-NADPH in a final volume of 1 mL in 0.1 *M* potassium phosphate buffer (pH 7.4). The samples were incubated for 15 min at 37°C, mixed with 0.5 mL of 17% perchloric acid (Sigma, St. Louis, MO), and centrifuged at 15,000*g* for 5 min. Formaldehyde was measured by the colorimetric method of Nash *(65)*. The samples were placed in a new tube and mixed with 0.4 mL of Nash reagent (0.02 *M* 2,4-pentanedione, 0.6% [v/v] glacial acetic acid, and 3.9 *M* ammonium acetate) and incubated at 70°C for 20 min. The final product was read on a spectrophotometer at 412 nm. Absorbance was compared to a standard curve generated from known concentrations of formal-

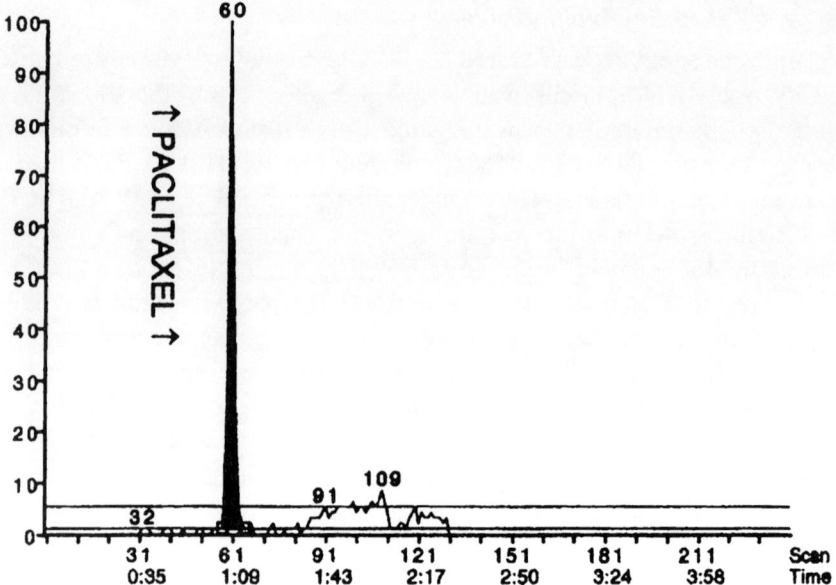

Fig. 3. Representative chromatogram depicting quantitative analysis of paclitaxel in plasma (*see* **Subheading 3.4.** for details).

dehyde. Activities were recorded as micromoles of formaldehyde per milligram of S9 protein per minute.

4. Results and Discussion

4.1. Plasma Paclitaxel Levels

Plasma concentrations (C_p) of paclitaxel were plotted against time, and the data are presented in **Fig. 3**. As expected, the curves had a biphasic appearance, with an initial distribution phase followed by an elimination phase. Mean values of paclitaxel C_p for the control PMO pretreatment group were typically somewhat lower than the vehicle pretreatment group. The lower C_p values were likely attributable to a wider error margin in that group. These differences were not statistically significant when subjected to one-way analysis of variance.

On the other hand, the differences in paclitaxel C_p in the antisense group were very interesting and significantly different from the vehicle group for the first 2 h following administration of paclitaxel. Pretreatment with the CYP3A2 antisense PMO AVI-4472 affected only the distribution phase of paclitaxel, leaving the elimination phase unaltered. At the first sampling point (20 min), the mean paclitaxel C_p was 18.9 ± 0.5 µg/mL for the antisense-treated group, compared with 6.3 ± 1.2 µg/mL for the oligomer vehicle-treated group ($p <$ 0.01). These differences in paclitaxel C_p got smaller with time, eventually overlapping between 2 and 3 h following the bolus administration of paclitaxel.

4.2. Pharmacokinetic Analysis of Plasma Paclitaxel Levels

The data were evaluated with the assistance of the computer program PKCALC as published previously *(66)*. The iv plasma data were fit to the following equation *(67)*:

$$C_p(t) = Ae^{-\alpha t} + Be^{-\beta t}$$

In which $C_p(t)$ is the plasma concentration at time t, A, and B are intercept terms, α is a distribution rate constant, and β is an elimination rate constant. All data are reported as the mean ± SD, and key parameters are presented in **Table 1**.

Paclixel's pharmacokinetics have been typically modeled with a two-compartment system: the drug is administered to and eliminated from a "central" compartment, and transport occurs into and out of a second "peripheral" compartment *(68)*. Although the compartments in pharmacokinetic models do not represent discrete anatomical or physiological entities, it may be estimated in rough terms that plasma constitutes the central compartment for paclitaxel. As it gets distributed, the drug is transported in the plasma to the liver, which is the major elimination site. The rest of the tissues in the body comprise the peripheral compartment.

Analysis of the paclitaxel C_p curves (**Fig. 4**) indicated that CYP3A2 antisense PMO altered only the distribution phase of the curve. Therefore, all analysis was focused on that phase, and the portion of the curve beyond 240 min was not taken into consideration for the purpose of this discussion.

4.2.1. Area Under the C_p Curve

Area under the C_p curve (AUC), which is an indication of total exposure to drug during the course of treatment, is perhaps the most important determinant of the success of this study because it underlines drug bioavailability. The paclitaxel $AUC_{20-240 \text{ min}}$ for the antisense PMO-pretreated group was 899.6 ± 35.6 µg·min/mL, compared with 265.8 ± 48.2 µg·min/mL ($p < 0.001$) for the vehicle-pretreated group. Note that this increase in AUC of approx 3.3-fold applies only to the distribution phase. Because the elimination phases of both antisense- and vehicle-pretreated groups were similar, the AUC difference for the total plasma C_p curves of paclitaxel is likely to be somewhat smaller.

Table 1
Select Plasma Pharmacokinetic Indices Following Bolus Intravenous Injection of Taxol in Male Rats Pretreated with PMO Targeted to CYP3A2

	PMO pretreatment[a]		
	Vehicle	Control	Antisense
Dose (mg/kg)	8.0	8.0	8.0
No. of animals	3	3	3
$T_{1/2}\,\alpha$ (min)[b]	12.0 ± 1.1	10.2 ± 5.3	31.5 ± 1.8[c]
$AUC_{20-240\,min}$ (μg·min/mL)	265.8 ± 48.2	169.9 ± 45.5	899.6 ± 35.6[d]
Plasma clearance (mL/min)	8.0 ± 1.5	11.7 ± 1.4	2.2 ± 0.1[c]
Volume of distribution (steady state) (L/kg)	2.2 ± 0.4	3.1 ± 0.5	0.5 ± 0.05[c]

[a]Data are expressed as the mean ± SD.
[b]$T_{1/2}\,\alpha$ represents the distribution half-life.
[c]$p < 0.01$ compared to vehicle group.
[d]$p < 0.001$ compared to vehicle group.

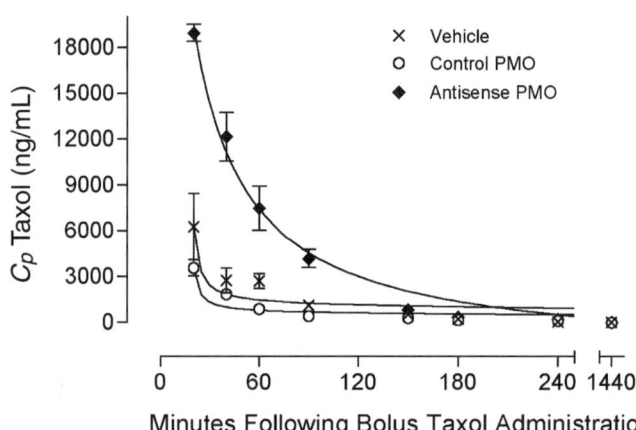

Fig. 4. Plasma concentrations of paclitaxel following administration of Taxol in PMO-pretreated rats. Vehicle, control, or antisense PMOs targeted to CYP3A2 mRNA were injected intraperitoneally into male rats at a dosage of 2.5 mg/(kg·d) for 3 d. Administration of PMO was started 48 h prior to administration of Taxol, which was administered as a single bolus iv injection at a dose of 8.0 mg/kg (*see* **Subheading 2.2.** for details).

4.2.2. Clearance

Clearance has become an important parameter over the last decade because of its extensive use in clinical pharmacokinetics. Clearance is defined as the volume of plasma from which the drug is removed per unit time. Although the definition is difficult to conceptualize, an important advantage of this parameter is that working equations are model independent and expressed in variables that are simple to obtain.

$$\text{Clearance} = F \times \text{Dose/AUC}$$

in which F is the absorbed fraction of the drug (essentially 1.0 for intravenously administered drugs).

The principal mechanism for clearance of paclitaxel from blood is considered to be hepatic metabolism, potentially from hepatic extraction and excretion into the bile, with resultant elimination in feces *(68)*. Renal elimination is generally not a major route. The plasma clearance of paclitaxel for the antisense PMO–pretreated group was 2.2 ± 0.1 mL/min, compared with 8.0 ± 1.5 mL/min ($p < 0.01$) for the vehicle-pretreated group. Because the doses were identical in both groups, the clearance term really reiterates the different C_p AUC values.

4.2.3. Distribution Half-Life

On iv administration, paclitaxel concentration in the blood decreases rapidly as the drug distributes to various tissues. The distribution half-life ($T_{1/2}\,\alpha$) has been previously reported in various studies to be between 20 and 40 min (human infusion studies) *(68)*. In the current study, we observed a plasma paclitaxel $T_{1/2}\,\alpha$ of 12.0 ± 1.1 min in the vehicle-pretreated group and 10.2 ± 5.3 min in the control PMO–pretreated group. The antisense-pretreated group had a remarkably longer $T_{1/2}\,\alpha$ value of 31.5 ± 1.8 min ($p < 0.01$). This substantial increase in $T_{1/2}\,\alpha$ is likely a direct result of reduced paclitaxel clearance from plasma. The change in half-life is a critical pharmacokinetic parameter because it determines the frequency of administration. An increase in half-life is likely to translate as decreased frequency of administration in the therapeutic context.

4.2.4. Volume of Distribution

The volume of distribution at steady state represents the volume in which a drug would appear to be distributed during steady state if the drug existed throughout that volume at the same concentration as that in the plasma. In previous studies (human), paclitaxel's volume of distribution has been reported variously between 45.8 and 182 L/m^2 (approx 1.1–4.5 L/kg of body mass) *(68)*.

Table 2
Enzyme Activities of Liver S9 Fraction for CYP3A2 and CYP1A1/1A2[a]

	Vehicle	Control PMO	Antisense PMO
CYP3A2	100.0 ± 5.5	103.7 ± 12.6	53.9 ± 8.2[b]
CYP1A1/1A2	100.0 ± 8.7	108.7 ± 8.4	106.3 ± 13.8

[a]CYP3A2 activity was measured by erythromycin demethylation in micromoles of formaldehyde per milligram of S9 protein per minute. CYP1A1/1A2 activity (used as a control) was measured by ethoxyresorufin O-dealkylation in picomoles of resorufin per milligram of S9 protein per minute. The data are expressed as percentage of control (mean ± SD), which are 82.7 ± 4.5 and 32.7 ± 2.8, respectively, for CYP3A2 and CYP1A1/1A2.
[b]$p < 0.01$ compared to vehicle group.

Because the volume of distribution is larger than total body water (0.7 L/kg), the implication is that paclitaxel is accumulated in tissues and blood components. It has also been previously reported that lung, liver, kidney, and spleen are the major accumulation sites for this drug, which is also highly bound to plasma proteins.

The value of the steady-state volume of distribution parameter is somewhat misleading in a study like this because it essentially represents an inverse function of the AUC number. It is presented nevertheless for sake of completeness. The value determined for paclitaxel in this study was 2.2 ± 0.4 L/kg for the vehicle-pretreated group and 0.5 ± 0.05 L/kg for the antisense-pretreated group. Clearly, the large decrease reflects the antisense-pretreated group's increased AUC. I speculate that a portion of decreased volume of distribution is attributable to decreased accumulation in the liver. However, such speculation cannot be corroborated in the absence of liver accumulation data.

4.3. Analysis of Liver CYP Enzyme Activity

4.3.1. CYP3A2

A decrease in liver CYP3A2 levels is clearly central to the testing of the previously stated hypothesis. Therefore, CYP3A2 levels were determined by two independent indices following termination of the study. First, a functional enzyme assay to determine erythromycin N-demethylation was done to measure the catalytic activity of the enzyme per unit of total protein content of the liver S9 fraction. As per approximate prediction *(60,61,63)*, the antisense-pretreated group had (53.9 ± 8.2)% activity of the vehicle-pretreated group **(Table 2)**. Second, the total CYP3A2 content was determined in the liver S9 fractions by immunoblot analysis of liver S9 fractions by a polyclonal anti-

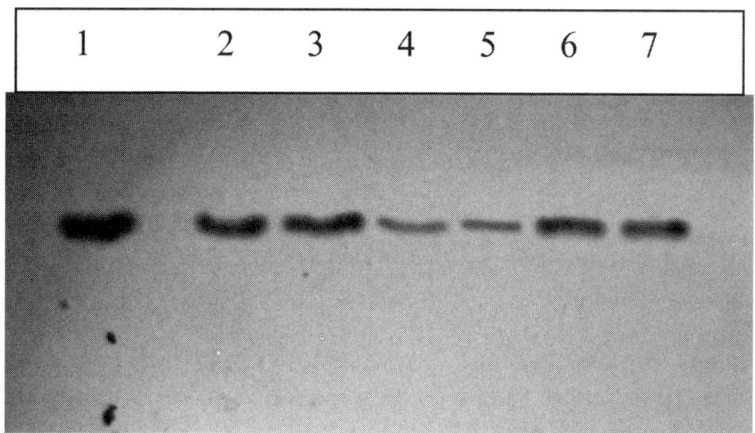

Fig. 5. Representative immunoblot analysis of rat liver S9 fractions for detection of 57-kDa CYP3A2 protein. All rats were treated with the same dose of paclitaxel (8 mg/ kg). PMO pretreatment varied and was as follows: *lane 1*, CYP3A2 protein positive control; *lanes 2 and 3*, 1-22-144 control PMO; *lanes 4 and 5*, AVI-4472 antisense PMO; and *lanes 6 and 7*, vehicle control.

body (Gentest, Woburn, MA). A representative blot, presented in **Fig. 5**, corresponds well with the erythromycin *N*-demethylation data.

4.3.2. CYP1A1/1A2

CYP1A1/1A2 levels were determined in the liver S9 fractions by ethoxyresorufin *O*-dealkylation assay. The levels of this related group of enzymes served as a control for the sequence specificity of the antisense PMO AVI-4472. CYP1A1/1A2 enzyme activity levels were statistically unchanged in all groups.

Unchanged levels of CYP1A1/1A2 activity highlight a key strength of the antisense approach—its exquisite sensitivity, which enables the investigator to target a single gene product. This is particularly important in the case of the large and closely related CYP family. All CYP catalytic activity consists of the following general steps *(69)*:

1. Substrate binding to the active site of the ferric form of enzyme.
2. Reduction of heme group from ferric to ferrous via NADPH/CYP reductase.
3. Binding of molecular oxygen.
4. Transfer of the second electron from CYP reductase.
5. Cleavage of the O–O bond.
6. Substrate oxygenation.
7. Product release.

Conventional CYP inhibitors are generally somewhat promiscuous. Specific clinically relevant inhibitors of CYP3A enzymes include grapefruit juice, azole antifungals, macrolide antibacterials, and human immunodeficiency virus (HIV) protease inhibitors. These inhibitors act via a variety of mechanisms that characterize the degree, specificity, and reversibility of their inhibition. Competitive inhibitors reversibly compete for the active site of the enzyme with other substrates and include azole antifungals and HIV protease inhibitors. Noncompetitive inhibitors bind reversibly to the enzyme on regions apart from the active site and include macrolide antibacterials *(22)*. Suicide inhibitors bind covalently to the enzyme's active site and include grapefruit juice metabolites *(70)*. In addition to the specificity issues related to the aforementioned types of inhibitors, the use of competitive inhibitors to occupy the active site of CYP enzymes can be associated with lipid peroxidation, generation of metabolites that react with the protein structure, or catalytic oxidation of substrates to intermediates that bind covalently to the prosthetic heme group *(71)*. The covalent binding of inhibitor substrates to CYP enzymes may also influence cellular responses that may contribute to drug toxicity or autoimmunity. For example, it has been reported that patients suffering from halothane hepatitis have autoantibodies that react with human CYP2E1. Such a break in immunological tolerance potentially occurred because CYP2E1 became trifluoroacetylated when it oxidatively metabolized halothane. It is possible that the covalently altered form of CYP2E1 may be able to bypass the immunological tolerance that normally exists against self-proteins *(72)*. Therefore, in terms of general strategy, the inhibition of specific CYP activity through the use of enzyme inhibitors can have significant pathological consequences.

5. Conclusion

Intraperitoneal administration of AVI-4472, antisense PMO targeted to the mRNA of CYP3A2, resulted in decreased activity of this enzyme in a sequence-specific manner in rats in vivo. Furthermore, AVI-4472 pretreatment resulted in a statistically significant alteration in various plasma pharmacokinetic parameters of bolus intravenously administered paclitaxel: increased AUC and distribution half-life combined with decreased plasma clearance. It is concluded that the hypothesis that using CYP3A2 inhibition by antisense PMO in rats will decrease the metabolism of paclitaxel and increase its bioavailability was proven.

What are the likely benefits that would result from such an approach? In the case of paclitaxel, the substantial decrease in clearance, increase in AUC, and increase in half-life suggest that its coadministration with a CYP3A4 antisense agent has clear therapeutic benefits. This approach is likely to allow for decreased dosage and increased administration interval. Taking into account the large variety of doses, infusion times, and treatment schedules commonly used for paclitaxel, it is also important to address whether the plasma concentrations observed in the present study are relevant to those achieved clinically. Despite the fact that this study consisted of bolus administration as opposed to the slow infusion administration used clinically, the values of key pharmacokinetic parameters observed in this study point to its potential therapeutic relevance.

It is also important to address the relevance of this rat study from the human perspective. Is it even feasible to extrapolate this concept to the human setting? The answer to this question would have to be an emphatic yes. In a recent article published from my laboratory *(59)*, human CYP3A4 inhibition studies were carried out in two distinct model systems: several primary cultures of freshly plated human hepatocytes and the human colon carcinoma cell line caco-2 transfected with CYP3A4 cDNA. Multiple antisense CYP3A4 PMO sequences were tested in these models in combination with three cytotoxic drugs: paclitaxel, cyclophosphamide, and cisplatin. Antisense PMOs targeted to CYP3A4 greatly increased the cytotoxicity of paclitaxel by decreasing its metabolism. By contrast, cells were rescued from the cytotoxicity of cyclophosphamide, which requires bioactivation by CYP3A4. Cisplatin was used as a control cytotoxic agent because it is not a substrate for CYP3A4. The same antisense sequence, targeted to the AUG region of the mRNA, was effective in primary hepatocytes from adult humans of multiple ethnic origins and either sex. Finally, the dose dependence and reversibility of the antisense approach make it a feasible choice in the metabolic redirection approach.

Acknowledgments

I would like to thank the PMO synthesis group at AVI, particularly Doreen Weller, for providing high-quality PMOs and coordinating with Kansas City Analytical Services. I also thank Pat Iversen and Gayathri Devi for helpful discussions. Melissa Cate provided technical assistance, and Derek Knapp arranged for Bristol-Myers Squibb Co to supply the Taxol.

References

1. Arora, V. and Iversen, P. L. (2001) Redirection of drug metabolism using antisense technology *Curr. Opin. Mol. Ther.* **3**, 249–257.
2. Rowinsky, E. K. (1997) The development and clinical utility of the taxane class of antimicrotubule chemotherapy agents. *Ann. Rev. Med.* **48**, 353–374.
3. Rowinsky, E. K. and Donehower, R. C. (1995) Paclitaxel. *N. Eng. J. Med.* **332**, 1004–1014.
4. Rowinsky, E. K. (1994) Update on the antitumor activity of Pacitaxel in clinical trials. *Ann. Pharmacol.* **28**, S18–S16.

5. Arbuck S. G. and Blaylock B. A. (1995) Taxol: Clinical results and current issues in development, in *Taxol Science and Applications* (Suffness, M., ed.), CRC Press, Boca Raton, FL, pp. 379–415.
6. Suffness, M. and Wall M. E. (1995) Discovery and development of Taxol, in *Taxol Science and Applications* (Suffness, M., ed.), CRC Press, Boca Raton, FL, pp. 3–26.
7. Rose, W. C. (1995) Preclinical antitumor activity of taxanes, in *Taxol Science and Applications* (Suffness, M., ed.), CRC Press, Boca Raton, FL, pp. 209–235.
8. Kumar, G. N., Oatis, J. E., Thornburg, K. R., Heldrich, F. J., Hazard, E. S., and Walle, T. (1994) 6-alpha-hydroxytaxol: isolation and identification of the major metabolite of taxol in human liver microsomes. *Drug Metab. Dispos.* **22,** 177–179.
9. Dorr, R. T. (1997) Pharmacology of the taxanes. *Pharmacotherapy.* **17,** 96S–104S.
10. Rahman, A., Korzekwa, K. R., Grogan, J., Gonzalez, F. J., and Harris, J. W. (1994) Selective biotransformation of taxol to 6 alpha-hydroxytaxol by human cytochrome P450 2C8. *Cancer Res.* **54,** 5543–5546.
11. Harris, J. W., Rahman, A., Kim, B. R., Guengerich, F. P., and Collins, J. M. (1994) Metabolism of Taxol by human hepatic microsomes and liver slices: participation of CYP3A4 and an unknown CYP enzyme. *Cancer Res.* **54,** 4026–4035.
12. Vuilhorgne, M., Gaillard, C., Sanderink, G. J., et al. (1995) Metabolism of Taxoid drugs, in *Taxane Anticancer agents,* (George, G. I., Chen, T. T., Ojima, I., Vyas, D. M., eds.), American Chemical Society, Washington, DC, pp. 98–110.
13. Monsarrat, B., Mariel, E., Cros, S., et al. (1990) Taxol metabolism: isolation and identification of three major metabolites of taxol in rat bile. *Drug Metab. Dispos.* **18,** 895–901.
14. Kostrubsky, V. E., Lewis, L. D., Wood, S. G., Sinclair, P. R., Wrighton, S. A., and Sinclair, J. F. (1997) Effect of Taxol on cytochrome P450 3A and acetaminophen toxicity in cultured rat hepatocytes: comparison to dexamethasone. *Toxicol. Appl. Pharmacol.* **142,** 79–86.
15. Schenkman, J. B. (1993) Historical background and description of the cytochrome P450 monooxygense system, in *Cytochrome P450* (Schemkman, J. B. and Greim, H., eds.), Springer-Verlag, London, UK, pp. 3–14.
16. Raucy, J. L. and Allen, S. W. (2001) Recent advances in P450 research. *Pharmacogenomics J.* **1,** 178–86.
17. Guengerich, F. P. (2002) Update information on human P450s. *Drug Metab. Rev.* **34,** 7–15.
18. Guengerich F. P. (1994) Catalytic selectivity of human cytochrome P450 enzymes: relevance to drug metabolism and toxicity. *Toxicol. Lett.* **70,** 133–138.
19. Backes, W. L. (1993) NADPH- Cytochrome P450 reductase: function, in *Cytochrome P450* (Schemkman, J. B. and Greim, H., eds.), Springer-Verlag, London, UK, pp. 3–14.
20. Roy, P., Yu, L. J., Crespi, C. L., and Waxman, D. J. (1999) Development of a substrate-activity based approach to identify the major human liver P-450 catalysts of cyclophosphamide and ifosfamide activation based on cDNA-expressed activities and liver microsomal P-450 profiles. *Drug Metab. Dispos.* **27,** 655–666.

21. Benet, L. Z., Kroetz, D. L., and Sheiner, L. B. (1996) Pharmacokinetics: the dynamics of drug absorption, distribution, and elimination, in *Goodman & Gilman's The Pharmacological Basis of Therapeutics* (Hardman, J. H., Limbird, L. E., Malinoff, P. B., Ruddon, R. W., and Golman, A. G. eds.), McGraw-Hill, New York, pp. 3–28.
22. Thummel, K. E. and Wilkinson, G. R. (1998) In vitro and in vivo drug interactions involving human CYP3A. *Annu. Rev. Pharmacol. Toxicol.* **38**, 389–430.
23. Guengerich, F. P. (1999) Cytochrome P450 3A4: regulation and role in drug metobolism. *Annu. Rev. Pharmacol. Toxicol.* **39**, 1–17.
24. Dresser, G. K., Spence, J. D., and Bailey, D. G. (2000) Pharmacokinetic-pharmacodynamic consequences and clinical relevance of cytochrome P450 3A4 inhibition. *Clin. Pharmacokinet.* **38**, 41–57.
25. Collins, J. M. (2000) Cytochrome P450 and other determinants of pharmacokinetics, toxicity and efficacy in humans. *Clin. Cancer Res.* **6**, 1203–1204.
26. Ekins, S., Vanden Branden, M., Ring, B. J., et al. (1998) Further characterization of the expression in liver and catalytic activity of CYP2B6. *J. Pharmacol. Exp. Ther.* **286**, 1253–1259.
27. Giancarlo, G. M., Venkatakrishnan, K. Granda, B. W., von Moltke, L. L., and Greenblatt, D. J. (2001) Relative contributions of CYP2C9 and 2C19 to phenytoin 4-hydroxylation in vitro: inhibition by sulfaphenazole, omeprazole, and ticlopidine. *Eur. J. Clin. Pharmacol.* **57**, 31–36.
28. Ball, S. E., Ahern D., Scatina, J., and Kao, J. (1997) Venlafaxine: in vitro inhibition of CYP2D6 dependent imipramine and desimipramine metabolism; comparative studies with selected SSRIs, and effects on human hepatic CYP3A4, CYP2C9 and CYP1A2. *Br. J. Clin. Pharmacol.* **43**, 619–626.
29. Yang, T. J., Krausz, K., Sai W. Y., Gonzalez, F. J., and Gelboin, H. V. (1999) Eight inhibitory monoclonal antibodies define the role of individual P-450s in human liver microsomal diazepam, 7-ethoxycoumarin, and imipramine metabolism, *Drug Metab. Dispos.* **27**, 102–109.
30. Kirkwood, L. C., Nation, R. L., and Somogyi, A. A. (1997) Characterization of the human cytochrome P450 enzymes involved in the metabolism of dihydrocodeine. *Br. J. Clin. Pharmacol.* **44**, 549–555.
31. Rasmussen, E., Eriksson, B., Oberg, K., Bondesson, U., and Rane, A. (1998) Selective effects of somatostatin analogs on human drug-metabolizing enzymes, *Clin. Pharmacol. Ther.* **64**, 150–159.
32. Caraco, Y., Tateishi, T., Guengerich, F. P., and Wood, A. J. (1996) Microsomal codeine N-demethylation: cosegregation with cytochrome P4503A4 activity. *Drug Metab. Dispos.* **24**, 761–764.
33. de Wildt, S. N., Kearns, G. L., Leeder, J. S., and van den Anker, J. N. (1999) Cytochrome P4503A: ontogeny and drug description. *Clin. Pharmacokinet.* **37**, 485–505.
34. Schuetz, J. D., Kauma, S., and Guzlian, P. S. (1993) Identification of the fetal liver cytochrome CYP3A7 in human endometrium and placenta. *J. Clin. Invest.* **92**, 1018–1024.

35. Wrighton, S. A., Schuetz, E. G., Thummel, K. E., Shen, D. D., Korzekwa, K. R., and Watkins, P. B. (2000) The human CYP3A subfamily: practical considerations. *Drug Metab. Rev.* **32,** 339–361.
36. Guengerich, F. P. (1993) Metobolic reaction: types of reaction of cytochrome P450 enzymes, in *Cytochrome P450* (Schemkman, J. B. and Greim, H., eds.), Springer-Verlag, London, UK, pp. 89–100.
37. Shimada, T., Yamazaki, H., Mimura, M., Inui, Y., and Guengerich, F. P. (1994) Interindividual variations in human liver cytochrome P-450 enzymes involved in the oxidation of drugs, carcinogens and toxic chemicals: studies with liver microsomes of 30 Japanese and 30 Caucasians. *J. Pharmacol. Exp. Ther.* **270,** 414–423.
38. Lown, K. S., Kolars, J. C., Thummel, K. E., et al. (1994) Interpatient heterogeneity in expression of CYP3A4 and CYP3A5 in small bowel: lack of prediction by the erythromycin breath test. *Drug Metab. Disposition* **22,** 947–955.
39. Paine, M. F., Khalighi, M., Fisher, J. M., et al. (1997) Characterization of inter- and intra-intestinal differences in human CYP3A-dependent metabolism. *J. Pharmacol. Exp. Ther.* **283,** 1552–1562.
40. Zhang, Q. Y., Dunbar, D., Ostrowska, A., Zeisloft, S., Yang, J., and Kaminsky, L. S. (1999) Characterization of human small intestinal cytochromes P-450. *Drug Metab. Dispos.* **27,** 804–809.
41. Wandel, C., Witte, J. S., Hall, J. M., Stein, C. M., Wood, A. J., and Wilkinson, G. R. (2000) CYP3A activity in African American and European American men: population differences and functional effect of the CYP3A4*1B 5'-promoter region. *Clin. Pharmacol. Ther.* **68,** 82–91.
42. Rebbeck, T. R., Jaffe, J. M., Walker, A. H., Wein, A. J., and Malkowicz, S. B. (1998) Modification of clinical presentation of prostate tumors by a novel genetic variant in CYP3A4. *J. Natl. Cancer Inst.* **90,** 1225–1229.
43. Felix, C. A., Walker, A. H., Lange, B. J., et al. (1998) Association of the CYP3A4 genotype with treatment-related leukemia. *Proc. Natl. Acad. Sci. USA* **95,** 13,176–13,181.
44. Parkinson, A. (1996) Biotransformation of xenobiotics, in *Casarett and Doull's Toxicology The Basic Science of Poisons*, 5th ed. (Klaassen, C. D., ed.), McGraw-Hill, New York, NY, pp. 113–186.
45. Gonzalez, F. (1989) The molecular biology of the cytochrome P450s. *Pharmacol. Rev.* **40,** 243–287.
46. Imaoka, S., Terano, Y., and Funae, Y. (1988) Constitutive testosterone 6 beta-hydroxylase in rat liver. *J. Biochem.* **104,** 481–487.
47. Kato, R. and Yamazoe, Y. (1992) Sex-specific cytochrome P450 as a cause of sex- and species-related differences in drug toxicity. *Toxicol. Lett.* **64-65,** 661–667.
48. Waxman, D. J., Ram, P. A., Pampori, N. A., and Shapiro, B. H. (1995) Growth hormone regulation of male-specific rat liver P450s 2A2 and 3A2: induction by intermittent growth hormone pulses in male but not female rats rendered growth hormone deficient by neonatal monosodium glutamate. *Mol. Pharmacol.* **48,** 790–797.

49. Kawai, M., Bandiera, S. M., Chang, T. K., and Bellward, G. D. (2000) Growth hormone regulation and developmental expression of rat hepatic CYP3A18, CYP3A9, and CYP3A2. *Biochem. Pharmacol.* **15,** 1277–1287.

50. Sonnischen, D. S., Liu, Q., Schuetz, E. G., Schuetz, J. D., Pappo, A., and Relling, M. V. (1995) Variability in human CYP paclitaxel metabolism. *J. Pharmacol. Expt. Therap.* **275,** 566–575.

51. Walle, T., Walle, U. K., Kumar, G. N., and Bhalla, K. N. (1995) Taxol metabolism and disposition in cancer patients. *Drug Metab. Dispos.* **23,** 506.

52. Summerton, J. and Weller, D. (1997) Antisense properties of morpholino oligomers. *Nucleosides Nucleotides* **16,** 889–898.

53. Hudziak, R. M., Barofsky, E., Barofsky, D. F., Weller, D. L., Huang, S. B., and Weller, D. D. (1996) Resistance of Morpholino phosphorodiamidate oligomers to enzymatic degradation. *Antisense Nucleic Acid Drug Dev.* **6,** 227–272.

54. Summerton, J. (1999) Morpholino antisense oligomers: the case for an RNase H-independent structural type. *Biochim. Biophys. Acta* **1489,** 141–158.

55. Giles, R. V., Spiller, D. G., Clark, R. E., and Tidd, D. M. (1999) Antisense morpholino oligonucleotide analog induces missplicing of c-myc mRNA. *Antisense Nucleic Acid Drug Dev.* **9,** 213–220.

56. Arora, V., Knapp, D. C., Smith, B. L., et al. (2000) c-Myc antisense limits rat liver regeneration and indicates role for c-Myc in regulating cytochrome P-450 3A activity. *J. Pharmacol. Exp. Ther.* **292,** 921–928.

57. Arora, V. and Iversen, P. L. (2000) Antisense oligonucleotides targeted to the p53 gene modulate liver regeneration in vivo. *Drug Metab. Dispos.* **28,** 131–138.

58. Mata, J. E., Ricker, J. L., Gattone, V. H., and Iversen, P. L. (2000) Treatment of C57BL/6J-*cpk/cpk* mice with c-myc antisense oligomer slows cystic disease and modulates protein expression in the kidney. *Proc. Int. Soc. Nephrol. Gene Ther. Symp.* (abstract).

59. Arora, V., Cate, M. L., Ghosh, C., and Iversen, P. L. (2002) Phosphorodiamidate morpholino antisense oligomers inhibit expression of human cytochrome P450 3A4 and alter selected drug metabolism. *Drug Metab. Dispos.* **30,** 757–762.

60. Arora, V., Knapp, D. C., Reddy, M. T., Weller, D. D., and Iversen, P. L. (2002) Bioavailability and efficacy of antisense morpholino oligomers targeted to c-Myc and cytochrome P-450 3A2 following oral administration in rats. *J. Pharm. Sci.* **91,** 1009–1018.

61. Arora, V., Hannah, T. L., Iversen, P. L., and Brand, R. M. (2002) Transdermal use of phosphorodiamidate morpholino oligomer AVI-4472 inhibits cytochrome P450 activity in male rats. *Pharm. Res.* **19,** 1465–1470.

62. Summerton, J. and Weller, D. (1997) Morpholino antisense oligomers: design, preparation, and properties. *Antisense Nucleic Acid Drug Dev.* **7,** 187–195.

63. Desjardins, J. P. and Iversen, P. L. (1995) Inhibition of rat cytochrome P450 3A2 by an antisense oligonucleotide in vivo. *J. Pharmacol. Expt. Ther.* **275,** 1608–1613.

64. Wrighton, S. A., Schuetz, E. G., Watkins, P. B., et al. (1985) Demonstration in multiple species of inducible hepatic cytochromes P-450 and their mRNAs related to the glucocorticoid-inducible cytochrome P-450 of the rat. *Mol. Pharmacol.* **28,** 312–321.

65. Nash, T. (1953) The colorimetric estimation of formaldehyde by means of the Hantzsch reaction. *Biochem. J.* **55,** 416–421.

66. Shumaker, R. C. (1986) Pkcalc: a basic interactive computer program for statistical and pharmacokinetic analysis of data. *Drug Metab. Rev.* **17,** 331–348.

67. Gibaldi, M. and Perrier, D. (1982) *Pharmocokinetics*, 2nd ed., Marcel Dekker, New York.

68. Straubinger, R. M. (1995) Biopharmaceutics of paclitaxel (Taxol): formulation, activity, and pharmacokinetics, in *Taxol Science and Applications* (Suffness, M., ed.), CRC Press, Boca Raton, FL, pp. 237–258.

69. Guengerich, F. P. (1993) Metabolic reactions: types of reactions of cytochrome P450 enzymes, in *Cytochrome P450* (Schemkman, J. B. and Greim, H., eds.), Springer-Verlag, London, UK, pp. 89–104.

70. Schmiedlin-Ren, P., Edwards, D. J., Fitzsimmons, M. E., et al. (1997) Mechanisms of enhanced oral availability of CYP3A4 substrates by grapefruit constituents: decreased enterocyte CYP3A4 concentration and mechanism-based inactivation by furanocoumarins. *Drug Metab. Dispos.* **25,** 1228–1233.

71. Ortiz de Montellano, P. R. and Correia, M. A. (1983) Suicidal destruction of cytochrome P-450 during oxidative drug metabolism. *Ann. Rev. Pharmacol. Toxicol.* **23,** 481–503.

72. Bourdi, M., Chen, W., Peter, R. M., et al. (1996) Human cytochrome P450 2E1 is a major autoantigen associated with halothane hepatitis. *Chem. Res. Toxicol.* **9,** 1159–1166.

VII

GASTROINTESTINAL

15

Antisense Oligonucleotide Treatment of Inflammatory Bowel Diseases

Bruce R. Yacyshyn

1. Introduction

Our expanding knowledge of inflammatory mediator pathways has created new options for the management of Crohn's disease and ulcerative colitis (*1*). In the past decade, the differential immune activation of intestinal and peripheral immune systems has been identified, and these observations have rapidly moved basic science observations to the bedside, as biological therapies continue to increase.

In general, inflammatory bowel diseases (IBDs) are treated with broad-spectrum antiinflammatory drugs, including 5-ASA, corticosteroids, and cytotoxic or noncytotoxic immunosuppressives, usually with only partial control of disease and often at a cost of substantial actual or potential toxicity. The recent identification of genetic mutations, including IBD1 and NOD2, has introduced new areas for potential drug development. However, these mutations are not yet available in clinical research (*2,3*). Currently, the approval of six new biological agents gives solid evidence as to the direction of current and future drugs, which is the specific targeting of select cytokines and their receptors. These agents include two antibodies to the interleukin-2 (IL-2) receptor for prophylaxis of acute renal transplant rejection (daclizumab, Zenapax®; basiliximab, Simulect®), an IL-2 diphtheria toxin fusion protein for lymphoma (denileukin diftitox, Ontak®), a tumor necrosis factor-α (TNF-α) chimeric antibody (infliximab, Remicade®) for Crohn's disease, a TNF-α p75 receptor-IgG$_1$ Fc fusion protein (etanercept, Enbrel®) for rheumatoid arthritis and psoriatic arthritis, and IL-1ra (anakinra, Kineret®) for rheumatoid arthritis.

From: *Methods in Molecular Medicine, Vol. 106: Antisense Therapeutics, Second Edition*
Edited by: I. Phillips © Humana Press Inc., Totowa, NJ

These new agents offer the potential for more effective and usually less toxic therapy. There is a great need for more specific or targeted therapies across the spectrum of immune-mediated and inflammatory diseases.

Antisense drugs can offer the potential for unprecedented specificity, owing to selective targeting of mRNA via Watson–Crick binding, and therefore tolerability *(4)*. By targeting intercellular adhesion molecule-1 (ICAM-1), an adhesion molecule that is fundamentally implicated in the inflammatory process, ISIS 2302 (alicaforsen) offers the potential for broad effectiveness in the treatment of immune-mediated and inflammatory diseases *(5,6)*. This antisense oligonucleotide (AS-ODN) hybridizes to a sequence in the 3' untranslated region of human ICAM-1 mRNA.

2. Phase 1–2 Trial Experience of Alicaforsen (ISIS 2302 Antisense to ICAM-1) in Crohn's Disease

In 1997, Glover et al. *(7)* reported on a phase I study of ISIS 2302 in healthy volunteers. These investigators showed that single iv doses of ISIS 2302 from 0.06 to 2.0 mg/kg and four doses every other day ranging from 0.5 to 2.0 mg/kg given by 2-h infusion were safe. The pharmacokinetic data realized by this study were the same as those found in primates. The only drug-related adverse event was a dose-related (approx 1.5-fold at a dose of 2.0 mg/kg) and transient (for 2–4 h after dosing) increase in activated partial thromboplastin time (aPTT). This event reflects transient plasma protein binding by phosphorothioate oligonucleotides *(8–10)*.

We have reported a small, placebo-controlled, double-blind, dose-escalating study of ISIS 2302 for Crohn's disease *(11)*. This phase I–II study was conducted in 20 patients between the ages of 18 and 80 with moderately active Crohn's disease (Crohn's Disease Activity Index [CDAI] of 200–350) despite stable background doses of steroids (maximum prednisone = 40 mg/d) with or without 5-ASA drugs. There were no differences in the baseline characteristics between ISIS 2302–treated and placebo groups. Corticosteroid dosages remained stable for the 26-d infusion period and then were adjusted by the investigator according to blinded clinical judgment. Patients received a total of 13 doses of ISIS 2302 or placebo by 2-h iv infusion. Four patients each were assigned to the 0.5- and 1.0-mg/kg dose cohorts, and the remaining 12 patients were assigned to the 2.0-mg/kg cohort. All patients were followed for a total of 6 mo. At the end of treatment on d 33, 7 of 15 (47%) ISIS 2302–treated patients and 1 of 5 (20%) placebo-treated patients were in remission by clinical criteria (CDAI < 150) ($p = 0.054$) (**Fig. 1A**). The placebo patient was in remission at baseline and had been enrolled in error. At mo 6, five of the seven ISIS 2302–treated remitters remained in remission (one had a brief increase in prednisone), and a sixth had a CDAI of 156. No clinically significant adverse events were reported, and there was no effect on routine laboratory safety indices. As

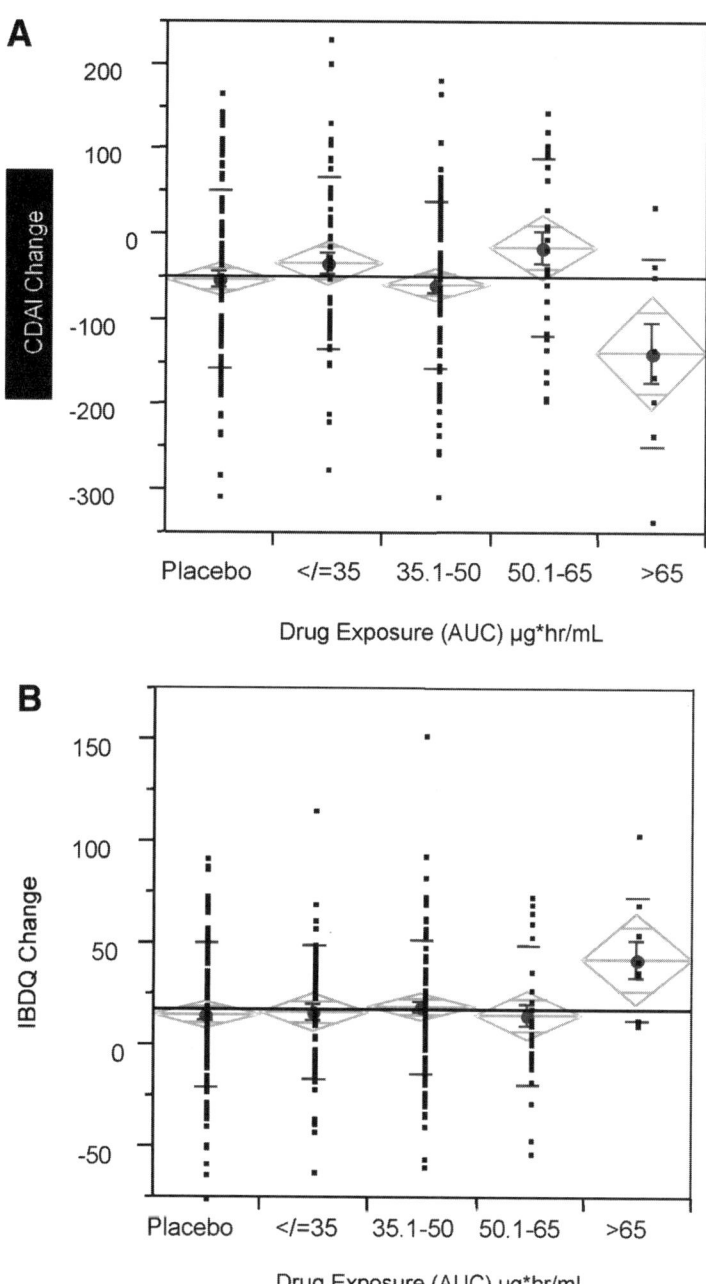

Fig. 1. Change (**A**) in CDAI and (**B**) IBDQ scores at wk 14, segregated by AUC cohort. Diamonds reflect the 95% confidence intervals and bars signify 1 or 2 SDs. CDAI and IBDQ significantly improved for the high-AUC subgroup vs placebo ($p = 0.027$ for both). (Reproduced with permission from **ref. *12*.)

anticipated, transient aPTT elevations of approx 3, 5, and 10 s occurred after doses of 0.5, 1.0, and 2.0 mg/kg, respectively. Moreover, we found that the clinical outcome changes were corroborated by an overall reduction in intestinal mucosal ICAM-1 expression after treatment with the 2 mg/kg dose.

3. Retreatment of Crohn's Disease Patients with ISIS 2302

Eight patients in the above study were retreated using the same treatment protocol. Four patients treated with alicaforsen were remitters to the first course of therapy: one was a responder, two were nonresponders, and one was a placebo-treated nonresponder. These patients were offered open-label retreatment with ISIS 2302 according to their originally assigned dosage, and the original placebo subject received 2 mg/kg. Baseline CDAI for patients entering the retreatment protocol ranged from 221 to 474. The ISIS 2302 retreatment protocol produced a remission in three of the four patients who were remitters to the first course of therapy. The patient who originally received placebo became a responder once treated with alicaforsen. This drug was only mildly or transiently effective again in the two previously nonresponding patients. The only drug-related laboratory effect found in the retreated patients was the expected dose-related and transient increase in aPTT (9.4 ± 5.6 s at the 2.0 mg/kg dose). The most common adverse event was headache.

4. Phase 2b Study of ISIS 2302 in Crohn's Disease (ISIS 2302-CS9)

Patients ages 14 to 80 were enrolled with moderately active (CDAI of 200–350), steroid-dependent Crohn's disease (active disease for ≥ 3 previous months despite 10–40 mg of prednisone or equivalent, with at least one unsuccessful taper attempt) *(12)*. Corticosteroids were stable for at least the 2 wk prior to study entry. Patients were allowed stable doses of aminosalicylates, but immunosuppressives were excluded within the prior 4 wk.

Patients were stratified by their baseline steroid dose (prednisone equivalent of 10–19 or 20–40 mg/d) for randomization. The protocol was based on pilot studies and the North American Methotrexate Study *(11,13)*. Patients were randomized to saline placebo or one of two ISIS 2302 regimens: 2 mg/kg intravenously three times a week for 2 wk or for 4 wk. Infusions were administered over 2 h and two treatment courses were administered in mo 1 and 3. The maximum permitted ISIS 2302 dose was 200 mg; the few patients weighing >100 kg were therefore administered slightly less than 2 mg/kg.

The primary end point was corticosteroid-free remission (CDAI < 150, corticosteroid dose = 0) at wk 14. Other secondary end points included corticosteroid utilization and quality of life, assessed by the Inflammatory Bowel Disease Questionnaire (IBDQ). Additional response levels included low-dose steroid-dependent remission (CDAI < 150, corticosteroid dose $\leq$ 10 mg/d), remission (CDAI < 150, corticosteroid dose ≤ baseline level), response (>100-point decrease in CDAI with corticosteroids ≤ baseline dose), and partial response (>70-point decrease in CDAI with corticosteroids ≤ baseline dose).

As part of the prospective design of this study, plasma for ISIS 2302 concentrations was obtained at 0, 2, 3, and 5 h after the start of the study drug infusion on d 1, and at the end of infusion (h 2) on d 10 and 26, for both treatment cycles. A population pharmacokinetic analysis was performed, which allowed for the analysis of sparse pharmacokinetic data yielding a mathematical model of drug behavior. The model is used to describe and predict drug concentrations in plasma and obtain reliable estimates of overall drug exposure in the treated population *(14–16)*. The population approach in this study was used to extract potentially useful pharmacokinetic/pharmacodynamic information from a complex clinical trial. Sixteen demographic and laboratory covariate terms were screened for their effects on the preliminary model, including age, gender, height, weight, steroid dose, liver function, renal function, and C3a and C5a levels. The covariate model was built using stepwise addition and backward elimination procedure *(14–16)*.

The model was established in a sample population of 70% of the total number of patients (index data set) and was tested against the remaining 30% of patients (validation data set). Bias and model predictions were evaluated in both the entire population and the population as stratified by important covariates using plots of predicted vs actual values and plots of random effects vs covariates. One of the outcomes measured by design was the relationship between efficacy and drug exposure, assessed using various statistical methods. Two measures of drug exposure per dose were computed for each ISIS 2302–treated patient: the predicted maximum concentration (C_{max}) and a predicted area under the concentration–time curve (AUC). Correlation of patient drug exposure (AUC) with remission rates, change in CDAI, or change in IBDQ was analyzed using logistic regression, linear regression, and descriptive statistics *(14–16)*. To determine whether there were AUC thresholds above which response rates improved, patients were categorized by 15-µg(h·mL) AUC increments over the dose per exposure range (groups ≤35, 35.1–50, 50.1–65; AUC > 65 µg/[h·mL]). The high AUC group was defined as those patients achieving an AUC >65 µg/(h·mL).

Overall, study completion rates through wk 14 for the 6-mo trial were 59.1% for the ISIS 2302–treated patients and 52.5% for placebo patients; most discontinuations were for lack of efficacy or disease progression (26 and 35% for ISIS 2302 and placebo, respectively). Eight ISIS 2302 subjects (4%) withdrew due to adverse events, including an ISIS 2302 drug reaction consisting of facial flushing, chest tightness, and dyspnea during the second course, which was reinduced on rechallenge infusion. A history of previous use of anti-TNF monoclonal antibody was reported in 5.7% of patients, the majority having discontinued this agent due to lack of effect. Interestingly, patients who had previously used immunosuppressives had a lower response rate (14%) than those who had not previously used immunosuppressives (26%). Even in the subgroup of patients with high AUC, only 40% (2/5) of patients who had previously used immunosuppressives achieved a response, whereas 75% (3/4) of patients who had not previously used immunosuppressives achieved a response.

Pharmacokinetic data were collected for 181 of 199 ISIS 2302–treated patients. The ISIS 2302 $T_{1/2}$ was 1.11 (±0.2) h and average C_{max} was 14.8 (±3) μg/mL. ISIS 2302 pharmacokinetics were well described by the one-compartment linear model, and evaluation of the model demonstrated good fit, without evidence of bias *(12)*.

Analysis identified two important covariates altering the pharmacokinetics of ISIS 2302: weight and gender (**Fig. 1B**). The average drug exposure per dose (AUC) was 38.0 (±10.2) μg/(hr·mL) in men and 44.7 (±12.7) μg/(h·mL) in women. For a constant weight of 70 kg, the AUC and C_{max} for females were higher than for males. For the average study weights of 80 (males) and 67 kg (females), there was still a trend toward increased AUC in females.

The effects of gender on drug exposure were primarily mediated through effects on drug clearance rates. Plasma clearance was 37% higher in males, resulting in a shorter $T_{1/2}$ of 0.94 h for 70-kg males vs 1.14 h for 70-kg females. The volume of distribution was also gender dependent, explained partially by differences in body fat composition, since ISIS 2302 has limited uptake by adipose tissue *(17)*. Metabolism was not affected by gender or exogenous estrogen use.

Of the 191 patients completing wk 14, only 64% (40/63) taking placebo had successfully discontinued steroids, compared with 78% (100/128) for the ISIS 2302–treated groups ($p = 0.032$). Cumulative steroid use through wk 14 did not differ among the intent-to-treat cohorts but was clearly lower for steroid-free remitters ($p < 0.001$). Similar proportions of patients in the three treatment groups achieved the primary end point, steroid-free remission at the end of wk 14: 19.2% in the 2-wk group, 21.2% in the 4-wk group (20.2% for combined ISIS 2302 groups), and 18.8% in the placebo-treated patients. Rates of remission (steroid-free, low-dose steroid-dependent, and steroids ≤ baseline) and response (response and partial response) over time were similar among treatment arms at various time points throughout the trial. Mean values of CDAI over time, steroid dose over time, duration of disease, prior immunosuppressive use, and IBDQ over time were similar among the three treatment groups (data not shown). Baseline steroid use did predict steroid-free remission, with remitters having an average corticosteroid dosage of 19.0 (±8.0) mg/d at baseline and nonremitters starting at 23.5 (±10.8) mg/d ($p < 0.005$).

There is a strong correlation between drug exposure and response to ISIS 2302 treatment. Following the 2 mg/kg iv ISIS 2302 dose, plasma AUC levels ranged from 13.49 to 96.09 µg/(h·mL), although all patients received the same 2 mg/kg dose. The AUC range was divided for analysis into 15 µg/(h·mL) dose increment groups: < 35 µg/(h·mL) (n = 54); 35.1–50 µg/(h·mL) (n = 90); 50.1–65 µg/(h·mL) (n = 28); and > 65 µg/(h·mL) (n = 9). Although the number of patients in higher-exposure groups was small, steroid-free remission rates increased as AUC increased. Remissions rose from 13.0% (7/54) for the lowest AUC subgroup, to 21.4% (6/28) for the subgroup with an AUC of 50.1–65 µg/(h·mL), subgroup, and 55.6% (5/9) for the highest AUC subgroup (p = 0.023 for the highest AUC subgroup vs placebo). The logistic regression of steroid-free remission rates as a function of AUC demonstrated a significant trend (p = 0.0064).

Other measures of clinical response consistently demonstrated improvement for the highest-AUC subgroup. The same strong correlation between drug exposure and secondary disease response measures was observed for changes in CDAI and IBDQ (**Fig. 1A,B**) at the end of wk 14, both showing statistically significant improvements vs placebo (p < 0.03). Correlation of improvement in CDAI at wk 14 and AUC confirmed this significant association (p = 0.02).

The highest-AUC subgroup (> 65 µg/[h·mL]) was predominantly female (eight of nine), Caucasian (eight of nine), and had relatively high obesity indices (average weight of 98.6 kg), resulting in required slower clearance and higher plasma levels necessary to achieve a higher drug exposure level. The baseline CDAI scores did not significantly differ (278 vs 276), nor did the baseline steroid dose for this group (18.9 vs 22.5 mg).

Time to steroid-free remission was similar across three treatment groups. For patients who reached steroid-free remission (within the 6-mo protocol), mean time (± SD) to first day of steroid-free remission was 85.3 (±35.9) d for placebo, 78.3 (±22.0) d for 2-wk cohort, and 82.5 (±29.0) d for the 4-wk cohort. The steroid taper was predefined in the protocol; thus, the earliest that steroid-free remission could be achieved was at 64 d. The 1-yr follow-up response duration was completed by 34 (17%) of ISIS 2302 and 16 (18.8%) of placebo patients from a potential pool of 40 (20.2%) ISIS 2302 and 19 (18.8%) placebo patients in steroid-free remission at mo 6.

5. Laboratory Assessment and Safety

Of 91 patients studied, one male subject developed new anti-ISIS 2302 IgG/IgM antibody by d 82 at a 1:10 titer without clinical sequelae. ISIS 2302 was well tolerated (**Table 1**). Medically important adverse events were limited to hypersensitivity reactions in a small proportion of patients (3%, including urticaria in six patients and two drug reactions).

Table 1
Adverse Reactions of Alicaforsen Treated Patients

	Placebo N = 101	ISIS 2302 2 mg/kg N = 198	ISIS 2302 Subset: AUC >65 ug/(h·mL) N = 9
Vasodilation	4 (4.0)	23 (11.6)[a]	1 (11.1)
Rash	15 (14.9)	19 (9.6)	0
Urticaria	0	6 (3.0)	0
Anaphylactoid reaction	0	2 (1.0)	0
Myocardial Infarction	0	1 (0.5)	0
Schwannoma	0	1 (0.5)	0
Asthenia	22 (21.8)	45 (22.7)	0
Flu Syndrome	12 (11.9)	27 (13.6)	1 (11.1)
Headache	26 (25.7)	65 (32.8)	1 (11.1)
Dizziness	16 (15.8)	36 (18.2)	1 (11.1)
Infection	37 (36.6)	88 (44.4)	3 (33.3)
Pancreatitis	1 (1.0)	0	0
Enterocolitis	26 (25.7)	44 (22.2)	2 (22.2)
Nausea	29 (28.7)	49 (24.7)	4 (44.4)
Vomiting	12 (11.9)	28 (14.1)	2 (22.2)
Abdominal Pain	16 (15.8)	30 (15.2)	2 (22.2)
Average post-infusion aPTT Increase, seconds	0.8 (+/- 4.66)	8.66 (+/- 7.29)[b]	8.70 (+/- 7.40)[b]
Maximum post-infusion aPTT Value, seconds	30.4 (+/- 7.6)	41.5 (+/- 7.4)[b]	42.1 (+/- 4.4)[b]

[a] $p = 0.03$ vs placebo.
[b] $p < 0.0001$ vs placebo.
(Reproduced with permission from **ref. *12*.**)

6. The Use of an Enema Formulation of ISIS 2302 in Ulcerative Colitis

The research experience to date with iv ISIS 2302 in ulcerative colitis is limited to two patients, one of whom received placebo. The treated patient had a sustained remission (7 yr) after treatment with 13 doses of iv ISIS 2302 (0.5 mg/kg). Van Deventer et al. reported the largest clinical experience with ISIS 2302 in ulcerative colitis at the United European Gastroenterology Week (2001). Using a nightly ISIS 2302 enema for 1 mo, they studied patients age >18 with active ulcerative colitis extending 5–50 cm from the anal verge. Patients were selected to have a Disease Activity Index (DAI) of 3–10, including an abnormal endoscopy score, and were not taking medications or a stable dose of oral 5-ASA for >2 mo. Patients had a negative stool culture and no history

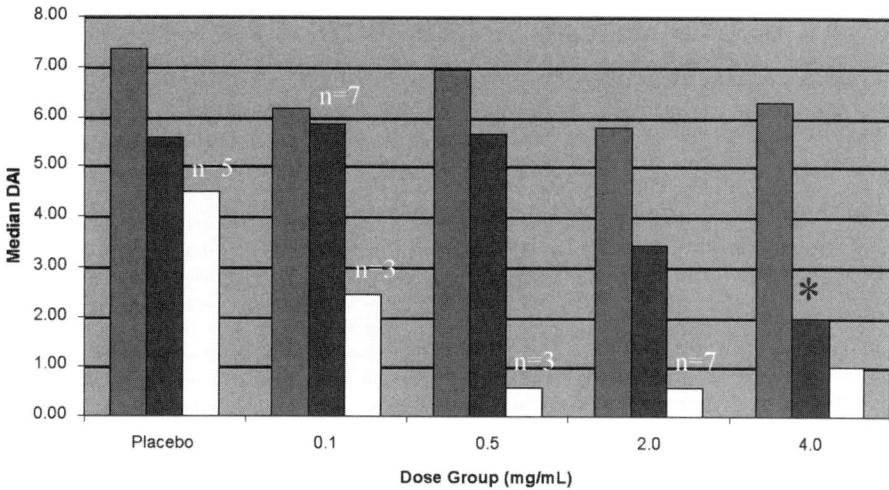

Fig. 2. Median DAI scores (0, mo 1, mo 3) by ISIS 2302 enema eoncentration. (Courtesy of ISIS.)

of bowel resection or stricture. The trial was dose escalated by dose cohort. Forty patients were studied, with 10 subjects (8 active/2 placebo) per dose cohort. The four doses administered were 0.1, 0.5, 2, and 4 mg/mL in a 60-mL enema. Enema retention time was recorded and patients received a 5-mo follow-up. Patient assessments included DAI at baseline and at mo 1, 3, and 6 and monthly erythrocyte sedimentation rate.

Colonoscopic biopsies were performed for ICAM-1 levels, histology, and pharmacokinetics. Primary end points were the change in DAI from baseline compared to placebo at the end of the treatment month (**Fig. 2**) *(18)*.

Overall, there was no difference in patient characteristics between the dose groups. The male/female ratio was 3:2, mean age was 56, and mean duration of disease was 6 yr. Of the patients studied, 45% had prior steroid use, 10% had received immunosuppressives (with or without steroids), and 45% had no history of either. Ninety-three percent were receiving concomitant therapy with oral 5-ASA drugs. The median change from baseline in DAI after 1 mo of enema administration for the 4 mg/mL cohort was significant compared to placebo ($p = 0.034$). These responses persisted and reduced the need for additional therapies during the 6-mo follow-up period. The drug was well tolerated and safe. One patient was hospitalized for anemia at mo 2, and two had a flare-up of ulcerative colitis at mo 2 and 5, respectively. No severe adverse events occurred. Minor adverse events included reversible elevation in GGT; reversible transaminitis; rectal cramping with enema administration; as well as mild nausea, asthenia, and dyspepsia. Additional trials of ISIS 2302 enema in patients with ulcerative colitis and pouchitis are planned in North America.

7. Conclusion

The current total number of patients that have received systemic ISIS 2302 safely, including volunteers; patients with Crohn's disease, rheumatoid arthritis, and psoriasis; and renal transplant recipients is over 600 persons.

The majority were receiving corticosteroids and frequently immunosuppressives (azathioprine, 6-mercaptopurine, methotrexate, cyclo-sporine). Safety in combination with drugs routinely administered for these diseases has been good. Monitoring for infections has not demonstrated increased rates compared to placebo with the exception of more herpes simplex in the renal transplant trial, in which ISIS 2302 was initiated concurrently with corticosteroids and cyclosporine.

The clinical benefit of an AS-ODN targeting ICAM-1 for Crohn's disease or ulcerative colitis remains under investigation, but the evidence for further studies and development of newer-generation molecules is compelling. The utility of antisense technology as a form of drug targeting has been well validated, and we have gained substantial knowledge using these molecules in cell lines, animal models of disease, and human gut disorders in the pathogenesis of these systems. To date, drug toxicity has not been a substantial issue. Patients tolerate the medication well, including retreatment protocols. For Crohn's disease, our understanding of drug pharmacokinetics has allowed for the current phase III trials of iv alicaforsen in Crohn's disease (CS20 and 21). These 150 patient studies are currently in the recruitment phase. Newer-generation AS-ODNs, which allow sc or oral administration and different targets (including TNF-α), should further contribute to our understanding of these compounds in gut inflammation.

References

1. Elson, C. O. (2002) Genes, microbes, and T cells—new therapeutic targets in Crohn's disease. *N. Engl. J. Med.* **346(8),** 614–616.
2. Ogura, Y., Inohara, N., Benito, A., Chen, F. F., Yamakoa, S., and Nunez, G. (2001) A Nod1/Apaf-1 family member that is restricted to monocytes and activates NF-kappaB. *J. Biol. Chem.* **276(7),** 4812–4818.
3. Cho, J. H., Nicolae, D. L., Gold, L. H., et al. (1998) Identification of novel susceptibility loci for inflammatory bowel disease on chromosomes 1p, 3q, and 4q: evidence for epistasis between 1p and IBD1. *Proc. Natl. Acad. Sci. USA* **95(13),** 7502–7507.
4. Baker, B. F. and Monia, B. P. (1999) Novel mechanisms for antisense-mediated regulation of gene expression. *Biochimica. Biophys. Acta* **1489,** 3–18.

5. Wu, H., Lima, W. F., and Crooke, S. T. (1999) Properties of cloned and expressed human RNase H1. *J. Biol. Chem.* **274**, 28,270–28,278.
6. Crooke, S. T. (1999) Molecular mechanisms of antisense drugs: human RNase H. *Antisense Nucleic Acid Drug Dev.* **9**, 377–379.
7. Glover, J., Leeds, J. M., Mant, T. G. K., et al. (1997) Phase 1 safety and pharmacokinetic profile of an intercellular adhesion molecule-1 antisense oligonucleotide (ISIS 2302). *J. Pharmacol. Exp. Ther.* **282**, 1173–1180.
8. Henry, S. P., Templin, M. V., Gillett, N., Rojko, J., and Levin, A. A. (1999) Correlation of toxicity and pharmacokinetic properties of a phosphorothioate oligonucleotide designed to inhibit ICAM-1. *Toxicol. Pathol.* **27**, 95–100.
9. Henry, S. P., Giclas, P. C., Leeds, J., et al. (1997) Activation of the alternative pathway of complement by a phosphorotioate oligonucleotide: potential mechanism of action. *J. Pharmacol. Exp. Ther.* **281**, 810–816.
10. Sheehan, J. P. and Lan, H. C. (1998) Phosphorothioate oligonucleotides inhibit the intrinsic tenase complex. *Blood* **92**, 1617–1625.
11. Yacyshyn, B. R., Bowsen-Yacyshyn, M. B., Jewell, L., et al. (1998) A placebo-controlled trial of ICAM-1 antisense oligonucleotide in the treatment of Crohn's disease. *Gastroenterology* **114**, 1133–1142.
12. Yacyshyn, B. R., Chey, W. Y., Goff, J., et al. (2002) Double blind placebo controlled trial of the remission inducing and steroid sparing properties of an ICAM-1 antisense oligodeoxynucleotide, alicaforsen (ISIS 2302), in active steroid dependant Crohn's disease. *Gut* **51**, 30–36.
13. Feagan, B. G., Rochon, J., Fedorak, R. N., et al. (1995) Methotrexate for the treatment of Crohn's disease. *N. Engl. J. Med.* **332**, 292–297.
14. Nedelman, J. R., Karara, A. H., Chang, C. T., et al. Inferring systemic exposure from a pharmacokinetic screen: model-free and model-based approaches. *Stat. Med.* **14**, 955–968.
15. Statistical Consultants (1986) PCNONLIN and NONLIN84: software for the statistical analysis of nonlinear models. *Am. Stat.* **40**, 52.
16. Yu, R. Z., Su, J. Q., Grundy, J. S., et al. (2003) Prediction of clinical responses in a simulated phase 3 trial of Crohn's patients administered the antisense phosphorothioate oligonucleotide ISIS 2302: comparison of proposed dosing regimens. *Antisense Nucleic Acid Drug Dev.* **13(1)**, 57–66.
17. Henry, S. P., Giclas, P. C., Leeds, J., et al. (1997) Activation of the alternative pathway of complement by a phosphorothioate oligonucleotide: potential mechanism of action. *J. Pharmacol. Exp. Ther.* **281**, 810–816.
18. Schroeder, K. W., Tremaine, W. J., and Ilstrup, D. M. (1987) Coated oral 5-aminosalicylic acid therapy for mildly to moderately active ulcerative colitis: a randomized study. *N. Engl. J. Med.* **317**, 1625–1629.

VIII

HEPATITIS

16

Optimizing Electroporation Conditions for Intracellular Delivery of Morpholino Antisense Oligonucleotides Directed Against the Hepatitis C Virus Internal Ribosome Entry Site

Ronald Jubin

1. Introduction

Hepatitis C virus (HCV) contains a positive-sense, single-stranded RNA viral genome that encodes viral structural and nonstructural proteins *(1)*. Initiation of translation is under control of the internal ribosome entry site (IRES) located within the viral 5'-nontranslated region (NTR) that is both highly conserved and structured *(2)*. In contrast to cellular mRNAs that initiate translation at the extreme 5' terminus, IRES elements direct the translational machinery directly to the initiator AUG codon. Mutations to various regions of the HCV IRES have proven deleterious to translation in numerous in vitro studies (reviewed in **ref**. *3*). In addition, several antisense studies have demonstrated dramatic reduction in IRES activity (reviewed in **ref**. *3*). Therefore, downregulation of IRES function has emerged as a valid therapeutic target for controlling HCV infection.

One general class of antisense oligonucleotides (AS-ODNs) are those that elicit cleavage of the mRNA-antisense complex by the cellular enzyme RNase H. These include unmodified DNA and phosphorothioate antisense (S-DNA). An alternative approach has been the use of AS-ODNs that contain modified backbone chemistries that do not elicit RNase H–mediated cleavage but, instead, block translation by steric mechanisms (RNase H independent). Both classes of oligonucleotides have been reported to be active against the HCV IRES *(3–6)*. One representative RNase H–independent type is the morpholino antisense (M-AS) that contains a phosphorodiamidate-morpholino chemistry in place

From: *Methods in Molecular Medicine, Vol. 106: Antisense Therapeutics, Second Edition*
Edited by: I. Phillips © Humana Press Inc., Totowa, NJ

of the phosphodiester-ribose sugar backbone. These oligonucleotides have been shown previously to be highly specific compared to S-DNA *(6)*. Because the M-AS does not lead to target degradation, it can be used to identify regions involved in RNA:RNA and RNA:protein interactions. This has been observed in HCV IRES antisense studies. AS-ODNs targeting stem loop (IIId) approx 60 nucleotides (nt) 5' of the viral open reading frame (ORF) inhibited translation at levels comparable with those directly targeting the initiator AUG *(4,5)*. Subsequent studies identified this region as critical for IRES:40S ribosome subunit interaction *(7,8)*.

Another advantage of RNase H–independent AS-ODNs is their high degree of specificity. This can be attributed to their lack of negatively charged backbone. However, this property makes it difficult to introduce M-AS into cells using common methods such as cationic lipid transfer.

Evaluation of AS-ODN activity in cell culture systems is inherently more difficult with many antisense chemistries. Cellular uptake, intracellular distribution, and stability can be factors that dramatically affect intracellular vs in vitro efficacy profiles. Moreover, extremely high levels of AS-ODNs can often lead to nonspecific inhibition of gene expression. Finally, cytotoxicity, either directly attributable to the antisense levels or indirectly through nonspecific interactions with cellular components, can adversely affect inhibition profiles and data interpretation. These factors underscore the importance of having the proper controls whenever performing AS-ODN studies (*see* **Note 1**). In an effort to monitor further for nonspecific effects of AS-ODNs directed against the HCV IRES, we have developed a dual-luciferase bicistronic plasmid that takes advantage of the differences between cellular and HCV translation initiation.

Plasmid pCMVR(1B/408)P contains the cytomegalovirus immediate/early promoter (CMVi/e) and two separate reporter cistrons (or genes) *(5)*. Transcription from the CMVi/e promoter produces a single mRNA transcript. Therefore, a single transcription element regulates production of both reporter genes. However, translation of each reporter gene is independent of one another owing to their distinct mechanisms. Translation of the first or upstream reporter *Renilla* luciferase (RL) is regulated by the *Xenopus* β-globin 5'-NTR. The RL reporter is expressed in a cap-dependent manner. The second or downstream reporter firefly luciferase (FL) is under translational control of the HCV IRES element. Expression of the FL reporter occurs independently of the RL reporter since the IRES directs the translational machinery directly to the initiator codon. Because this ORF is internally located within the transcript, translation of FL occurs by a cap-independent fashion. A principle advantage of this system is that target and control reporters are part of a single transcription product and therefore are present at stoichiometric levels. The use of two luciferase reporters is preferred because they have similar signal-to-noise ratios, linear range, and decay rates *(9)*. In addition, the use of the Dual Luciferase Assay Kit™ simplifies procedures by assaying both FL and RL within a single well. Nonspecific antisense activity that affects cellular transcription or cap-dependent translation will result in inhibition of the RL control reporter. A truly specific IRES inhibition profile should demonstrate only a reduction in FL reporter. Thus, dual luciferase bicistronic plasmids can be of great use in determining antisense specificity. Finally, as an additional control, we performed a 3-(4,5-dimethylthiazolyl-2)-2,5-diphenyl tetrazolium bromide (MTT) cytotoxicity assay to further control for nonspecific cytotoxic effects *(10)*.

The in vitro activity of M-AS directed to the HCV IRES has been recently described *(5)*. We have developed an effective electroporation protocol for the introduction of M-AS into cells in culture (*see* **Note 2**). This chapter outlines methods for the evaluation of M-AS directed to the HCV IRES in HeLa cell transfection assays. The goal of our study design is to determine several parameters: First, does cotransfection of the M-AS with the bicistronic plasmid result in specific inhibition of the HCV IRES? Direct comparison of M-AS-treated and nontreated reporter signals within the same transfection set determine this. Second, what transfection conditions are the least toxic to cells? This can be achieved by directly comparing MTT values in each transfection set with one another. Third, are the concentrations of M-AS and/or DNA (the H_2O controls) toxic to cells? DNA toxicity can be determined by direct comparison of each no-treatment sample with the no-treatment mock-transfection MTT values. The effect of the addition of M-AS can be determined by comparison of M-AS-treated and nontreated RL and MTT values within the same transfection set. Fourth, does antisense activity correlate with cell cytotoxicity? This can be determined by directly comparing the percentage of cytotoxicity and percentage of IRES inhibition within each transfection set. The obvious goal is to identify transfection conditions that produce the greatest specific IRES inhibition with the least amount of cytotoxicity.

Owing in part to the large number of transfections required for initial M-AS optimization studies, the mismatch controls M-AS have been omitted. However, it is critical to repeat the optimal transfection conditions and include appropriate controls for increased confidence of results. Finally, the use of fluorescent-labeled AS-ODN or an equivalent means to visually monitor cellular uptake should also be determined with optimal transfection conditions.

2. Materials

1. Purified, endotoxin-free plasmid DNA (*see* **Note 3**).
2. Morpholino AS-ODNs prepared in cell culture–grade water and filter sterilized (AVI BioPharma, Corvallis, OR).
3. Humidified 37°C (5% CO_2) incubator.

4. Tissue culture biosafety cabinet.
5. Adherent HeLa cell cultures (*see* **Note 4**).
6. Cell culture medium components: Eagle's minimum essential medium supplemented with 10% fetal bovine serum (FBS), 2m*M* L-glutamine, 100 IU/mL of penicillin, 100 µg/mL of streptomycin; trypsin (0.25%)-EDTA (0.1%) solution; Dulbecco's phosphate-buffered saline (PBS) without calcium and magnesium (Mediatech).
7. Sterile plasticware for passaging cells (T-175 flasks), for plating assays (96-well plates), for cell pelleting (250-mL conical, polypropylene), and for dilutions (5.0-mL polypropylene round-bottomed tubes) (Falcon).
8. Tabletop centrifuge for pelleting cells (Beckmann Allegra 25R).
9. Gene Pulser II Electroporator with capacitance extender and 0.4-cm gene pulser cuvets (Bio-Rad, Hercules, CA).
10. MTT assay reagents: MTT, *N,N*-dimethylformamide (NN-DMF), sodium dodecyl sulfate, PBS (with Mg and Ca).
11. Shaker unit with microplate platform (Labline).
12. Microplate reader equipped with a 570 n*M* filter (MRX; Dynex).
13. Microplate injectable luminometer (Dynex MLX luminometer).
14. Microlite 1$^+$ 96-well round-bottomed white microtiter plates (Dynex).
15. Dual-Luciferase Assay Kit™ (Promega, Madison, WI).

3. Methods

3.1. Preparation of AS-ODNs

1. Resuspend morpholino AS-ODN HCVm330-354 (5'-GAUUUGUGCUCAU GAUGCACGGUCU-3') that exactly complements HCV 1b genotype IRES sequences spanning nt 330–354 and a random mismatch oligonucleotide, RDMm330-354 (5'-UUGGGCCUGUAGUCCAUAUCAGGUU-3'), were distilled deionized sterile (cell culture–grade) H$_2$O at a final concentration of 100 µ*M*, filter sterilize, aliquot, and freeze at –20°C.
2. Thaw the aliquots on ice and dilute in cell culture–grade H$_2$O for assays. Cell culture–grade H$_2$O is used as a no-treatment control and to keep assay volumes constant.

3.1.1. Maintenance and Passage of Cell Cultures

All cell culture manipulations are performed with sterile reagents under aseptic conditions.

1. Maintain adherent HeLa cell monolayers in T-175 flasks in a humidified 37°C (5% CO$_2$) incubator and routinely passage prior to confluency.
2. Twenty-four to 36 h prior to electroporation studies, determine the number of transfection points to be carried out, and split an adequate number of flasks at a 1:3 ratio (approximately one flask will provide enough cells for one transfection).

3.1.2. Harvesting of Cells

1. Wash the (approx 80–90% confluent) T-175 flasks with 50 mL of PBS without Ca/Mg.
2. Aspirate and replace with 5 mL of (0.25%) Trypsin-EDTA and transfer to a 37°C incubator for 3–5 min.
3. When the majority of the cells have rounded up (monitor using a microscope), transfer back to the tissue culture hood and neutralize the trypsin by adding 5.0 mL of complete medium (with 10% FBS).
4. Vigorously rinse the walls of the flasks repeatedly with a 10-mL pipet to break up cell clumps.
5. Pool the collected cells from all the flasks into a sterile 250-mL conical polypropylene centrifuge tube placed on ice.
6. Transfer 200 μL of the cell mix to a microfuge tube containing an equal volume of 0.4% Trypan blue solution.
7. Pellet the 250-mL cell suspension in a tabletop centrifuge by centrifuging (800g) for 5 min at 4°C.
8. While the cells are pelleting, transfer the cell/trypan blue suspension to a hematocytometer and perform a cell count.
9. Following centrifugation, wash the cells in a 10X cell pellet volume of ice-cold PBS without Ca/Mg (*see* **Note 5**).
10. Repeat **step 7**.
11. Recover cells the by aspirating all the PBS without Ca/Mg wash from the pellet.
12. Resuspend the cells in the ice-cold PBS without Ca/Mg to a final concentration of 1.0×10^7 cells/mL and keep on ice. *See* **Subheading 3.1.6.** for instructions on medium preparation prior to starting transfections.

3.1.3. Transfection Setup

Since the experiment involves varying several different parameters, the use of a spreadsheet is advisable (**Table 1**). Transfections are assembled on ice. Each transfection set will consist of two samples, the first containing 0.1 mL of a 15 μM M-AS stock and the second no-treatment control with an equal volume of cell–culture grade H_2O. The volume of cells is held constant at 0.4 mL/transfection (4.0×10^6 total cells). Plasmid DNA is at a constant 10 μg/transfection level in all transfections. Before assembling transfections, remove a 0.4-mL aliquot of cells and mix with 12.0 mL of complete medium. Place on ice. This will be used as a "cells only" control. The preparation of master stocks is preferred to minimize variability owing to pipeting. A total of 14 transfections (7 sets) is to be performed. We routinely set up for 16 transfections to allow for spill volume. Transfer 6.4 mL of cell suspension (0.4 mL cells/transfection × 16 transfections) to a 50-mL conical tube. Remove 3.2 mL of the cells/DNA mix and transfer to a separate 50-mL conical tube. To the first tube, add 0.8 mL of 15 μM M-AS stock (0.1 mL × 8 transfections). The final concen-

Table 1
Transfection Worksheet[a]

Set no.	Transfection no.	Volts (kv)	Resistance (W)	Transfection components	Time constant (ms)
1	1	0.25	250	Cells/DNA/M-AS	7.8
1	2	0.25	250	Cells/DNA/H$_2$O	7.4
2	3	0.25	500	Cells/DNA/M-AS	13.0
2	4	0.25	500	Cells/DNA/H$_2$O	13.2
3	5	0.25	960	Cells/DNA/M-S	26.0
3	6	0.25	960	Cells/DNA/H$_2$O	27.0
4	7	0.5	25	Cells/DNA/M-AS	0.8
4	8	0.5	25	Cells/DNA/H$_2$O	0.8
5	9	0.75	25	Cells/DNA/M-AS	0.8
5	10	0.75	25	Cells/DNA/H$_2$O	0.8
6	11	1	25	Cells/DNA/M-AS	0.8
6	12	1	25	Cells/DNA/H$_2$O	0.8
7	13	NP	NP	Cells/DNA/M-AS	NP
7	14	NP	NP	Cells/DNA/H$_2$O	NP

[a]The spreadsheet shows the electroporator parameters and components for each individual transfection. The time constant values were recorded following each electroporation.

314

tration will be 3 μM in all samples containing M-AS. Add 0.8 mL of cell culture–grade H_2O to the second tube.

3.1.4. Electroporator Setup

The electroporator-controlled parameters include both the voltage and the resistance applied to the cells. Varying either parameter can affect the overall length of time that the cells are exposed to current (measured in time constant [millisecond] values). The assay parameters selected in this protocol were based on preliminary cell culture studies. Two approaches were undertaken. First, a constant low voltage and varied resistance profile was performed (sets 1–3). Second, a constant low resistance and varied voltage was analyzed (sets 4–6). If cell lines other than HeLa are to be assayed, the exact conditions (e.g., DNA concentration, diluent, voltage, resistance) need to be determined empirically.

1. Under sterile conditions, transfer 0.5 mL of cells/morpholino mixture with a sterile pipet into a 0.4-cm electroporation cuvet and keep on ice.
2. Repeat with no-treatment controls.
3. Visually check to make sure that no air bubbles are trapped (if necessary dislodge by repeated pipeting).
4. Immediately proceed with electroporation (*see* **Note 6**).

3.1.5. Electroporation

1. Set the electroporator to the desired voltage and resistance parameters for each sample set.
2. Prior to placing the cuvet in its holder, ensure that all ice/moisture has been removed from outside the cuvet (*see* **Note 7**).
3. Transfer the dry cuvet into its holder and slide into the unit.
4. Pulse until a beep is heard, signaling the end of the transfection.
5. Note the time constant value on the spreadsheet.
6. Transfer the cuvet back to ice and repeat the procedure for all transfections (varying voltage and resistance as necessary).
7. Retain the final set (set 7) of transfections (with and without M-AS) but do not electroporate. These will serve as mock-transfection controls.

3.1.6. Plating Transfections

1. Prior to electroporation, fill 14 separate 50-mL conical tubes with 11.0 mL of complete medium.
2. Following transfection, transfer the transfected cells from the cuvet to the 11.0 mL medium using a sterile pipet (*see* **Note 8**).
3. Following transfer, immediately rinse the cuvet with 1.0 mL of complete medium and combine with the respective transfection. Repeat for each transfection. Each transfection/medium mix contains a final concentration of 3.33×10^5 cells/mL.

4. Plate out 150 μL of each transfected cell population (5.0 × 10⁴ cells/well) in replicates of six in column format (transfections 1–12).

5. Reserve the bottom two rows for the nonelectroporated controls.

6. Plate transfections 13 and 14 (set 7) at the same volumes except in replicates of six arranged in row format. The bottom row should be plated with untreated cells ("cells-only" control). This setup enables all transfections to be assayed on the same plate while providing for maximum replicate numbers. A minimum of two plates should be setup for each experiment (*see* **Note 9**).

7. Label the plates accordingly for analysis in either the reporter assay or cytotoxicity assay.

8. Transfer both plates to a humidified 37°C (5% CO₂) incubator overnight for 21–24 h.

3.2. MTT Cytotoxicity Study

3.2.1. Preparation of Reagent

1. Prepare the MTT solution by dissolving powdered MTT into PBS to a final concentration of 1 mg/mL. Store at 4°C protected from light (*see* **Note 10**).

2. Prepare the solubilization buffer by making a 1:1 500-mL stock solution using NN-DMF and cell culture–grade H₂O.

3. Dissolve 100 g of sodium dodecyl sulfate (SDS) into the NN-DMF/H₂O solution, and adjust the final volume to 500 mL for a final 20% SDS concentration (*see* **Note 11**).

3.2.2. MTT Assay

1. Following overnight incubation, aspirate media from the cells and wash with 100 μL/well of PBS.

2. Aspirate the PBS and replace with 100 μL/well of MTT stock solution, and return the plates to the 37°C incubator (proceed directly with **Subheading 3.3.**).

3. Approximately 4–6 h postincubation (following the luciferase assays), examine the cells-only control lane in a light microscope. A strong field of purple crystal-like structures within the cells demonstrates that MTT reduction has occurred.

4. Aspirate the MTT stock solution and replace with 100 μL/well of solubilization buffer.

5. Transfer the plate to a microplate shaker platform and mix vigorously for ≥15 min.

6. Measure the absorbance in a microplate reader equipped with a 570-nm wavelength filter.

3.3. Dual Luciferase Assay

1. Prior to harvesting cells, thaw dual luciferase assay buffers in a room temperature water bath (keep substrates on ice).

2. After complete thawing, assemble the buffer components.

3. For the firefly luciferase substrate (Luciferase Assay Reagent™ II [LARII]), transfer 10 mL of buffer directly into a glass bottle containing lyophilized sub-

strate and mix gently by repeated inverting. For the Renilla substrate (Stop & Glo™), transfer 200 μL of substrate solvent into the lyophilized substrate vial followed by a brief, gentle vortex.

4. Transfer the entire mix to 10 mL of the thawed Stop & Glo buffer.
5. Mix briefly by repeated inverting.
6. Equilibrate the substrate components to room temperature before use in the assay (*see* **Note 12**).
7. Prepare a 1X stock of passive lysis buffer.
8. Remove the reporter assay plate from the incubator and process for luciferase assays.
9. Aspirate the medium and wash the cells with 100 μL/well of PBS.
10. Aspirate and replace with 60 μL/well of passive lysis buffer.
11. Transfer the plate to a microplate shaker and shake vigorously for 10 min to promote cell lysis.
12. Confirm lysis by microscopic evaluation.
13. Transfer 50 μL/well to a microlite 1$^+$ microtiter plate.

3.3.1. Luminometer Setup

The Dynex MLX is used with the dual injector setup (*see* **Note 13**).

1. Wash the injectors with 3X strokes of cell culture–grade water.
2. Remove the dead volume remaining in the tubes by 3X strokes on the recover injectors setting (capture the recovered water in a waste container and discard).
3. Prime the injectors with 3X strokes of complete LARII (injector A) or complete Stop & Glo (injector C) substrate mixes (*see* **Note 14**).

3.3.2. Injection and Read Parameters

Each well is assayed by the sequential injection of each luciferase substrate (LARII followed by Stop & Glo).

1. Using the Assay wizard function, setup the following assay parameters:
   a. Calculation mode: ratio of first to second integral.
   b. Gain setting: autogain (0-1000 relative light units).
   c. Dispense volume A: 100 μL (LAR II).
   d. Dispense volume C: 100 μL (Stop & Glo).
   e. Total read time: 24.0 s.
   f. Start time of first integration: 2.0 s.
   g. Start time of second integration: 14.0 s.
   h. Scaling factor: scale data by 1.
   i. Heating plate: temperature disabled.
   j. Delay between A and C: 12 s.
   k. Dispense method: normal refill.
   l. Shaking: enabled for each well.
   m. Stop time of first integration: 12 s.
   n. Stop time of second integration: 24 s.

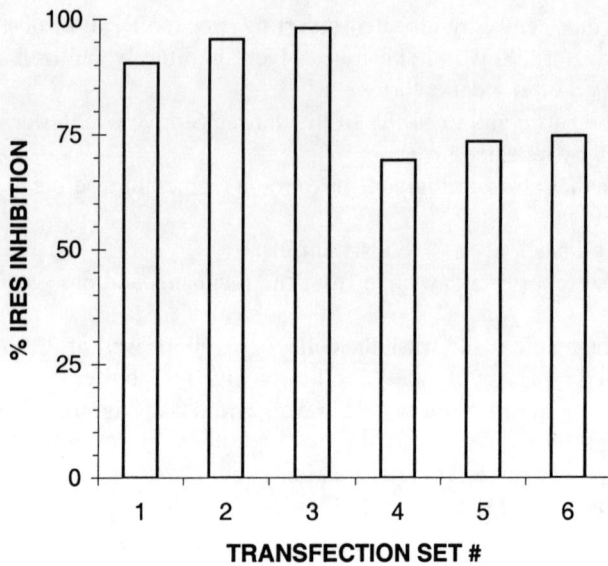

Fig. 1. Inhibition of IRES translation by M-AS. Transfection reporter values were averaged, and the FL/RL ratio was determined for each individual transfection. The percent inhibition of FL (IRES) reporter activity was determined by direct comparison of transfections containing M-AS with those that did not (within the same transfection set).

2. Place the microtiter plate containing cell lysates into the plate carrier and select the plate run. The luminometer will read each well individually, then report all results on completion of the run.
3. Following the assay, recover the buffers and wash the tubing with 5X strokes of a mild detergent followed by 10X strokes of sterile H_2O (leave remaining H_2O in the lines).

3.4. Data Analysis

The addition of the RL (Cap-mediated) signal in the bicistronic system allows for the simultaneous normalization of transfection efficiency and M-AS specificity within each transfection set. The relative translation efficiency of IRES-mediated FL values is determined by comparing the averaged FL/RL ratios of M-AS-treated to nontreated controls for each transfection set (e.g., transfection 1 vs transfection 2). Values obtained are plotted as percent inhibition for each transfection set (**Fig. 1**). Comparison of inhibition profiles showed that M-AS reduced FL levels in all transfection sets, but to varying levels.

MTT data were analyzed individually for each transfection. Values for each transfection were averaged. Comparison of cells-only controls to the mock-

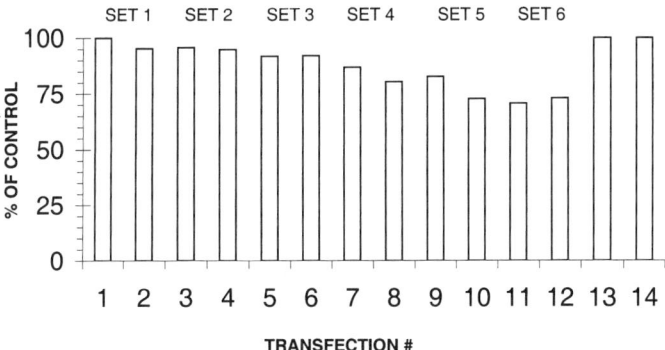

Fig. 2. Inhibition of cell viability. MTT values were averaged for each transfection, and cellular toxicity was determined by direct comparison of transfections with the mock-transfection controls.

transfected controls (transfections 13 and 14) showed no difference in MTT values, suggesting that the addition of transfection components to cells was well tolerated (data not shown). Subsequently, these mock-transfected values (arbitrarily set to 100%) were directly compared with actual transfection values. Values are plotted as a percentage of the mock-transfected control values (**Fig. 2**). Comparison of results within each transfection set showed no differences in MTT values, demonstrating that the 3 μM M-AS test level was not toxic to cells. However, comparison of transfection sets with each other showed varying levels of inhibition. Sets 1–3 showed little or no inhibition, whereas sets 4–6 exhibited a more dramatic reduction in MTT levels compared to controls. Comparison of the time constant values showed an inverse relationship between time constant levels and cytotoxicity (**Table 1**). Moreover, this increase in time constant values was directly associated with increased M-AS efficacy (**Table 1** and **Fig. 1**). The transfection parameters used for set 3 produced the greatest inhibition levels (>95%) with only minor (<10%) cytotoxicity.

Use of the optimal transfection procedures can subsequently be used in more elaborate assays including controls. Therefore, an expanded assay was performed using increasing concentrations of HCVm330-354 or the randomized control, RDMm330-354, and 0.25-kV/960-µF electroporator settings. Values were plotted in a fashion identical to **Fig. 1**. The results demonstrated that HCVm330-354 produced a dose-dependent inhibition of FL (IRES) values (**Fig. 3**). Inhibition of the samples containing RDMm330-354 showed minor but not dose-dependent inhibition, further supporting the specificity of HCVm330-354 toward the HCV IRES.

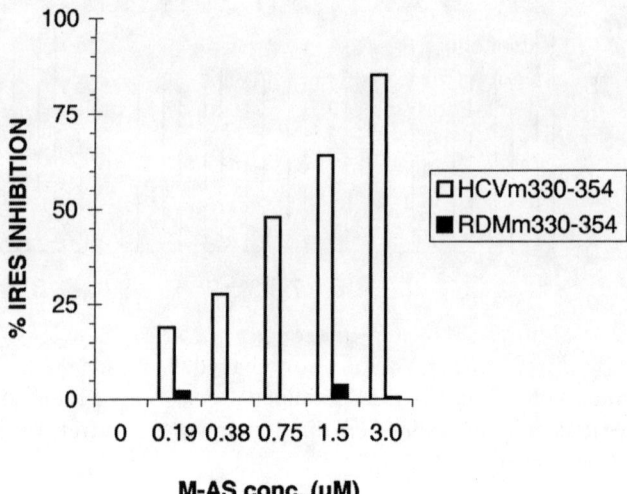

Fig. 3. Dose-dependent inhibition of the HCV IRES. The authentic antisense HCVm330-354 (open bars) or randomized control RDMm330-354 (shaded bars) were added to separate transfections at increasing concentrations. Transfection conditions were identical for all assay conditions (0.25kV/960 μF). Percent inhibition profiles were calculated in a fashion identical to those in **Fig. 1**.

These data demonstrated that electroporation can be a valuable approach for introducing nonionic M-AS into HeLa cells in culture. The use of a bicistronic plasmid simplified the assay format while providing a valuable control for transfection efficiency and specificity. Furthermore, the concomitant analysis of cytotoxicity by MTT assay demonstrated that limiting cytotoxicity should greatly enhance the inhibition profile of a given M-AS, which is a critical aspect of selecting candidate AS-ODNs for further development.

4. Notes

1. Whenever reporter assays are used, it is essential to check for nonspecific activity against the specific reporter gene before performing elaborate studies. The use of reporter controls (e.g., pCMV-FLUC and pCMV-RLUC) should be analyzed prior to elaborate transfection assays to ensure that M-AS does not produce undesirable nonspecific effects. Furthermore, the use of random or mismatch controls will lend greater confidence of M-AS specificity.
2. This described protocol uses M-AS. However, it should also be useful for transfection studies using other nonionic antisense chemistries including peptide nucleic acids.

3. Cell-based assays require plasmids that possess a promoter element that is active within the cell line of choice. Plasmid DNA should be prepared by a method such as column purification (Maxi-kit; Qiagen, Valencia, CA) and eluted into endotoxin-free sterile H_2O. Poorly prepared DNA can have extremely adverse effects on transfection.

4. The HeLa cells used in this study were Texas Hela cells received from Dr. C. Guantt, University of Texas Health Science Center, San Antonio. Substitution of another adherent Hela line should produce comparable results.

5. The cell pellet should first be resuspended in a minimal volume of ≤5.0 mL. Trying to resuspend the cell pellet in a large volume will result in cell clumps remaining in solution.

6. Work quickly. Cells that remain on ice for extended periods of time can settle out into the bottom of the cuvet, which can adversely affect transfection efficiency.

7. Excess moisture on electrodes can result in arcing, potentially damaging the electroporator.

8. Be sure to select a pipet that is long enough to reach the bottom of the cuvet and has an adequate capacity (e.g., 1.0 mL serological pipet [7521; Falcon]).

9. Additional remaining cell volumes can be used to setup additional plates for time course experiments if desired.

10. MTT is very toxic. Handle carefully only with gloved hands and dispose of used solutions properly.

11. To hasten dilution of SDS into the NN-DMF/H_2O solution, heat the solution to 37°C.

12. The firefly luciferase/luciferin substrate reaction is strongly influenced by temperature. Preequilibration to room temperature ensures that the substrate will not change temperature during the course of the assay.

13. We used the dual injection system to prevent adverse effects of reporter decay when assaying an entire plate. As an alternative, samples can be assayed individually in a single-format luminometer. In addition, newer substrate mixes that have increased stability are commercially available, which may alleviate the need for injectors (Promega).

14. The dual luciferase assay must be performed sequentially. FL must be assayed first, followed by RL. Be sure that injectors are primed properly before starting the assay.

References

1. Choo, Q. L., Kuo, G., Weiner, A. J., Overby, L. R., Bradley, D. W., and Houghton, M. (1989) Isolation of a cDNA clone derived from a blood-borne non-A, non-B viral hepatitis genome. *Science.* **244,** 359–62.
2. Brown, E. A., Zhang, H., Ping, L. H., and Lemon, S. M. (1992) Secondary structure of the 5' nontranslated regions of hepatitis C virus and pestivirus genomic RNAs. *Nucleic Acids Res.* **20,** 5041–5045.
3. Jubin, R. (2001) Hepatitis C IRES: translating translation into a therapeutic target. *Curr. Opin. Mol. Ther.* **3,** 278–287.

4. Hanecak, R., Brown-Driver, V., Fox, M. C., et al. (1996) Antisense oligonucle-otide inhibition of hepatitis C virus gene expression in transformed hepatocytes. *J. Virol.* **70,** 5203–5212.

5. Jubin, R., Vantuno, N. E., Kieft, J. S., et al. (2000) Hepatitis C internal ribosome entry site (IRES) contains a phylogenetically conserved GGG triplet essential for translation and IRES folding. *J. Virol.* **74,** 10,430–10,437.

6. Summerton, J., Stein, D., Huang, S. B., Matthews, P., Weller, D., and Partridge, M. (1997). Morpholino and phosphorothioate antisense oligomers compared in cell-free and in-cell systems. *Antisense Nucleic Acid Drug Dev.* **7,** 63–70.

7. Kolupaeva, V. G., Pestova, T. V., and Hellen, C. U. (2000) An enzymatic footprinting analysis of the interaction of 40S ribosomal subunits with the internal ribosomal entry site of hepatitis C virus. *J. Virol.* **74,** 6242–6250.

8. Kieft, J. S., Zhou, K., Jubin, R., and Doudna, J. A.. (2001) Mechanism of ribo-some recruitment by hepatitis C IRES RNA. *RNA* **2,** 194–206.

9. Sherf, B. A., Navarro, S. L., Hannah, R. R., Wood, K. V. (1996) Dual-luciferase reporter assay: an advanced co-reporter technology integrating firefly and Renilla luciferase assays. *Promega Notes* **57,** 2–9.

10. Gerlier, D. and Thomasset, N. (1986) Use of MTT colorimetric assay to measure cell activation. *J. Immunol. Methods* **94,** 57–63.

Index